AF739775

INNOVATION IN CRISIS MANAGEMENT

This book deals with how to measure innovation in crisis management, drawing on data, case studies, and lessons learnt from different European countries.

The aim of this book is to tackle innovation in crisis management through lessons learnt and experiences gained from the implementation of mixed methods through a practitioner-driven approach in a large-scale demonstration project (DRIVER+). It explores innovation from the perspective of the end-users by focusing on the needs and problems they are trying to address through a tool (be it an app, a drone, or a training program) and takes a deep dive into what is needed to understand if and to what extent the tool they have in mind can really bring innovation.

This book is a toolkit for readers interested in understanding what needs to be in place to measure innovation: it provides the know-how through examples and best practices. The book will be a valuable source of knowledge for scientists, practitioners, researchers, and postgraduate students studying safety, crisis management, and innovation.

Chiara Fonio is Research Associate at the Vrije Universiteit (VU, Amsterdam), Faculty of Social Sciences, Organization Sciences. She has been working in academia and European institutions (Joint Research Centre, European Commission, where the work on the trial guidance methodology was designed and developed). She has extensive experience in European-funded projects and authored a number of publications on crisis management and surveillance (including, among others, *Big Data, Surveillance, and Crisis Management*, co-edited with Kees Boersma, 2017). Her research interests cover disaster governance, crisis management and surveillance studies.

Adam Widera is Managing Director of the Competence Center for Crisis Management at the European Research Center for Information Systems (ERCIS), hosted by the University of Münster, Germany. After graduating in political science, philosophy, and political economy, he pursued his doctorate in information systems. His research areas cover modelling, simulation and performance measurement in humanitarian and disaster relief logistics and design and evaluation of information systems for humanitarians. He has published over 50 research papers in international journals and conferences and has been involved in various international research projects.

Tomasz Zwęgliński is Colonel of the Polish State Fire Service working as an academic at the Main School of Fire Service in Warsaw, with above 20 years of experience. He has a PhD in security sciences and a master's in fire engineering, is a qualified expert in the Union Civil Protection Mechanism, and is an expert in crisis and disaster management, CBRN, and occupational safety. He is also a manager of research and international civil protection capacity-building projects (e.g. in Ukraine). His key research interest focuses on the evaluation of disaster management exercises, including new solutions. EU Research Executive Agency expert. His recent publications include *Exercising in a Radioactive Environment–a Case Study from CBRNE Exercise in Chernobyl Exclusion Zone* (2019), *Cascading Effect as a Contemporary Challenge for Crisis Management* (2020), and *Polish Approach to Sharing Resources Deployable for the EU Civil Protection Mechanism* (2021).

Routledge Studies in Science, Technology and Society

The Cultural Authority of Science
Comparing across Europe, Asia, Africa and the Americas
Edited by Martin W Bauer, Petra Pansegrau, Rajesh Shukla

Blockchain and Web 3.0
Social, Economic, and Technological Challenges
Edited by Massimo Ragnedda and Giuseppe Destefanis

Understanding Digital Events
Bergson, Whitehead, and the Experience of the Digital
Edited by David Kreps

Big Data – A New Medium?
Edited by Natasha Lushetich

The Policies and Politics of Interdisciplinary Research
Nanomedicine in France and in the United States
Séverine Louvel

Apocalyptic Narratives
Science, Risk and Prophecy
Hauke Riesch

Distributed Perception
Resonances and Axiologies
Edited by Natasha Lushetich and Iain Campbell

Questing Excellence in Academia
A Tale of Two Universities
Knut H. Sørensen and Sharon Traweek

Public Communication of Research Universities
'Arms Race' for Visibility or Science Substance?
Edited by Marta Entradas and Martin W. Bauer

Innovation in Crisis Management
Edited by Chiara Fonio, Adam Widera and Tomasz Zwęgliński

For the full list of books in the series: www.routledge.com/Routledge-Studies-in-Science-Technology-and-Society/book-series/SE0054

INNOVATION IN CRISIS MANAGEMENT

Edited by Chiara Fonio, Adam Widera and Tomasz Zwęgliński

Routledge
Taylor & Francis Group
LONDON AND NEW YORK

Cover image: © Henry Browne

First published 2023
by Routledge
4 Park Square, Milton Park, Abingdon, Oxon OX14 4RN

and by Routledge
605 Third Avenue, New York, NY 10158

Routledge is an imprint of the Taylor & Francis Group, an informa business

© 2023 selection and editorial matter, Chiara Fonio, Adam Widera and Tomasz Zwęgliński; individual chapters, the contributors

The right of Chiara Fonio, Adam Widera and Tomasz Zwęgliński to be identified as the authors of the editorial material, and of the authors for their individual chapters, has been asserted in accordance with sections 77 and 78 of the Copyright, Designs and Patents Act 1988.

All rights reserved. No part of this book may be reprinted or reproduced or utilised in any form or by any electronic, mechanical, or other means, now known or hereafter invented, including photocopying and recording, or in any information storage or retrieval system, without permission in writing from the publishers.

Trademark notice: Product or corporate names may be trademarks or registered trademarks, and are used only for identification and explanation without intent to infringe.

British Library Cataloguing-in-Publication Data
A catalogue record for this book is available from the British Library

Library of Congress Cataloging-in-Publication Data
Names: Fonio, Chiara, editor. | Widera, Adam, editor. | Zwęgliński, Tomasz, editor.
Title: Innovation in crisis management / edited by Chiara Fonio, Adam Widera, Tomasz Zwęgliński.
Description: New York, NY: Routledge, 2023. | Series: Routledge studies in science, technology and society | Includes bibliographical references and index.
Identifiers: LCCN 2022038497 (print) | LCCN 2022038498 (ebook) | ISBN 9781032189154 (paperback) | ISBN 9781032189178 (hardback) | ISBN 9781003256977 (ebook)
Subjects: LCSH: Technological innovations–Economic aspects. | Industrial management–Technological innovations. | Crisis management.
Classification: LCC HC79.T4 I546554 2023 (print) | LCC HC79.T4 (ebook) | DDC 338/.064–dc23/eng/20220811
LC record available at https://lccn.loc.gov/2022038497
LC ebook record available at https://lccn.loc.gov/2022038498

ISBN: 978-1-032-18917-8 (hbk)
ISBN: 978-1-032-18915-4 (pbk)
ISBN: 978-1-003-25697-7 (ebk)

DOI: 10.4324/9781003256977

Typeset in Bembo
by Apex CoVantage, LLC

CONTENTS

FIGURES

TABLES

CONTRIBUTORS

Angelos J. Amditis, PhD, Research and Development Director of ICCS, has a doctorate from the Electrical and Computer Engineering School of the National Technical University of Athens. In 2002, he founded the I-SENSE Research Group. He is active in ICT for environmental and security applications, critical infrastructure, secured smart cities, risk and crisis management, sensors and information fusion, and so on. He has been the scientist responsible for more than 150 research projects, currently coordinating the projects INGENIOUS, NIGHTINGALE, EIFFEL, 5G-IANA, CitySCAPE, eCharge4Drivers, DIONE, HYPERION, CYBER-MAR, and RESIST, and he has also been the coordinator of the IN-PREP Project.

Funda Atun is Assistant Professor at the University of Twente in the Netherlands (Department of Urban and Regional Planning and Geo-Information Management). She is an urban planner with a track record in the field of societal resilience, disaster risk reduction, and climate change adaptation. She has been teaching in Italy, Switzerland, and the Netherlands and has been involved in many research projects revolving around, *inter alia*, disaster and migration and disaster risk reduction and climate change.

Karolina Bralewska has a PhD in environmental engineering and a master's in civil safety engineering and is a specialist in the area of combining the results of environmental research on indoor and atmospheric air quality and water and soil pollution with risk assessment methods developed as part of safety engineering. She is an academic, working at the Safety Engineering Institute at the Main School of Fire Service (SGSP) in Warsaw, Poland. She is an author of more than 20 publications at the international level and a contractor of scientific projects in the area of safety, crisis management, and methods of using information systems in safety & security.

Claas Cassens is a student in the information systems master's programme at the University of Münster. He previously received a bachelor's degree in information systems at the same university. He worked as a graduate student assistant at the Chair of Information Systems and Supply Chain Management at the University of Münster. His primary fields of research are software engineering and simulation modelling in the context of humanitarian supply chains, with a focus on agent-based modelling in disaster evacuation.

Carsten Dalaff is working as Senior Project Manager and Researcher within the institute for transportation systems in the German Aerospace Center (DLR). For many years, he has been in charge of the topic transportation systems in various responsibilities. During this time, he successfully led various national and international projects as project coordinator and manager. In recent years, he focused his work on traffic management in crisis situations and major events. In this book, he brings his experience and knowledge from the role of DRIVER+ trial owner.

Anna Foks-Ryznar, Deputy Chief of the Crisis Information Centre at the Space Research Centre of the Polish Academy of Sciences (Warsaw, Poland). She holds a PhD in technical sciences and specialises in the development and evaluation of new technologies for crisis management, rescue services, and humanitarian aid. Her recent research includes the assessment of new technologies using the DRIVER+ methodology published in *Trial as a pragmatic and systematic approach for assessing new solutions in crisis management and rescue operations* (2019).

Max Friedrich holds an MSc in psychology from the Technische Universität Braunschweig, Germany. Currently he is working as a researcher at the Institute of Flight Guidance of the German Aerospace Center (DLR) and acts as coordinator of ResponDrone. His research focusses on human-machine interface design for supervisory control of multiple unmanned aircraft systems. In 2018, he spent a semester abroad at NASA Langley Research Center as a visiting researcher.

Veronika Gstaiger studied geography as well as disaster management and risk governance and has been working as a research assistant at DLR since 2008. In addition to research work on automated image analysis in the context of national and international projects related to disaster management, she is involved in the development of aerial image acquisitions systems for disaster management and coordinates the planning and operation of research flight campaigns.

Jonathan Hall is a fire chief, a public sector executive director, and an emergency/disaster professional, specialising in national and international civil protection arrangements. He is the founder of the Resilience Advisors Network (RAN), working worldwide to improve emergency and disaster preparedness. A qualified expert in the European Union Civil Protection Mechanism (UCPM), he now manages an immediate network of crisis and disaster experts and provides a broader network for more than 1,500 European crisis and innovation practitioners supporting organisations, such as the UN, the EC, and humanitarian NGOs in research projects.

Denis Havlik is a researcher at the Safety and Security Department of the AIT Austrian Institute of Technology GmbH. His main field of expertise is in design and conceptualisation of sensor- and data-centric decision support and knowledge management systems for crisis management, climate change adaptation, and climate change mitigation domains. Denis is the principal architect of the DRIVER+ portfolio of solutions (https://pos.driver-project.eu/en).

Bernd Hellingrath serves as Chair for Information Systems and Supply Chain Management at the University of Münster, Germany, and he is Academic Director of the Competence Center for Crisis Management at the European Research Center for Information Systems (ERCIS). He published over 200 research papers and is continuously involved as track chair and reviewer in various conferences and programs. Research interests lie in the area of supporting logistic operations in crisis situations with information systems.

Martijn Hendriks, Software Development Manager at Merlin Software BV, holds an MSc in media and knowledge engineering from Delft University of Technology (NL). Based in the Netherlands, he has experience in both graphics and web-based software development in service of crisis and emergency management. Within the DRIVER+ project, Martijn was coordinator of the technical staff for the trial.

Dražen Ignjatović is a researcher in the field of digital safety and security at the Austrian Institute of Technology. His area of expertise is the design and development of software solutions for data-centric decision support and information management, with a focus on crisis and pandemic management and on the military and police sectors. Ignjatović's recent works include *Applying data mining techniques in the context of social media to improve situational awareness at large-scale events* (2021) and *Use of social media and open-source data to enhance situational awareness in the Austrian crisis and disaster management* (2021).

Konstanze Lechner holds a PhD in physical geography. Since 2008, she has been a scientific employee at the German Aerospace Center for the Center for Satellite-Based Crisis Information. As Senior Researcher, she has been coordinating national and international research projects on the use of earth observation data for crisis management, humanitarian relief and in the field of civil security. Main topics are the earth-observation-based monitoring of crisis situations and linking research with operations through close interaction and cooperation with users. Since 2018, she has been deputy head of DLR's Center for Satellite-Based Crisis Information.

Joonas Lieb graduated with a master's degree in aerospace engineering (2017) from the RWTH in Aachen, Germany. Since 2017, he has been working as a scientific researcher for the German Aerospace Center (DLR) in Braunschweig in the UAS group of the Institute of Flight Guidance. Currently, he is working in a variety of international projects for DLR, including the integration of unmanned aircraft systems into the existing airspace. In 2019, Joonas Lieb obtained a drone pilot license for Germany and Italy.

Nina Merkle received BSc and MSc degrees in mathematics from the University of Applied Science Regensburg, Germany, in 2011 and 2013, respectively. In 2018, she received the PhD degree in computer science and geoinformatics from the University of Osnabrück, Germany. Since 2014, she has been working as a research associate with the Remote Sensing Technology Institute (IMF), German Aerospace Center (DLR), Oberpfaffenhofen, Germany. Her current research interest includes the automatic information extraction from of aerial and satellite data for disaster management and support of humanitarian organisations after natural disasters.

Panagiotis Michalis, BEng (Hons), MSc, PhD, is Senior Researcher at the Institute of Communication and Computer Systems. He has significant international experience in the application of novel technological solutions at industrial level to assess critical infrastructure. He has been appointed as an external evaluator of EU projects and a reviewer in journals and has delivered invited talks to universities, companies, and media. Panagiotis is currently lead author of more than 80 publications in journals and international conferences. His research interests include the digital transformation of systems and the development and application of ICT solutions to enhance disaster risk management and infrastructure performance.

Michael Middelhoff is a member of the Competence Center for Crisis Management at the European Research Center for Information Systems and a research assistant at the Chair for Information Systems and Supply Chain Management at the University of Münster. His research activities focus on the design and implementation of information systems in organisational environments to provide resilience against malicious actions and crises. He is an expert on enterprise architecture and business process management and their use to implement security. He is further active in the assessment of solutions for crisis and pandemic management in Europe.

Fay Misichroni, innovation engineer at I-SENSE Group of ICCS and CIBOS Innovations. From 2006 to 2010 she worked as a telecommunication engineer at Siemens. Since 2010, she has been a researcher at ICCS, involved in several FP7 and H2020 European Research Projects as a software developer and technical manager. Since 2021, she has also been an innovation engineer at CIBOS Innovations, focusing on turning high TRL research prototypes to industrial products. Her research interests include, among others, interoperability, standardisation, research data management, real-time data manipulation and transportation, statistical and mechanistic modelling and simulations, telecommunication encryption, and multilayer software architecture design.

Georg Neubauer is Senior Researcher at the Safety and Security Department and Deputy Head of the Competence Unit Cooperative Digital Technologies at the AIT Austrian Institute of Technology GmbH. His main field of expertise encompasses data-centric decision support and knowledge management systems for crisis and pandemic management and interoperability of systems with a focus on the syntactic and semantic levels. Moreover, he is involved in national and international standardisation and is a lecturer at the Technical University of Vienna.

Christian Niermann has a master's degree in human factors from the Technical University of Berlin. He has worked as a scientist at the German Aerospace Center (DLR) in the Pilot Assistance Department since 2014. In the DRIVER+ project, he was the DLR project manager and was primarily responsible for the flights in Trial Austria. His own research focuses on the human-centred design of software in helicopter cockpits and the evaluation of single-pilot operations in airliner operations, which he leads in a variety of projects. He is particularly interested in spatial audio for guiding pilots in highly critical situations.

Paweł Ogrodnik is Associate Professor of Technical Sciences in the discipline of environmental engineering, mining, and energy working at the Main School of Fire Service (SGSP) in Warsaw, Poland. His scientific interest covers issues related to the circular economy as well as safety and security. His recent scientific publications concern the use of waste materials in concrete production (2021), personal protection equipment (2021), and problems related to the fire safety of photovoltaic farms (2022).

Eleftherios Ouzounoglou, PhD, is Senior Researcher at the Institute of Communication and Computer Systems, holding a BSc in informatics and telecommunications and an MSc in Bioinformatics. He has obtained a PhD degree from the National Technical University of Athens in 2018. His research interests include, among others, machine learning, software development for desktop and mobile devices, data and communications security, systems biology, and multiscale cancer modelling. He has participated in several EC-funded research projects as a software developer, bioinformatician, and project manager in the secure societies and health domain.

Camilo Palacio Ramirez is a crisis and disaster manager with over 15 years of both academic and practical experience in emergency services and risk management work. He has a Master's degree in risk and emergency management and in the last years has worked at the department of national disaster management and research of the Austrian Red Cross where he is performing as a project manager for different European projects related to disaster management.

Leonidas Perlepes, PhD, R&D Project Manager, has been involved as a project manager in a number of commercial and EU-funded projects, focusing on the public safety domain and the systems that could facilitate the related operations, such as incident management and command and control systems. He is a co-author of more than 20 papers in international journals and conferences. His research interests include command and control systems, GIS systems, wireless sensor networks, and embedded systems.

Niclas Rotering is an information systems master student at the University of Münster where he priorly received his bachelor's degree in information systems. He currently writes his master thesis about combining simulation and optimisation methods for blood supply chains and works as a graduate student assistant at the Chair for Information Systems and Supply Chain Management. His primary fields are simulation modelling and implementation for various scenarios of the humanitarian supply chain, including disaster evacuation and blood supply chains.

Alexander Scharnweber graduated from Technical University Braunschweig in 2005 with a diploma in aerospace engineering. Since 2006, he has been employed as a researcher with the Institute of Flight Guidance at the German Aerospace Center (DLR) in Braunschweig. His current project and research activities include airspace integration for unmanned aircraft systems (UAS) use in crisis management, as well as concept and system development for next-generation cockpits focusing on single-pilot operations.

Elisa Schröter holds an MSc in geomatics from the HafenCity University Hamburg, Germany. From 2018 to 2022, she has been a research associate at the German Aerospace Center (DLR) for the Center of Satellite-Based Crisis Information (ZKI), where she was engaged in research activities on the application of emerging methods in remote sensing for civil security and crisis management, besides working as a cartographer and GIS analyst in ZKI's rapid mapping service. Being an advocate of user-centred design, her interest includes the development of new 3D-mapping applications for the benefit of civil security and the public sector.

Gunnar Schwoch has been working as a scientist at the German Aerospace Center (DLR) in the Institute of Flight Guidance since 2008. The main topics of his research activities are the integration of unmanned aircraft into the airspace and the deployment of unmanned aircraft and drones in crisis situations to enable real-time situational awareness. In DRIVER+, he acted as a solution coordinator for the trial in the Netherlands and as a solution provider and operator of the ground control station during the trial in Austria.

Evangelos Sdongos, Managing Director of ASTRIAL GmbH, specialised in command and control systems for public safety agencies. He holds an MEng in electrical and computer engineering and an MBA. He has extensive experience in international, EU, and national security projects. He is active in the field of security and defence under several domains focusing on uptake of novel technologies. He is chairing in various crisis management and security-related conferences (e.g. ISCRAM) and working groups of standardisation bodies and associations (e.g. CEN-CENELEC) and has published more than 30 articles in book chapters, journals, and conference proceedings.

Marcin Smolarkiewicz is Associate Professor of Safety at the Main School of Fire Service in Warsaw, Poland. His research areas include crisis management, risk analysis, the use of matrix methods in public safety, CBRN threat analysis, and methods for validation of innovation for crisis management. His recent papers include *Exercising in a radioactive environment – a case study from CBRNE exercise in Chernobyl Exclusion Zone* (2019), *Effect of enhanced supply air filtration in buildings on protecting citizens from environmental radioactive particles* (2020), and *Testing selected personal protection items of firefighters in combined conditions of mechanical loads and temperatures occurring during gas LNG leaks* (2021).

Todor Tagarev is Professor at the Institute of Information and Communication Technologies of the Bulgarian Academy of Sciences and Head of its Centre for Security and Defence Management. He is an experienced security and defence policy-maker with sound theoretical knowledge and background in cybernetics, complexity, and security studies. Professor Tagarev is the author of "A Taxonomy of Crisis Management Functions" (with Valeri Ratchev, 2020), a co-editor of *Digital Transformation, Cyber Security and Resilience of Modern Societies* (2021), and one of the lead authors of the European Union's volume *Science for Disaster Risk Management 2020: Acting Today, Protecting Tomorrow.*

Steven van Campen, a senior crisis management trainer, consultant, and exercise manager, holds an MSc in aerospace engineering and in industrial design engineering from the Delft University of Technology. He acquired several certifications as senior incident command instructor. Based in the Netherlands, he managed several Dutch, international, and European research and development projects, implemented simulation and training innovations, and coordinated multiple joint-agency exercises and train-the-trainer courses. Steven was involved in all DRIVER+ trials for scenario creation and role-player coordination, and he led the development of the training module. He is also passionate about cooking, theatre, woodworking, and travelling.

Richard van Oorschot holds an MSc in theoretical biology. After graduating, he left academics to go into business, where he has performed a variety of roles in technical management. He is Technical Coordinator of Respondrone, after having co-written the proposal.

Cor-Jan Vermeulen, Senior Advisor at HKV Consultants on water and safety. He is a specialist in water management and crisis/disaster management related to floods, with more than 20 years of experience. He specialised in the operational control of water systems, often related to implementation of information systems with decision support or automated control. Since 2005, he has been active in the field of crisis management, especially on floods and evacuation. This involves flood and risk maps for different sources (pluvial, river, coastal), flood preparedness plans, and training professionals in flood crisis management and exercises on flood crisis.

Erik Vullings is Senior System Integrator at TNO, the largest research organisation in the Netherlands. He holds a PhD in electrical engineering and an MSc in mechanical engineering from Delft University of Technology. His experiences include project and programme management of large research projects for Philips, in Australia, and at TNO, and his main passion is in finding elegant solutions for complex technological challenges. In DRIVER+, he was the technical lead in developing the testbed and was responsible for the realised architecture.

Bernard Wiśniewski, Professor of Social Sciences, is Director of the Centre for Research Methodology in Security Sciences at the WSB University in Dąbrowa Górnicza (Poland). He also works at the Main School of Fire Service (SGSP) in Warsaw in the Department of Safety Engineering. He is a retired officer of the Polish Army. He has published around 300 scientific and popular science papers, published in Poland and several other European countries. He conducts research in the fields of security theory, state security, internal security, military security, crisis management, and security research methodology.

ACKNOWLEDGEMENTS

This book would not have seen the light without the institutions, resources, and people that made it possible. DRIVER+ received funding from the European Union's Seventh Framework Programme for Research, Technological Development and Demonstration under Grant Agreement (GA) Nr. #607798. Thanks to this EU research funding programme, a wide variety of stakeholders (some of whom are the authors of *Innovation in Crisis Management*) worked and grew together towards the success story that DRIVER+ has become. Our heartfelt thanks also go out to the DRIVER+ project officer, Guillaume Lapeyre, who gave feedback and advice that impacted enormously on our work and stirred it in the right direction.

We are grateful for the engagement and the support of all crisis management practitioners who were involved in the DRIVER+ trials.

We would like to thank all authors who contributed to this book. They all worked under tight deadlines, with dedication and in a spirit of creative collaboration.

Sebastian Henke (University of Munster) deserves a special thanks for supporting this book throughout the process: his work was key for us.

FOREWORD

Dear reader,

As the former technical coordinator of the DRIVER+ project, it is a great pleasure to welcome you to this book about innovation in crisis management. The book is largely based on the experiences we have gained during our exciting DRIVER+ journey, but luckily not exclusively. I am more than happy to see that our outputs have been adopted and further developed by many other projects and organisations, who are sharing their insights in this book as well.

We are confronted with natural disasters, which are more frequent and/or intense due to climate change, and increasing cascading effects. Terrorist attacks are a growing concern for the protection of public spaces. The COVID-19 crisis taught us about the devastating impact of a worldwide epidemic. Moreover, with the rise of hybrid threats, new types of crises are emerging, affecting the physical and cyber domains simultaneously, leading to less clarity about the origin, intention, and consequences of crisis situations and how to best face these. It is evident there is a constant need for society at large to adapt and to prepare, respond, and recover.

However, it is not a secret if I tell you that innovation in crisis management can be rather difficult and complex. This is in particular the case for practitioner organisations. Innovation takes time, and practitioners do not always have time as they are focused on today's crises. It is not clear how they can realistically assess the value of new solutions, especially when these are not yet integrated in current operations. Access to knowledge and research results is not easy, let alone how to best implement these. And lastly, the innovation ecosystem involves many different stakeholders, and it is simply not easy getting connected with these other stakeholders.

Within DRIVER+, we therefore explicitly put the practitioners in the driver seat. We made a promise to develop products enabling the crisis management practitioners to take the lead in assessing and validating solutions that are addressing their operational needs. DRIVER+ had the ambition to introduce a new practitioner-centred approach to boost capability development and innovation in crisis management, facilitating its use beyond the project. The trials we have conducted each involved between 44 and 155 participants, coming from 8 to 18 countries. Throughout the project, we reached out to and actively involved 137 practitioner organisations, 64 solution providers, 41 policy representatives, 36 projects, and 19 platforms/networks. The centre of expertise network has grown to 13 full and associate members from throughout Europe. Over 25 EC-funded and national projects and multiple practitioner organisations throughout Europe apply and further develop the DRIVER+ outcomes.

I am extremely proud that DRIVER+ has been officially labelled by REA as a success story and was selected as one of the ten most impactful innovation breakthrough projects of 2020 funded by the EC. And I am very proud this book is published. We have learnt a lot, and we are very pleased to share this body of expertise and experiences with the crisis management community. Nevertheless, this book should not be regarded as the final stage of the DRIVER+ journey but merely as a next step. New innovations need to be constantly developed and implemented by both practitioners, policy-makers, researchers, solution providers, and citizens. It requires a whole of community approach to face the current and emerging challenges. The DRIVER+ methods and tools boost this joint capability development and innovation in crisis management, increasing Europe's resilience. I sincerely hope this book inspires many to continue the journey.

Enjoy reading the book!

Marcel van Berlo, TNO, the Netherlands

ABBREVIATIONS

3C	command, control, and communication
AAR	after-action review
AC/CT	action centre / crisis team
AC/CT	action team / crisis team
AI	artificial intelligence
ANC	airborne network controller
BRK	Bavarian Red Cross
CAP	Common Alerting Protocol
CD&E	concept development and experimentation
CIS	common information space
CM	crisis management
CMINE	Crisis Management Innovation Network Europe
COP	common operational picture
COVID-19	coronavirus disease 2019
CSS	common simulation space
DED	data exchange diagram
DLR	German Aerospace Center
DR1/DR2	dry run 1 / dry run 2
DRB	disaster resistance business
DRIVER+	Driving Innovation in Crisis Management for European Resilience project
DRS	disaster-resilient society
DSS	decision support system
EDA	European Defence Agency
EMT	emergency medical technician
ERCC	Emergency Response Coordination Centre
EUCPT	European Union Civil Protection Team

FFFVM	Forest Fire Fighting with Vehicles Module
FP7	Seventh Framework Programme
FSX	full-scale exercise
FTP	file transfer protocol
GCS	ground control stations
GIS	geographic information system
IPCC	Intergovernmental Panel on Climate Change
ISOK	the Polish IT system dedicated to facilitate natural disasters risk analyses
JRC	European Commission's Joint Research Center
KPI	key performance indicator
LCMS	the Dutch national crisis management system
LEMA	Local Emergency Management Authority
OST	observer support tool
PCR	payload communications relay
PFA	psychological first aid
PIUEM	Performance Indicators Uncertainty Estimation Method
PoS	portfolio of solutions
REA	Research Executive Agency
REI	result-effort indicator
RESILOC	Resilient Europe and Societies by Innovating Local Communities
ROT	regional operational team
RPAS	remotely piloted aircraft systems
SitRep	situational report
SLR	systematic literature review
SME	small- and medium-sized enterprises
SotA	state of the art
SRH	Safety Region Haaglanden
TGM	trial guidance methodology
TGT	trial guidance tool
TIM	trial integration meeting
TMM	traffic and mission management system
TMT	trial management tool
TNO	Netherlands Organization for Applied Scientific Research
TRL	technology readiness level
TTX	tabletop exercise
UAV	unmanned aerial vehicle
UCPM	Union Civil Protection Mechanism
WHO	World Health Organisation
WWU	University of Muenster, Germany
XVR	XVR simulation, the Netherlands
ZKI	Center for Satellite-Based Crisis Information, Germany

INTRODUCTION

Chiara Fonio, Adam Widera, and Tomasz Zwęgliński

Over the last years, disaster management continuously faces an enormous challenge of increasing number and severity of natural and man-made disasters with their broad and deep negative impact on widely understood sustainability of a human being (Aifadopoulou et al., 2018; Halkos et al., 2015; Sholoiko, 2017; Telizhenko et al., 2017; Yezer, 2010). Consequently, there is a need to increase capability building upon new techniques, technologies, and other new approaches. Current and future challenges require a development and uptake of innovative solutions that are tailored to meet operational needs of practitioners dealing with disaster management due to the severe impact of disasters of different origin (Zwęgliński, 2020). Efficiency of decisions taken prior to a disaster, or in its very initial phase, are more and more dependent on supportive technologies which are used by disaster managers (Izumi et al., 2019; Shaw et al., 2016).

The nature and impact of natural hazards and man-made disasters have changed dramatically while societies and citizens rely increasingly more on interconnected infrastructures. Societal and technological evolutions affect people's ability to cope with crises and impact our vulnerabilities. According to the EM-DAT database, Europe faced 1,170 natural hazards and 506 technological crises between 2000 and 2022. Climate change is exacerbating the intensity of disasters. The latest IPCC report (2022) states that even temporarily exceeding global warming with 1.5°C will result in severe risks and impacts for society. Recent floodings in Europe and wildfires worldwide are induced by climate change (e.g. WWA, 2021). The impact of human-induced climate change on fire weather has already emerged above the range of natural variability (Jolly et al., 2015). These changes have created acute challenges for crisis management in Europe. As a result, a market for innovative and integrative solutions has originated. This is a rapidly developing global market with a projected value of USD 122.94 billion by 2023 (Marketsandmarkets, 2018).

DOI: 10.4324/9781003256977-1

In this challenging context, this book provides key insights into how potentially innovative solutions can be assessed by crisis management organisations before committing resources. Drawing on the rich experiences and data collected in one of biggest and most ambitious crisis management project ever funded in Europe (DRIVER+,[1] Driving Innovation in Crisis Management for European Resilience, 2014–2020), the book provides an in-depth overview of the pan-European effort that was carried out in the frame of the project by a wide variety of stakeholders to assess innovation. Researchers, practitioners, solution providers, and IT developers sit at the same table to measure the impact of solutions in realistic settings through crisis management (CM) trials. This resulted in an impressive, instructive, and intense journey of learning, experiences, and growth, which is now captured in all its colourful and complex nuances in the pages of this book.

This edited collection is structured along multiple perspectives. The first and second parts describe the methodological and technical approaches to assess CM innovations. In the third and fourth parts several dedicated applications and lessons learnt are presented and discussed. The fifth part, the annex, provides insights into specific issues: terminology, literature, and additional information.

- A *methodological perspective*. Part 1 outlines the main rationale behind the development of the trial guidance methodology (TGM) designed in the project and describes the backbone of the methodological approach by focusing on what is needed to carry out systematic and robust trials. Furthermore, Part 1 focuses on why trials are suitable to measure the extent to which solutions (e.g. training, apps, common operational picture tools, simulation tools, etc.) can make a difference within a specific context. The TGM is a practitioner-centred approach which was designed with and for crisis management practitioners. The co-creative and iterative nature of the TGM was essential to improve it in the course of the project and beyond since it was applied and adapted in other contexts and projects. The first part of the book offers all the background knowledge needed to understand the other perspectives described in the next parts, and it is the "conceptual glue" that keeps the core elements of the edited collection together.
- A *technical perspective*. Part 2 revolves around the technical infrastructure and the toolset to connect solutions (e.g. current operational systems with new solutions) and simulate crisis environments in a realistic way. It provides essential information on all the technical components (as well as supporting tools and services) of the so-called DRIVER+ testbed, conceived to facilitate trials through a set of open-source software tools. The latter enable the exchange and the collection of data and provide an immersive simulation environment in which both trials and CM exercises can be carried out. Furthermore, Part 2 describes the online management and inference toolset consisting of a repository of CM solutions (the portfolio of solutions, or PoS), an inventory of CM gaps and the trial guidance tool (TGT) which supports the application of the TGM.

- A *practical perspective*. Part 3 dives deep into the DRIVER+ trials through an emphasis on the trials themselves by providing examples, analyses, results of the trials, and considerations on simulation and situational awareness tools and methods. It is worth considering that the DRIVER+ trials are the main sources of inspiration for this part, and the same trial is sometimes referred to in more than one article (e.g. the trial hosted by the Safety Region Haaglanden, SRH, in the Netherlands). The reason behind it is that experiences and the amount of data collected were, respectively, so rich and large that allowed for many reflections. These range from descriptions and evaluations of all the solutions involved in trials to the implementation of new approaches drawing on the TGM, from the use of flood modelling in disasters to practitioner-driven simulation approaches for managing mass evacuation in urban areas, and from the added value of 3D modelling and orthophotomapping to support decision-making processes to the added value of airborne camera systems to obtain high-resolution images of damaged areas.
- A *glimpse into the future*. Part 4 describes how the DRIVER+ outputs have been applied in other projects and hence provides the basis for a paradigm shift in assessing innovation in the CM domain. As shown, the outputs, specifically the TGM, have the potential to go well beyond the successful yet limited timeframe of a project since they have already been adapted and applied in other projects and by different stakeholders. This part tackles the benefits but also the potential for improvements and the results in applying the TGM. Moreover, it outlines how the methodological approach is flexible and scalable enough to be adapted: for example, it is currently being used in a project focused on pandemic management.
- *Annex*. Part 5 offers additional resources to the readers, both to better understand the book (the glossary) and facilitate the reading (the list of abbreviations) and to explore further literature (references to systematic literature review, described in Part 1) and information (technical and non-technical).

Overall, the book takes the readers through a journey that did not end at the end of DRIVER+. On the contrary, the paradigm shift mentioned earlier has just started: the findings described in the book can be considered as catalysts for changing the way in which innovation is conceived, understood, and measured in the crisis management domain.

Note

1 www.driver-project.eu, https://cordis.europa.eu/project/id/607798.

References

Aifadopoulou, G., Chaniotakis, E., Stamos, I., Mamarikas, S., & Mitsakis, E. (2018). An intelligent decision support system for managing natural and man-made disasters. *International Journal of Decision Support Systems*, 3(1–2), 91–105. https://doi.org/10.1504/IJDSS.2018.094263.

EM-DAT: The International Disaster Database (https://emdat.be). *Hosted by the Centre for Research on the Epidemiology of Disasters (CRED) within the Université Catholique de Louvain (UCLouvain)*, Belgium.

Halkos, G., Managi, S., & Tzeremes, N. G. (2015). The effect of natural and man-made disasters on countries' production efficiency. *Journal of Economic Structures*, 4(10). https://doi.org/10.1186/s40008-015-0019-2.

Intergovernmental Panel on Climate Change/IPCC. (2022). *Sixth Assessment Report – Technical Summary*. www.ipcc.ch/report/ar6/wg2/downloads/report/IPCC_AR6_WGII_FinalDraft_TechnicalSummary.pdf.

Izumi, T., Shaw, R., Djalante, R., Ishiwatari, M., & Komino, T. (2019). Disaster risk reduction and innovations. *Progress in Disaster Science*, 2. https://doi.org/10.1016/j.pdisas.2019.100033.

Jolly, W. M., Cochrane, M. A., Freeborn, P. H., Holden, Z. A., Brown, T. J., Williamson, G. J., & Bowman, D. M. J. S. (2015). Climate-induced variations in global wildfire danger from 1979 to 2013. *Nature Communications*, 6, 7537. https://doi.org/10.1038/ncomms8537.

Marketsandmarkets. (2018). *Incident and Emergency Management Market – Global Forecast to 2023*. www.marketsandmarkets.com/Market-Reports/incident-emergency-management-market1280.html.

Shaw, R., Izumi, T., & Shi, P. (2016). Perspectives of science and technology in disaster risk reduction of Asia. *International Journal of Disaster Risk Science*, 7, 329–342. https://doi.org/10.1007/s13753-016-0104-7.

Sholoiko, A. (2017). Financing losses from natural and man-made disasters by use of crowdfunding. *Investment Management and Financial Innovations*, 14(2–1), 218–225. https://doi.org/10.21511/imfi.14(2-1).2017.07.

Telizhenko, O., Mashyna, Y., & Opanasyuk, Y. (2017). Organizational and economic basis of natural and man-made disasters consequences management. *Journal of Environmental Management and Tourism*, 8(1), 270–278. https://doi.org/10.14505//jemt.v8.1(17).25.

World Weather Attribution/WWA. (2021). *Rapid Attribution of Heavy Rainfall Events Leading to the Severe Flooding in Western Europe during July 2021*. www.worldweatherattribution.org/wp-content/uploads/Scientific-report-Western-Europe-floods-2021-attribution.pdf.

Yezer, A. M. (2010). Expectations and unexpected consequences of public policy toward natural and man-made disasters. In W. Kern (Ed.), *The Economics of Natural and Unnatural Disasters* (pp. 39–64). Kalamazoo, MI: W.E. Upjohn Institute for Employment Research. https://doi.org/10.17848/9781441678812.ch3.

Zwęgliński, T. (2020). The use of drones in disaster aerial needs reconnaissance and damage assessment – Three-dimensional modeling and orthophoto map study. *Sustainability*, 12(15), 6080. https://doi.org/10.3390/su12156080.

PART 1

Current and New Methodological Approaches to Assess Innovation in CM

This part revolves around both the foundations of the trial guidance methodology (TGM) and the TGM itself that was applied in the DRIVER+ trials and beyond. It provides an in-depth overview of the rationale behind the TGM by focusing on the points of departures for developing the TGM, as well as the systematic literature review that was carried out in the frame of the project to design a new methodological approach for trialling solutions in the CM domain (Chapter 1) In doing so, this part has the ambition of setting the conceptual boundaries of the trial guidance methodology. Furthermore, it takes the readers by the hand, through a step-by-step explanation of the trial guidance methodology. Chapter 2 describes the TGM anatomy so that phases, steps, tools, and more general considerations on who should be involved, how and when in a trial journey are clarified. This part paves the ground to the rest of the book: the TGM, in fact, enables trialling thanks to its systematic and robust approach. Put simply, it is an essential reading to understand why the technical infrastructure (Part 2) was developed in a certain way and to be familiar with the trial-like setting and concepts before taking deep dives into them (Part 3).

DOI: 10.4324/9781003256977-2

1

MEASURING INNOVATION

The Current State of the Art

Chiara Fonio, Adam Widera, and Funda Atun

Introduction

This chapter provides an overview of the foundations of the methodological approach which guided all trials in the frame of the DRIVER+ project and has been implemented in other contexts, as outlined in Part 4 of this book. The development of the trial guidance methodology was far from being easy. It required years of trial-and-error approaches, brainstorming and lessons learnt based on the application of the concept development and experimentation (CD&E), which resulted into being an excellent source of inspiration, as will be shown, but needed to be adjusted to serve the purposes of the trials carried out in DRIVER+ (Labbé et al., 2006; Pikner, 2015). We had a point of departure (the CD&E) which helped us to design the first experiments, but we acknowledged that we had to adapt it to ensure a more realistic assessment of potentially innovative solutions in the context of crisis management organisations. This is why, in order to craft a more suitable methodological approach, a systematic literature review was carried out. The findings were key to defining the conceptual boundaries and the core principles of the TGM. In the following pages, we offer a glimpse into the long journey that led us to develop the TGM.

The Foundations of the Trial Guidance Methodology (TGM)

The design process that led to the development of the TGM implied exploring a specific approach that was pioneered in military defence research: the CD&E (Labbé et al., 2006). The latter was selected for some features which are considered relevant also for carrying out trials:

- In the CD&E, new concepts and ideas are evaluated in realistic settings and before spending resources or before organisational changes are implemented.

DOI: 10.4324/9781003256977-3

- The CD&E is a creative process, meaning that concepts are developed through brainstorming, evaluation, and analyses combined with inputs from experiments.
- The starting point is the identification of a capability gap or a new solution that has some potentials to deal with challenges and gaps within an organisation.
- The concept matures until they can be trialled in an operational setting. Ideally, it would be assessed in experiments so that evidence-based recommendations on the implementation readiness level of concepts or solutions can be provided.

If, on the one hand, this approach can be extremely useful to evaluate solutions, on the other, two main shortcomings were identified. It is worth considering that the shortcomings refer to the application of the CD&E for carrying out trials, not the approach in general:

1. The crisis management domain differs from the military area as roles and responsibilities are defined in a less "rigid" way, meaning that there is not a command-based decision chain and the reality of CM organisations are more complex, uncertain, and dynamic.
2. Experiments cannot serve the purpose of assessing potentially innovative solutions since they are designed, planned, and carried out in "laboratory-like" settings which would not allow for researching and analyses that must take into account the reality of practitioners. Those realities, their daily work, and the way in which gaps, challenges, and needs are perceived and must be addressed are dynamic and require a more realistic – yet robust – methodological process.

For these reasons, it was decided to carry out a systematic literature review. This led us to explore a bulk of knowledge which provided the basis for the TGM.

Applied Systematic Literature Review (SLR) Methodology

Next to the previously mentioned TGM roots, the consortium decided to execute a systematic literature review (SLR) focused on trial-like events conducted in the past. The SLR approach was chosen to reduce the bias of study selection, data extraction, and presentation and ensure a higher quality and being reproducible because of a systematic and well-documented procedure. There are guides for most research domains from medicine (Mulrow, 1987, 1994; Higgins & Green, 2011) to social sciences (Petticrew & Roberts, 2006) to software engineering (Kitchenham, 2004; Kitchenham & Charters, 2007). From different available SLR procedures, the approach presented by Thomé et al. (2016) was followed. The main advantage of this approach is the fact that it considers peer reviews in selecting and analysing the literature, and hence, a high quality of the SLR can be achieved. As there are no crisis-management-specific guides, the approach of Thomé et al. (2016) was chosen because the tasks and procedures of both are much alike.

In general, the SLR process according to Thomé et al. (2016) can be summarised as follows: In the first step, planning and formulating the problem, a review team needs to be set up, the scope needs to be clarified and set in context. Additionally, a protocol is set up that describes how the next steps are to be executed. This protocol is the main differentiation between conventional literature reviews and SLRs. Before the actual execution, the search and selection, data gathering, and quality evaluation, analysis and interpretation procedures are defined. For the search process, the databases, keywords, and time span are defined. The selection process is done in two iterations by at least two members of the review team. First, the abstract, title, and keywords are reviewed, and predefined selection criteria are applied. A criterion stands for a decision to include or exclude a study, so one "applies" the criteria, meaning the reviewer includes or excludes a study by choosing an appropriate criterion for the study. Second, the full text is reviewed and included/excluded based on the chosen criterion that fits the research paper best. Data must be extracted in data extraction forms (codebooks) defined within the protocol. The codebook is a representation or summary of the relevant and most valuable information from the (selected) existing literature. Next to meta information of each paper (e.g. title, authors), the codebooks contain the following sections:

- Experiment, exercise, simulation, or trial objectives: This section provides a description and objectives of any trial-like events conducted as part of the study (e.g. simulation, serious games, etc.).
- Research questions: This field aims to state the specific question or objective of the presented work, answering "What is the paper presenting?" or "What is the paper's contribution?" This was decided because the majority of reviewed papers have not included their objective in the form of a question statement.
- Experiment planning and deviations: This section describes how the presented example of an experiment was planned, including specific considerations, steps, or phases which were taken in order to conduct such an event.
- Research methods: This field covers the methodology followed by the authors for their research.
- Metrics and key performance indicators (KPIs): Here, the metrics, measures, or indicators are summarised.
- Data collection plan: In this field, all available information about how data was collected are summarised (e.g. sample size).
- Data analysis: This field explains how the way data was analysed and which specific procedures or standards were followed.
- Ethical procedures: This field provides information on the protocols or specific procedures for obtaining permission to make use of the data collected.

- Results: Here, key findings of the described events are summarised.
- Methodological lessons learnt: This section explains any methodological learnings derived from the experience of conducting the research.

There was also additional space for comments as suggested in multiple SLR guidelines. In the next paragraph, the application of the methodology to trial-like events in CM is presented.

Step 1: Planning and Formulating the Problem

The aim of the SLR was twofold: First, as the CD&E was deemed not comprehensive enough for the purposes of DRIVER+, an overview of existing approaches for conducting trial-like events in the past decade in the crisis management domain was considered a necessary starting point. Second, a solid and robust knowledge base for trials needed to be provided. More pragmatically, the expected contribution from the SLR is to support the design of a dedicated CM innovation evaluation approach through the following:

1. The analysis of the state of the art (SotA) concerning the applied of methods over the course of the past decade (What kind of methods were used? What kind of research questions were asked?).
2. Knowledge base that can be used to support CM staff to plan, execute, and evaluate trials.

The application context was narrowed to the crisis management domain in general and considered contributions from practitioners, researchers, and solution providers. The review team consisted of members from four different organisations: the German Aerospace Center (DLR), the European Commission's Joint Research Center (JRC), the Netherlands Organization for Applied Scientific Research (TNO), and the University of Muenster (WWU). The SLR's objective was to "analyse the state of the art (SotA) concerning the use of evaluation methods and approaches for CM innovations." Accordingly, the overarching SLR research question was formulated as follows: "How to design and evaluate a space for trialling socio-technical innovations for crisis management in a realistic and multi-stakeholder setting?"

In order to contextualise the research question, the PICOC (population, intervention, control, outcome, and application context) characteristics, introduced by Kitchenham (2004) and recommended by Thomé et al. (2016), were chosen.

In order to compile a search string supporting the SLR objectives, the applied search string structure, including the search terms and synonyms, is depicted in the following table.

The keywords were collected and discussed within the SLR team, having dedicated experience in CM research, covering different disciplines, such as security research, engineering, information systems, sociology, psychology, or logistics. Thus, a high quality of the applied search string was ensured.

TABLE 1.1 Applied PICOC Criteria

Characteristic	*Description*
Population	CM practitioners, CM researchers, policy-makers.
Intervention	Exploration of trial-like approaches, which evaluate socio-technical solutions in the crisis management domain.
Control	Concept development and experimentation (CD&E) approach; lessons learnt of the first phase of the demonstration project Driving Innovation in Crisis Management for European Resilience (DRIVER+); contribution of the multidisciplinary DRIVER+ consortium.
Outcome	1. Answer if there are other "holistic" approaches like CD&E. 2. Specific elements of existing approaches which even cover only a small set on how to trial and evaluate solutions. 3. Knowledge base.
Application context	Crisis management practitioners, researchers, and solution providers.

TABLE 1.2 Compilation of the Search String

Field of Interest	*Keywords for Search Query*
Application context	crisis management, emergency management, disaster relief, humanitarian operation, disaster management, disaster response
Functional description	simulation, serious game, exercise, game, test, trial, experiment, training
Research object	innovation, software, algorithm, decision support, tool, solution, process, organisation, partnership

Step 2: Literature Search

The search scope was limited to peer-reviewed journal papers to ensure a minimum degree of quality behind the respective articles. To enable the use of a software solution it was important that the data could be exported as a .ris file. This led to the following source list:

> "EBSCO," "Google Scholar," and "ScienceDirect" (The idea of using JSTOR had to be dropped, as this one was not able to handle the long search query that came up.)

By combining the keywords defined before the following search string was created:

> ("crisis management" OR "emergency management" OR "disaster relief" OR "humanitarian operation" OR "disaster management" OR "disas-

> ter response") AND ("simulation" OR "serious game" OR "exercise" OR "game" OR "test" OR "trial" OR "experiment" OR "training") AND ("innovation" OR "software" OR "algorithm" OR "decision support" OR "tool" OR "solution" OR "process" OR "organisation" OR "partnership")

The search results were included for the initial screening if they were peer-reviewed, in English and published between 2007 and 2017. This resulted in 2,934 results. First, 320 duplicate papers were deleted. Subsequently, an initial filter was applied. Keywords, titles, or abstracts had to include at least one of the following words:

> "assessment" OR "evaluation" OR "generalizability" OR "method" OR "methodology," "procedure" OR "qualitative" OR "reliability" OR "validity"

These filter terms form the very core of the project as it aims at assessing and evaluating innovation by using a defined methodology that is a qualitative, reliable, and valid process that leads to generalisable (but case-driven) results. This step reduced the number of possibly relevant papers to 948. After screening the abstract and the full text and manually deleting undetected duplicates, a total of 239 studies were included for the next steps. Studies were included if the authors conducted some sort of experiment (simulation, case study, tabletop exercise, etc.) included communication or coordination between different organisations in a crisis management context, covered an interdisciplinary approach for emergency preparedness, had a training component, concerned crisis management decision-making, or included the test of a socio-technical solution. Studies were excluded if it did not contain any of the previously mentioned subjects. Additional 21 studies had to be excluded, either because the full text could not be retrieved or deemed irrelevant, when reading the entire content of the paper. Figure 1.1 shows the steps as a flow diagram based on the PRISMA standard (according to Moher et al., 2009).

Step 3: Data Gathering

The data gathering was done by creating a codebook for each included paper that was introduced for the lessons learnt from experiments conducted in the initial project phase to ensure reproducibility and comparability. The initially applied filter was crucial to reduce the number of possibly relevant papers quickly and start manual screening. Here mainly sources from Google Scholar were omitted because they often did not have an abstract and/or due to the poor citation export possibilities provided by the site. Additionally, the filter also excluded too theoretical papers and/or papers that did not assess innovations in trial-like experiments. During the manual screening the reasons to exclude a study were most often caused by the semantic interpretation of a crisis (e.g. economic crisis, mental health crisis, etc.). Other excluded papers also often did not test solutions in a CM setting or tested a non-socio-technical solution (mainly medical or psychological interventions).

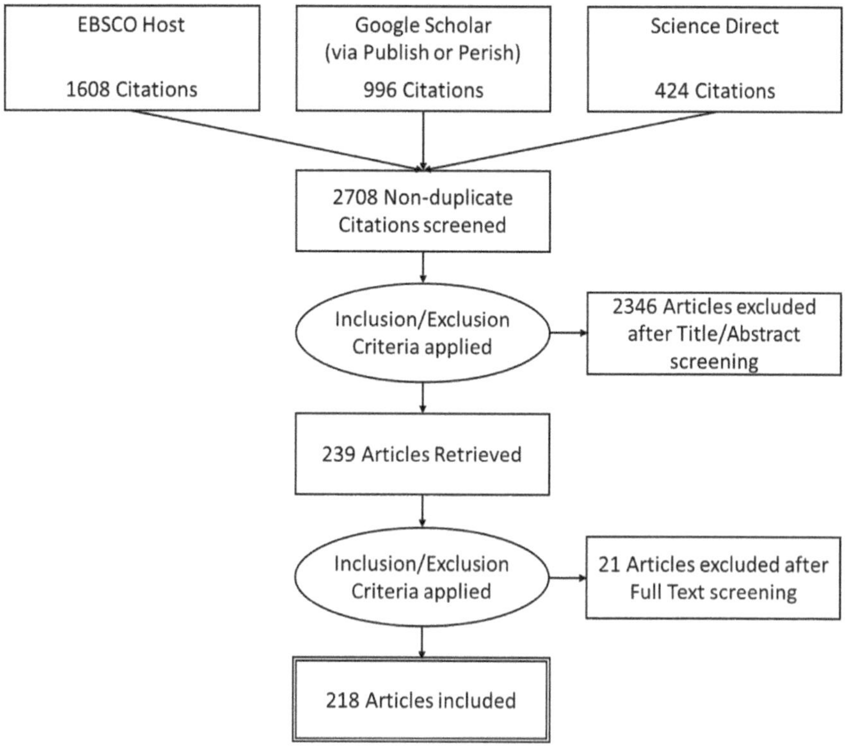

FIGURE 1.1 SLR Flow Diagram according to PRISMA.

Also, too specific and too theoretical, mathematical, or computational models were a major part of excluded studies. The review of the included papers was organised according to the following peer-review procedure.

Each paper was completely read and analysed and short summaries for each of these categories were saved in a codebook. The 218 relevant papers were split between the four involved organisations so that for each paper two codebooks would be created by two different team members. The results were then synthesised to one single codebook for each paper, which will be used as the foundation for the knowledge base.

Step 4: Quality Evaluation

One of the SLRs aims was to support the design of the TGM and to serve as a database of best practices on trial-like event methodologies. The latter objective requires a careful quality assurance of the usability of each potential best practice for TGM applicants. Therefore, only peer-reviewed studies were included; thus, the high publication standards served as quality assurance. Additionally, data gathering was done in two rounds, where first two SLR team members created decentralised

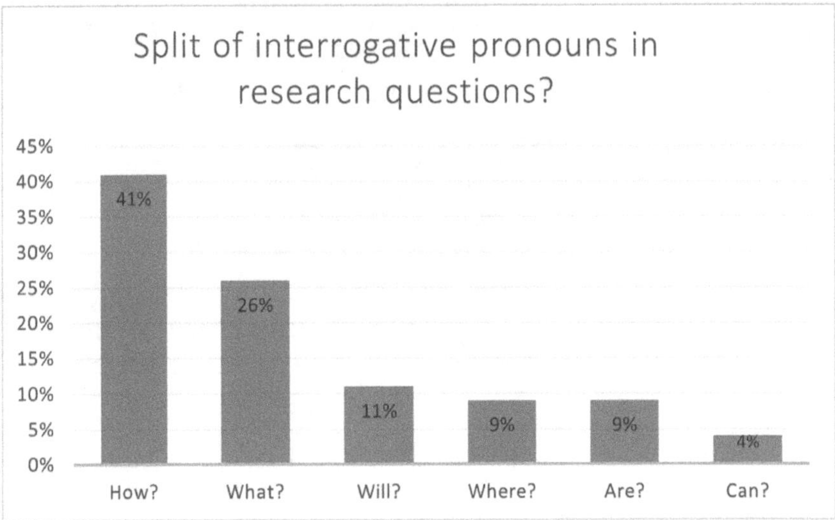

FIGURE 1.2 Data gathering procedure.

and independently a codebook, which was then synthesised in a second round. In consequence, the peer extraction of DRIVER+-relevant information on the best practices supported a certain inter-subjectivity of the created codebooks. This is not only relevant in order to double-check the codebooks in a proper way but also to support the reviewers in terms of the different scientific disciplines of the papers and the reviewers. This also reduced the risk of missing as much information as possible. Hence, the task to "follow up on missing information," as stated by Thomé et al. (2016), was not necessary.

Step 5: Data Analysis and Synthesis

A first data analysis was done by using basic descriptive statistics so that analysing the state of the art concerning the use of methods, research questions, experiment planning, KPI usage, data collection, and data analysis could be aggregated in order to interpret the current SotA.

Step 6: Interpretation

Interpreting data was discussed within the review team. It was decided to use graphics and charts to depict the findings. This will enable others to also interpret the data as quickly as possible and, furthermore, puts the data into the context of the formulated objective for the whole SLR.

Step 7: Presenting Results

The results are presented in the following section. It was decided to use a visual approach by using graphics and giving a small explanation text. In order to enable the use of the results for the target audience of DRIVER+, the knowledge base was created. As it is part of the trial guidance tool (see Part 2), it is online and free for everyone to use while preparing a trial or informing oneself about trials in general.

Step 8: Updating the Review

This step is planned to be an ongoing process. Every time a journal paper is identified by the consortium (during the project phase) and/or by the parties involved during the setup and execution of a trial (during and after the project phase), a codebook should be filled in by the authors and then be fed into the knowledge base. Furthermore, all consortium members are encouraged to fill the knowledge base with more codebooks of relevant peer-reviewed journal articles they encounter. This will be taken up by the trial guidance tool that allows users to suggest new entries into the knowledge base. Here, quality will be ensured, as all new entries will be peer-reviewed.

Tool Support

The chosen SLR approach by Thomé et al. (2016) is divided into three phases: preparation (step 1), execution (steps 2 to 4), and summarisation (step 5 to 8). This separation is also embedded in the software StArt (State of the Art through Systematic Review), which was developed by the Laboratory of Research on Software Engineering (LaPES) of the Computing Department of the Federal University of São Carlos (DC/UFSCar). It was decided to use this tool, as it is not only available as freeware but it also has extensive online tutorials. Both mentioned circumstances allowed every member of the SLR team to use the tool easily.

For data extraction, Microsoft Word was used. Here, the template for the codebooks was created and spread among the team members, who used it to create one codebook per paper assigned to him/her. According to the objective of exploring trial-like events the task was to extract the main information from the selected papers (e.g. which RQ was followed or which evaluation approach was chosen). Hence, no subjective text mining was needed to be executed, but a structured and comparable summary of the most relevant meta information of the selected papers was targeted. As one paper was always assigned to two reviewers, in the end the reviews could be easily combined. Because the number of relevant papers is significantly higher than average (13), manual synthesis was not feasible; thus, the software MAXQDA Analytics Pro 2018 (release 18.0.3) was used. Furthermore, all entries from the combined codebooks were inserted in Microsoft Excel, where some basic analysis could be performed (e.g. the number of empty fields).

Findings

We now turn to present the findings of the SLR by focusing mainly on data analysis and synthesis (step 5) and interpretation (step 6).

Research Questions

We first looked at the research questions in the selected literature. In the frame of the project DRIVER+, research questions have a prominent role as they guide the trial design and facilitate the evaluation phase, which revolves, inter alia, around potential answers to the initial questions identified by the practitioner. Our first expectation was to find the research questions in interrogative forms. Surprisingly, only 10% of the selected peer-reviewed papers contained a statement in the form of an interrogative question. Within this 10% of the papers, the interrogative research questions were formulated by the words "How" (40%), "What" (26%), "Will" (11%), "Where" (9%), "Are" (9%), and "Can" (4%) (Figure 1.3).

Considering the purposes of the trials, the "How" and "What" questions are the most fitting options. The former focuses mainly on analysing the potential impact of innovative solutions on the socio-technical setup of a crisis management organisation (how does solution XY impact on mobilising volunteers?), and it is used to define or quantify a particular subject or measure respectively. Moreover, "What" questions set a direction for the steps to be followed for a specific process and are used to widen to or narrow down a specific subject – for instance, "What produces a change in the Standard Operating Procedures (SOP) of our medical team when solution X is used?"

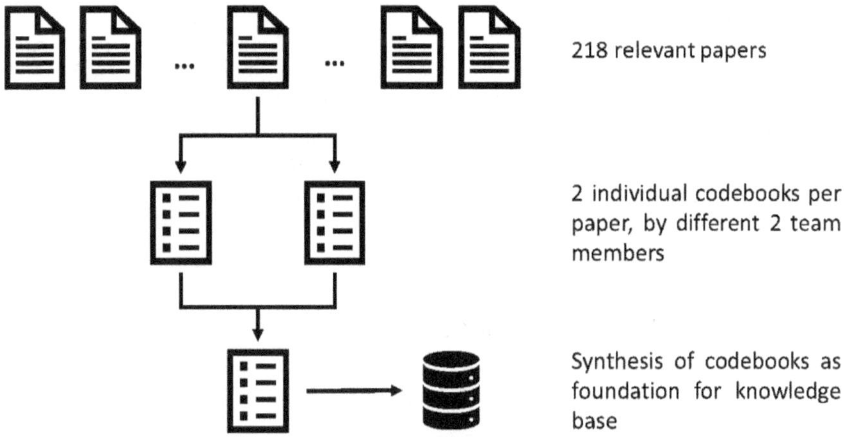

FIGURE 1.3 Split of interrogative words in research questions.

When we use "Can/Is/Are" constructions (13%), it is essential to ensure that the questions refer to a specific, measurable subject. Otherwise, the risk is having subjective or broad answers. There is a risk that the formulated questions lead to very subjective or general answers (e.g. "Is this a good solution?"). On the contrary, a question like "Is this solution usable on our fire truck?" will result in a measurable trial setup. The rest of the research questions were filled in with the aim or objective the papers meant to address. Recurring topics were the construction of frameworks or methodologies for designing and developing decision-support systems, simulation testbeds, or even training programs identifying critical skills. Another topic studied within this domain was training programs for which simulations are often used to evaluate a program's effectiveness or establish the training itself.

Experiment Planning and Deviations

The systematic literature review aimed to provide an overview of experiment planning and learn more about the deviations used often in the past decade. We identified six broad terms: questionnaire/survey, exercise/test, study, analysis/synthesis, interview, and workshop (Figure 1.4).

Furthermore, we found two alternatives to plan an experiment: (1) develop the experiment theoretically or (2) design and execute a practical experience. The former is 32% of the reviewed research. However, most cases use the experiential approach through questionnaires, exercises, or workshops. In the codebook, the purpose of the "experimental planning" attribute is to analyse the essential aspects of the proper scoping of experiments, such as "scenario planning," "roles assignment," or "protocol construction." This criterion provides us with information on

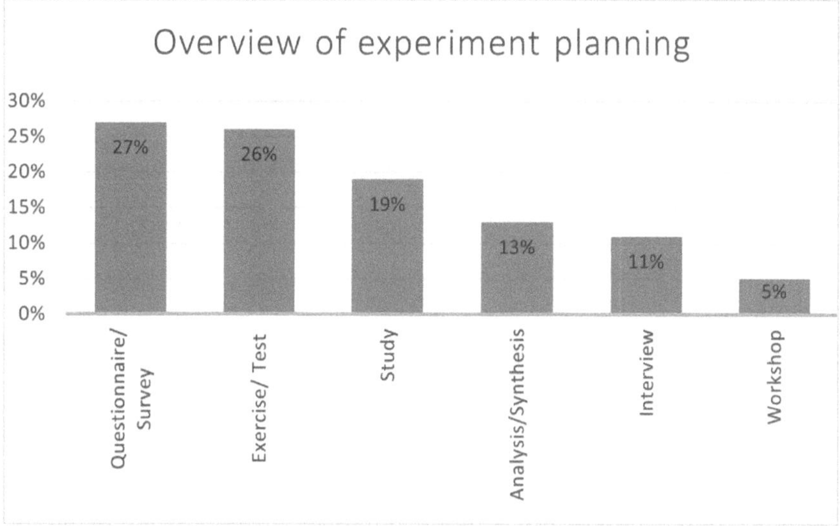

FIGURE 1.4 Overview of experimental planning.

selecting the scope of the experiment correctly, deciding on the roles of participants, and a proper physical simulation space (for example, furnishing, availability of equipment like audio and video recorders, space for providing instructions and teaching, etc.).

Research Methods

There are various research methods in the crisis management domain, and it is challenging to identify a specific trend in research (Figure 1.5). The findings show that with 31%, simulation is the most preferred research method. Modelling (21%) and review (19%) follow simulation. Interviews and surveys account for 29% of the methods used, and they support simulations by providing insights from various perspectives The most preferred approach is the mixed methods that combine quantitative and qualitative data gathering techniques to validate the trial outcomes, such as supporting virtual reality with surveys and focus group meetings.

Experiment, Exercise, Simulation, or Trial Objective

Over the last decade, the domain of crisis management has hosted various types of exercises, simulations, and trials, such as simulate (25%), develop (21%), study (21%), train (15%), plan (12%) and compare (6%) (Figure 1.6). We observed some participatory methods based on discussion sessions and tabletop exercises in this criterion. The latter works to validate plans and policies and identify potential gaps or weaknesses in the execution of an experiment.

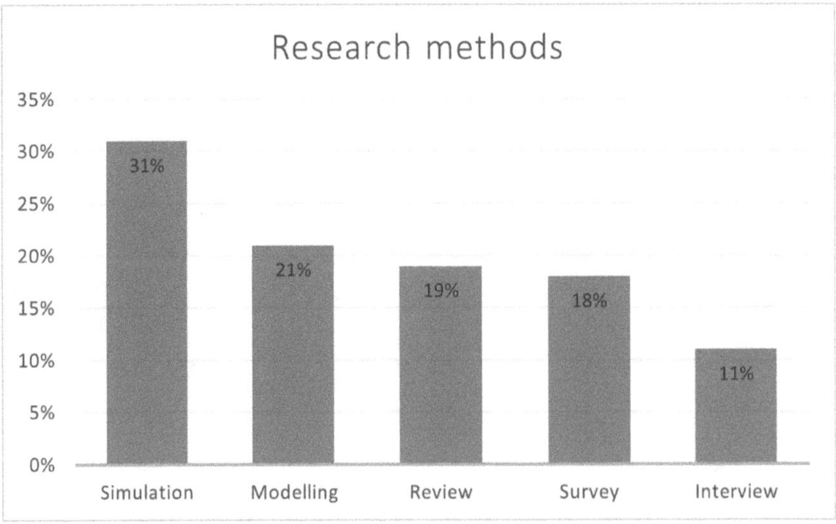

FIGURE 1.5 Research methods.

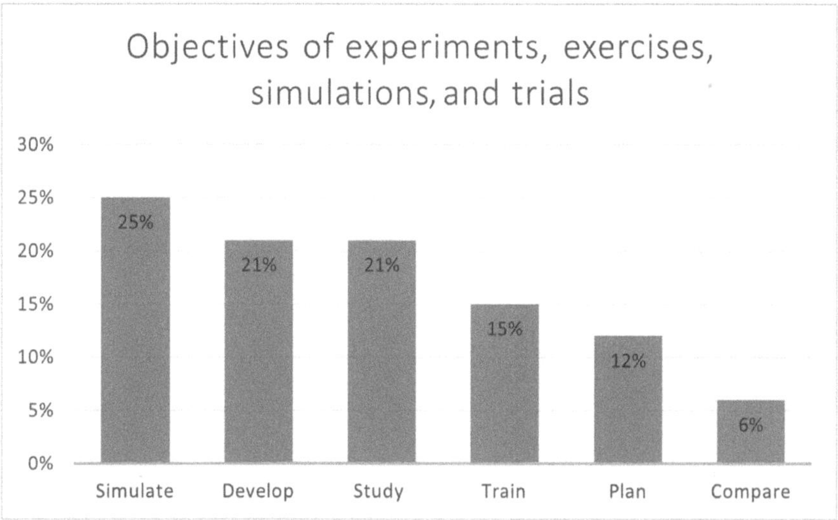

FIGURE 1.6 Objectives of experiments, exercises, simulations, and trials.

Simulation of different scenarios and crises, such as earthquakes or pandemics, helps to identify best practices to detect the relevant skills and decide on the best evacuation or response plan. Training is another purpose for conducting experiments. On the one hand, it helps develop new skills for practitioners in disaster and crisis management; on the other hand, it supports the improvement of training programs for crisis management.

Metrics and Key Performance Indicators

What we found is that 32% of the literature on trials does not mention the metrics and key performance indicators (KPIs). However, analysing a trial and assessing its performance cannot be done without KPIs. Performance, number/sum, quality, cost, and time/delay/speed are the main metrics and indicators that we have found in the literature. Time, cost, and quality are the prominent ones used for KPIs. Though "time" is mentioned in 29 codebooks, it cannot be considered a common metric or KPI. Not having a common metric is proof of the difficulty in establishing generic performance measurement approaches in crisis management. We did not find a pattern, as in each peer-reviewed article we identified some specific metrics that are directly related to the content of the article. Additionally, we found the use of a real metric, such as a number, in only 12 papers.

Data Collection and Analysis

Data collection is directly related to the research question. Most of the analysed papers followed a mixed-method approach that requires both quantitative (questionnaires or surveys) and qualitative (focus groups, interviews, or case studies) data collection methods (Figure 1.8). The former is analysed statistically; the latter requires hermeneutics or semiotics methods. Interviews, evaluations, surveys, and questionnaires are the most preferred methods within the analysed literature. Focus groups are also an important method, but trained personnel are required to conduct effective and efficient focus groups. Although it is the most effective

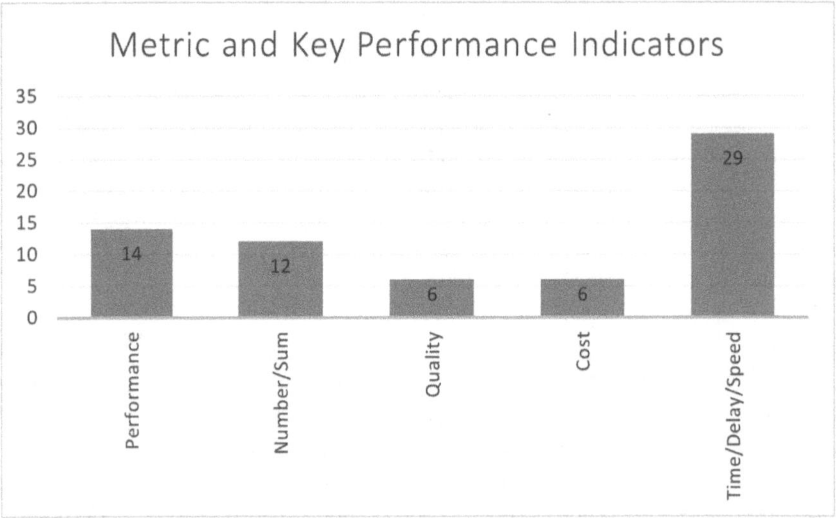

FIGURE 1.7 Metric and key performance indicators.

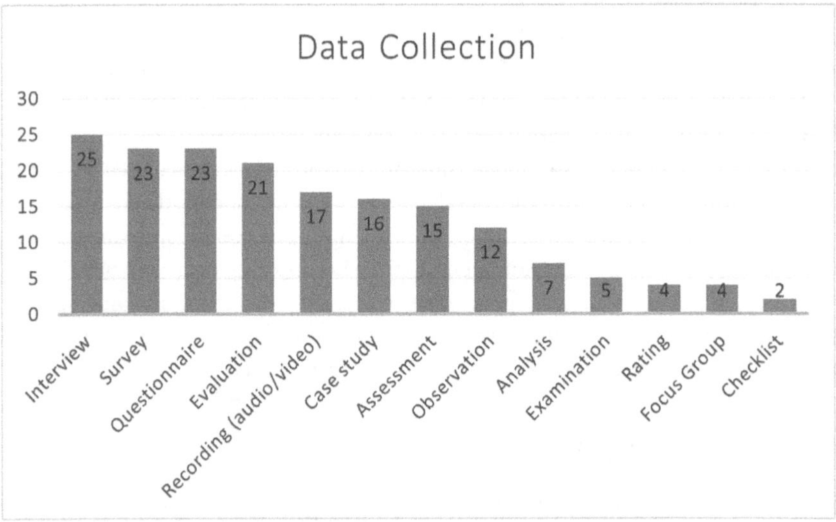

FIGURE 1.8 Data collection.

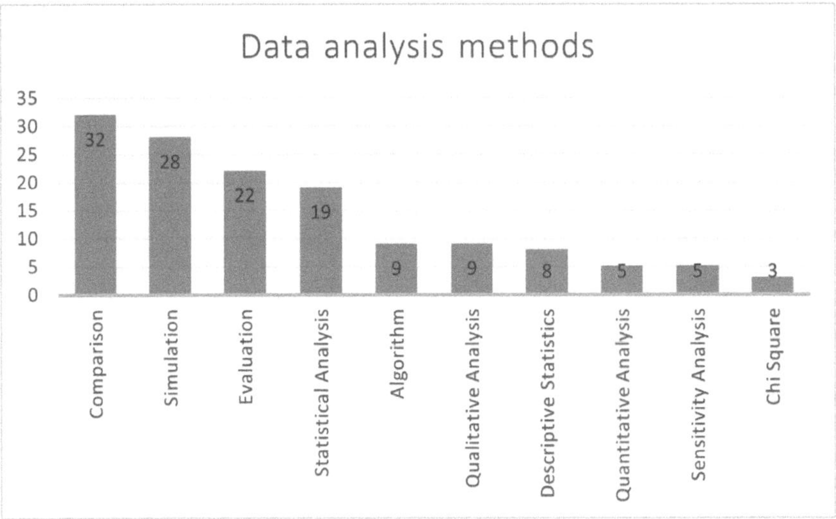

FIGURE 1.9 Data analysis methods.

method, it is one of the least preferred data collection methods due to the requirement of trained personnel. Regarding the data analysis methods, we investigated ten different data analysis methods (Figure 1.9). The most preferred methods were comparison, simulation, evaluation, and statistical analysis.

Ethical Procedures

The last section in the codebooks was about ethical procedures. Including ethical procedures has become obligatory in most scientific journals. However, at the time of the systematic literature review, only 21.1% of the analysed peer-reviewed papers mentioned the ethical procedures that were followed in the trials. We encountered three main ethical procedures based on (1) approval (30 papers), (2) informed consent and agreement (14 papers), and (3) anonymizing/coding the answers (4 papers) (Figure 1.10). Regarding the approval, 27 of the 30 papers mentioned exactly the board that approved their research, such as ethics committee, review board, and total quality council. Only a few of the papers state the followed protocol in detail. Obtaining signed informed consent forms is the second most common approach in the analysed literature. Only three papers include some details about the ethical considerations, such as "ethics-by-design approach" (Neville et al., 2016), "hard copies in a cabinet in a locked office" (Al Khalaileh et al., 2010), and "files are stored in a locked file cabinet located in a locked room" (Gamboa-Maldonado et al., 2012).

Turning the State of the Art (SotA) into a Knowledge Base

The purpose of the SotA was to make it available both in the frame of the project and for the wider crisis management community. This implied turning it into a

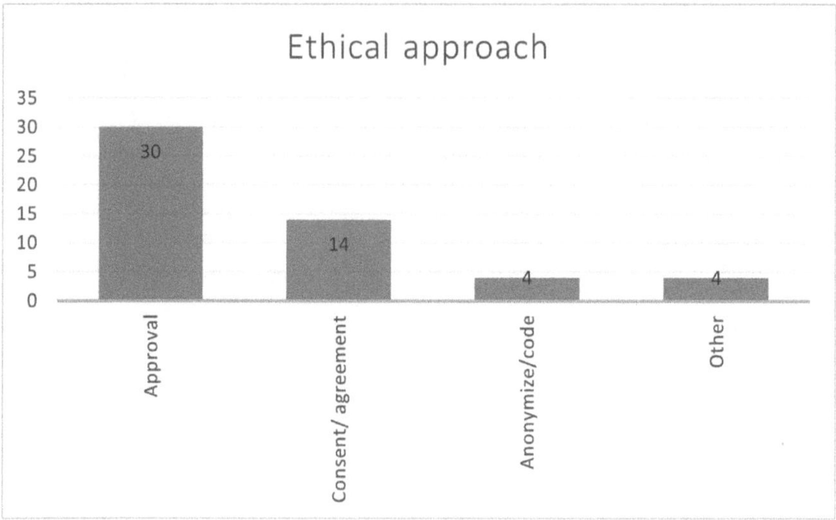

FIGURE 1.10 Ethical procedures.

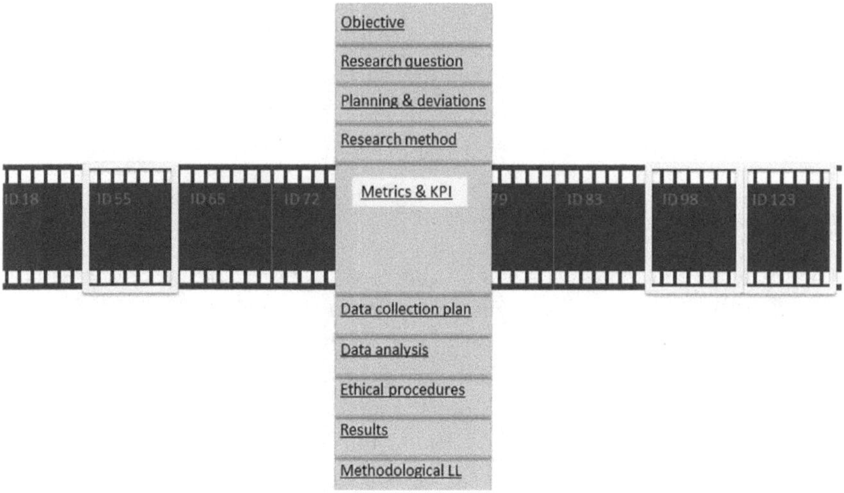

FIGURE 1.11 Utilising the SLR results as a knowledge base.

relational database that shows only the information one is looking for through a system of "tuples" and "attributes." A tuple is a dataset of a single item (e.g. our codebook), while an attribute is a specific element of the tuple (e.g. the topics in each code book: objectives, research questions, etc.). Moreover, the overall idea behind the relational database is provide two options for searches for future users:

- A horizontal search for every codebook that has information on, for example, serious games in the "metrics and KPIs" (attribute). The results will be all in the attribute "metrics and KPIs."

- A vertical search to explore more on a specific ID so that users can explore beyond the attribute they were looking for (e.g., not only metrics and KPIs but also data collection plan and data analysis).

The results of these efforts are available in the portfolio of solutions under the knowledge database here: https://pos.driver-project.eu/en/gt/knowledge.

Conclusions

Exploring the CD&E approach and carrying out the SLR were key for two reasons: the heterogeneous disciplines, methods, and trial-linked events confirmed the need for implementing a mixed-research approach in the TGM; the knowledge base can support and facilitate the application of the TGM. Furthermore, it provides access, through the trial guidance tool, to a considerable amount of knowledge that can be of benefit to practitioners and researchers alike.

The identification and analysis of 218 as relevant identified peer-reviewed journal papers directly contributed to two major objectives. The identified heterogeneous disciplines, research methods, and trial-like events confirmed the underlying DRIVER+ decision to consider a mixed-research approach for the TGM. Furthermore, the SLR enabled the creation of high-quality best practices in the TGM knowledge base that now supports every trial designer in applying the TGM. This knowledge base is available via the trial guidance tool (see also Part 2). Another finding is that the past decade did in fact struggle with considering ethical aspects. Within the TGM design, these are taken up, as described in this chapter.

The design of the TGM, its core principles and the application in the context of crisis management organisations are dependent on the foundations which are outlined in this chapter. Having this work in mind, in the next chapter we provide an in-depth description of the methodological approach that was applied in all DRIVER+ trials and beyond.

References

Al Khalaileh, M. A., Bond, A. E., Beckstrand, R. L., & Al-Talafha, A. (2010). The disaster preparedness evaluation tool: Psychometric testing of the Classical Arabic version. *Journal of Advanced Nursing*, 3, 664–672. http://doi.org/10.1111/j.1365-2648.2009.05208.x.

Gamboa-Maldonado, T., Marshak, H. H., Sinclair, R., Montgomery, S., & Dyjack, D. T. (2012). Building capacity for community disaster preparedness: A call for collaboration between public environmental health and emergency preparedness and response programs. *Journal of Environmental Health*, 75(2), 24–29.

Higgins, J. P. T., & Green, S. (Eds.). (2011). *Cochrane Handbook for Systematic Reviews of Interventions*. The Cochrane Collaboration. www.handbook.cochrane.org.

Kitchenham, B. (2004). Procedures for performing systematic reviews. *Keele, UK, Keele University*, 33(2004), 1–26.

Kitchenham, B., & Charters, S. (2007). *Guidelines for Performing Systematic Literature Reviews in Software Engineering*. https://www.researchgate.net/publication/302924724_Guidelines_for_performing_Systematic_Literature_Reviews_in_Software_Engineering.

Labbé, P., Bowley, D., Comeau, P., Edwards, R., Hiniker, P., Howes, G., Kass, R., Morris, C., Nunes Vaz, R., & Vaughan, J. (2006). Guide for understanding and implementing defense experimentation GUIDEx. *The Technical Cooperation Program*. http://opanalytics.ca/aus/pdf/GUIDExBookFeb2006.pdf.

Moher, D., Liberati, A., Tetzlaff, J., Altman, D. G., & for the PRISMA Group. (2009). Preferred reporting items for systematic reviews and meta-analyses: The PRISMA statement. *BMJ*, 339, 1–8. https://doi.org/10.1136/bmj.b2535.

Mulrow, C. D. (1987). The medical review article: State of the science. *Annals of Internal Medicine*, 106(3), 485–488. https://doi.org/10.7326/0003-4819-106-3-485.

Mulrow, C. D. (1994). Systematic reviews: Rationale for systematic reviews. *BMJ*, 309(6954), 597–599. https://doi.org/10.1136/bmj.309.6954.597.

Neville, K., O'Riordan, S., Pope, A., Rauner, M., Rochford, M., Madden, M., Sweeney, J., Nussbaumer, A., McCarthy, N., & Cian O'Brien, C. (2016). Towards the development of a decision support system for multi-agency decision-making during cross-border emergencies. *Journal of Decision Systems*, 25(Sup 1), 381–396. http://doi.org/10.1080/12460125.2016.1187393.

Petticrew, M., & Roberts, H. (2006). *Systematic Reviews in the Social Sciences*. Malden, MA: Blackwell Publishing.

Pikner, I. (2015). Concept development & experimentation as tool for capability development. *International Conference Knowledge-Based Organization*, 21(1), 88–93. De Gruyter Open.

Thomé, A. M. T., Scavarda, L. F., & Scavarda, A. J. (2016). Conducting systematic literature review in operations management. *Production Planning & Control*, 27(5), 408–420.

2

THE TRIAL GUIDANCE METHODOLOGY

Chiara Fonio and Adam Widera

Introduction

Did you ever ask yourself when and why you switch from a "running system" to a new approach? Think of buying a new smartphone or an electric bicycle as an innovation within your ecosystem. Reflect what drives you to switch from an old to a new system: Is your current smartphone broken so you need a new one? Did your friend offer you a ride with an e-bike, so you realised advantages to your traditional bike? While a new smartphone reveals a necessity because of a shortcoming (the old one is broken), the new e-bike is not related to a shortcoming or gap per se but appears as an opportunity or potential improvement you have recently learnt.

Triggers to innovate might have different sources, such as shortcomings or gaps and opportunities or solutions. Every person has an explicit and implicit set of criteria to decide for or against a potential innovation. Also, decision-makers have at least the feeling of evidence whether a change in the ecosystem might be beneficial or not in the light of the (explicit and/or implicit) decision criteria. This (perceived) evidence might be an independent report by an authority or an established domain expert, but decisions can also have simple anecdotal evidence, such as a recommendation of someone who did decide to innovate. Going back to the introductory examples, we can conclude that the decision to buy or not to buy a new smartphone or e-bike remains very much on the highly specific or individual perception of the decision-maker: the perception itself is influenced by factors in the target (such as the novelty of the innovation), factors in the situation (e.g. urgency of the gap), and also factors in the perceiver, such as attitude, expectations, or experiences (Robbins et al., 2010). In consequence, decisions about innovations might greatly vary even if the gaps and the solutions are identical: while early adaptors might perceive the need to buy new phones or e-bikes, laggards tend to

DOI: 10.4324/9781003256977-4

wait until the majority provides a broad database supporting the evidence about the intended advantages.

Innovation is also highly uncertain and innovation processes are complex, lengthy, cumulative, and non-linear (Mazzuccato, 2015, 2018). As argued by Mazzuccato, "it is better able to recognise the serendipity and uncertainty that characterise the innovation process" (2015, p. 43), which is apparent when looking, for instance, at the rise and fall of economic powers in history.

In addition, one should not underestimate that decisions in general, not just those triggered by innovation, involve human judgement and "whenever there is judgement, there is noise" (Kahneman et al., 2021, p. 12). This highlights the complexity of making decisions in all types of organisations: variability in judgement can cause serious problems, as illustrated by Kahneman et al. (2021).

Even if classifications of human behaviour, such as Roger's (1962) theory on diffusion of innovation, support anticipations on innovation attitudes, the areas in which the potential innovation take place cause very different requirements for decision-makers. While smartphone users with, let us say, "unlimited" resources are more likely to buy a new device, smartphone users with limited resources tend to be more critical towards the degree of evidence about the potential added value of a new device. In sum, it can be contrasted that the process of recognising a trigger to innovate (gap- or solution-driven), the wish to receive a sufficient level of evidence towards the potential added value (data collection), and context-dependent sense and decision-making (data analysis and evaluation) follows a recurring pattern: one must have a clear objective to reach, collect, and prioritise metrics (making the objective measurable) and gather and process the required data and evaluate the results.

Now, imagine that decisions about investments into innovations refer to critical ecosystems, such as the crisis management domain: What if the new smartphone is supposed to be used by first responders in complex emergencies? What if the e-bike is supposed to be used by blood services in urban settings with frequent congestions? And what if you – as an individual decision-maker – are responsible for the decision to spend a considerable amount of your organisation's budget to innovate from an established and routine way of working into an uncertain but maybe more efficient or more effective way of working in the future? What would be your requirements to make such a decision? The objective of this chapter is to present the foundations, mechanisms, and procedures of *one way* how crisis management organisations and authorities can be empowered and supported to *systematically* identify the potential *added value and risks* of innovations *driven by their very specific perception(s)* and context: the trial guidance methodology (TGM).

Before turning to describe how the TGM can support crisis management organisations, some considerations have to be made. First, this specific methodological approach was developed and applied, as described in the introduction, in the frame of an EU-funded research project. This implies that the amount of available resources and assets (e.g. human, economic, technical, time management, etc.) was set and known in advance. While uncertainties and unexpected challenges occur even in highly predictable contexts, all stakeholders are committed to work together, using the available resources, to achieve common objectives. In other words, resources were *a given*, and this is not always the case in crisis management organisations, which may have the willingness to explore the potential added value of innovations but may also have to face the reality of, for example, unexpected crises which hamper all work processes. The second aspect is interconnected with the first and refers to the stakeholders who were involved in developing, implementing, and working with the TGM. Researchers, practitioners, and solution providers sit at the same table, and while the decision-making process was far from being easy and implied many feedback loops, the richness of the perspectives, needs, and ideas was quite unique. The uniqueness of the circumstances should not be seen as a limitation, meaning that the TGM can only be applied under the exact same conditions. As shown in this chapter, we are dealing with a scalable and adaptable approach, by design. If researchers can be of support, practitioners and solutions providers are nonetheless essential in the process of co-creation that leads to trials. Third, the TGM was conceived as part of a bigger vision: developing a European testbed for crisis management capabilities development. The testbed consists of physical, methodological, and technical infrastructures to carry out the so-called trials. And it is from those that we will start outlining the conceptual boundaries of this methodological approach.

Why and How to Trial

If the TGM is *one way of* how crisis management organisations and authorities can be empowered and supported to *systematically* identify the potential *added value and risks* of innovations *driven by their very specific perception(s)* and context, trials define the contours and set the stage for the application of the TGM. As will be mentioned further, the "stage" metaphor is fitting since a trial is an event that takes place in a specific context, entails roles and responsibilities, and unfolds within a predefined timeframe.

Before using metaphors to explain concepts, it is key to describe what a trial is and why crisis management organisations may benefit from it. A trial is a systematic test to see how well something works under as realistic as possible circumstances. To frame it along the lines described earlier, a trial may serve the purpose to understand how, if, and to what extent a new smartphone or a new app can make a difference (i.e. can be truly innovative) in the context of a crisis management organisation and a specific scenario. To assess if investments are worth, trialling out solutions (e.g. a new software or work process) may be the most effective way to figure out the potential impact of a change (new solutions) on one organisation. Trialling means testing, using the solution(s) under specific conditions and within a scenario.

Response organisations are more familiar with crisis management exercises, training, and simulation games which revolve around the crisis management capability of *participants* within one or more organisations (Grunnan & Fridheim, 2017) than with trials. The differences between this set of valuable activities and trials lie in the *objective of analysis*. At the core of the trial, there is the operational performance and consequent measurement of a change brought about by potentially innovative solutions, while the scope of an exercise is to test the performance of the participants. Put simply, in exercises decision-making abilities of individuals or teams are tested, in trials decision-making as a whole (meaning, of an organisation) is assessed. The point is not testing abilities but rather understanding if new solutions can help deal with gaps and challenges via measuring the *operational performance of crisis management organisations*. This cannot be done without a structured approach which is called TGM. The first key element of the acronym has been outlined already. "Guidance" and "methodology" are also crucial as they define the nature of the approach. The TGM provides guidelines and recommendations to those interested in assessing innovation on the basis of a systematic methodology. It "guides" by providing practical recommendations during the whole trial experience, but the guidelines may have to be adapted to the contexts in which they are applied. For instance, there are actors and stakeholders with roles and responsibilities (from the trial owner to the evaluation coordinator, as will be discussed further), but the available resources and time within an organisation may not allow for the full spectrum of responsibilities that must be fulfilled. Or the solutions selection process will be carried out in a "lighter" way, especially if practitioners are already familiar with the solutions they want to test.

In practice, what does a trial look like? It looks like an event which unfolds in the context of one (or more than one at the same time) crisis management organisation. If we consider the latter as a "stage," there will be experts both behind the scenes and on that stage to ensure that everything runs smoothly. There will also be a proper "scenography": different pre-prepared rooms in which the trial participants use the solutions – for instance, a flood management tool or common operational picture (COP) tool. All actors will have to follow a script (the scenario), and what you will see on stage is only the execution of the event. To get there, a step-by-step preparation phase is needed to make sure that all actors are aligned around a common vision (the objective of the trial), the technical infrastructure is well integrated with existing legacy systems and new solutions can exchange data. The execution will be followed by the evaluation of the results. This is when the potential added value of the solutions is assessed through the data collected during the execution phase.

To do all this properly, "hats" have to be assigned, meaning that roles and responsibilities should be defined in advance. The question is, who should be involved? Who should take the most important decisions? In other words, who is really *needed* to carry out a trial, besides the trial participants?

There are at least *four* main roles to nourish the brain power needed for managing a trial (the so-called trial committee):

1. The *trial owner* – namely, the crisis management organisation which is responsible for the trial itself. Being responsible for a trial entails following the entire

trial experience and hosting it. More importantly, it entails having the "vision" of the trial (e.g. objectives and overall goal), which is captured in a proper scenario so that gaps and needs can emerge. Having the "right" facility is also important here, but there are no minimum requirements as it depends on the objective(s) of the trial and the resources (both human and not) that will be involved. Usually, having multiple rooms and a basic set of technologies (see: iso.org, 2018, ISO 9241-11:2018) may facilitate the execution phase.

2. The *technical coordinator*. As the name indicates, the technical coordinator deals with the technical setup of the scenario so that potentially innovative solutions can be properly assessed. This requires, *inter alia*, the integration of legacy systems with the trialled solutions. The technical coordinator is also in charge of the technical testbed infrastructure (see Part 2)) and is responsible for involving the solutions providers in key moments (e.g. ad hoc meetings before executing the trial). Moreover, he or she works closely with the evaluation coordinators to provide technical insights, as needed. Another aspect that should not be underestimated and falls under the responsibility of the technical coordinator is the training. When trialling new solutions, those who will be using the solutions should get an understanding of the main functions; hence, a training, provided by solution providers, must be put in place.
3. The *practitioner coordinator*. Practitioners (namely, crisis management stakeholders) are at the core of the TGM due to its inherently practitioner-driven nature. They do have the "last word" about the trial, specifically what should be assessed and how the results should be interpreted from a practitioner's perspective. In this frame, a practitioner coordinator should have a background crisis management (CM) as he or she is responsible for a wide set of CM-related activities: from the selection of the trial participants to the development, together with the trial owner, of a realistic scenario; from the identification of key performance indicators related to crisis management dimension to expectations management. Overall, a practitioner coordinator manages (contact management and communication) the pool of trial participants in a consistent and structured way so that they are engaged with all relevant activities.
4. The *evaluation coordinator*. The evaluation coordinator has the difficult task of facilitating an independent, multi-perspective, and robust assessment of the trialled solutions. It is a central role, considering the main rationale behind trials. It is a role that calls for strong neutrality and independence but also for strong collaboration with the other three coordinators and the trial participants, especially the CM practitioners and solution providers. A key recommendation for selecting a proper evaluation coordinator calls for having as little as possible biases towards the trial participants. Besides, a minimum level of analytical skills is needed in order to carefully elaborate, pre-evaluate, and apply the data collection plan. However, it must be noted that specific data analysis techniques might be outsourced to team members with appropriate qualifications and experiences.

After this short overview on key roles and responsibilities, we now turn to describe the TGM's anatomy in all its phases and steps.

The TGM Cycle

The trial guidance methodology consists of three main phases:

1. Preparation
2. Execution
3. Evaluation

While the three phases sound obvious (you have to prepare first, execute "on stage" when you are ready, and evaluate the performance at the end), they imply a rigorous step-by-step approach that ensures a proper design, execution, and evaluation of the results throughout the trial journey. As mentioned earlier, the TGM can be scalable, but the backbone of the methodology (the TGM cycle) shall remain the same.

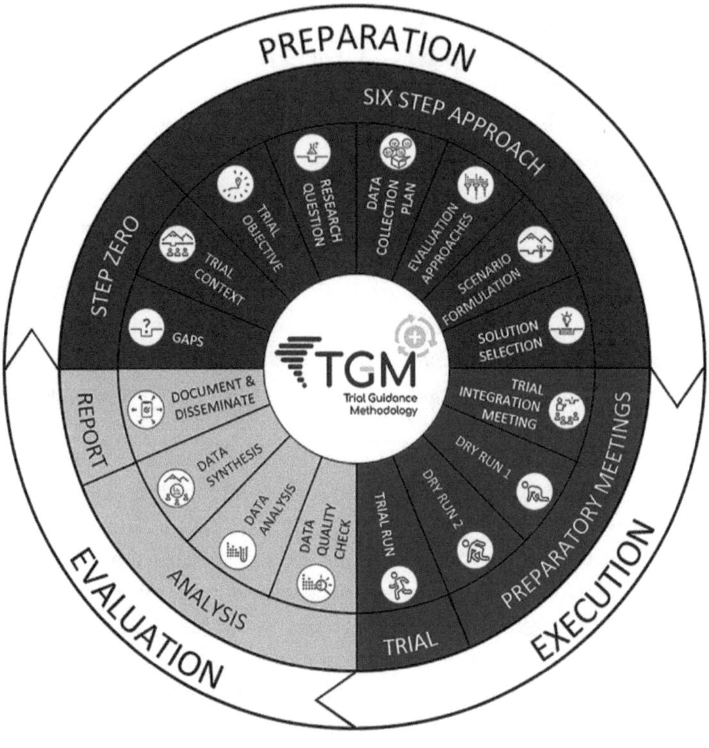

FIGURE 2.1 The trial guidance methodology cycle.

Preparation Phase

The preparation phase revolves around two parts: the so-called step-zero, which deals with two essential prerequisites and the six-step approach focused on the trial design.

Step Zero

This phase is crucial to enter the execution phase with a robust design. To prepare for a trial, there are some prerequisites (the so-called step zero) to consider. First, the potential added value and risks of innovations can be identified by crisis management organisations only if innovations are measured vis-à-vis a baseline in which there are *gaps* to address. Otherwise, proper measurement would not be possible: the potential added value of solutions emerges in relation to existing circumstances and problems that, for example, slow down work processes or limit the effectiveness of operations. Those problems are the gaps that CM organisations are facing which are familiar and well-known to them. They are the triggers for carrying out a trial: to do things better and to have an innovative approach to challenges: can we shorten the response phase using drones? Is this app for managing volunteers helpful to mobilise them faster and more effectively? Can we improve situational awareness through this new common operational tool? If yes, to what extent? As we will explain later, the question "to what extent" is crucial in the TGM.

The role of the TGM is not to identify those gaps on behalf of an organisation but rather to provide recommendations on how to do this. To start with, capability gaps must be (1) identified and (2) discussed with practitioners.

There are many ways to identify gaps. In this regard, there is not a one-size-fits all solution as it depends on (1) the existing knowledge on problems and gaps and (2) the resources available to specify internal gaps with relevant stakeholders. If the existing knowledge is enough (e.g. documented reports and lessons learnt), there is no need to "identify" new gaps but rather to collect, systematise, and discuss them in workshops, focus groups, and/or brainstorming with practitioners. If the knowledge is non-existent or outdated, ad hoc workshops with proper facilitation would be needed. The aim of the workshops is to discuss the current situation and focus on what can be improved or what is missing. The gaps should also be validated with relevant crisis management stakeholders in follow-up meetings.

Gaps do not happen in a vacuum. They are embedded within a context and experienced by many stakeholders. Mapping the trial context is the second prerequisite and involves the identification of all key aspects surrounding the context of a crisis management organisation. The trial context "frames" the environment in which the gaps occur. To be more specific, the practitioner coordinator has to work closely with CM stakeholders to depict a baseline which is the "as-is process": who, what, and how operations and procedures take place. This involves depicting all processes and actors involved in (1) in the trial context and (2) in the specific crisis management operation in which the gaps emerge. The trial context gives a zoom-in glimpse into what a crisis management organisation wants to address in a trial: who is involved, when, how, and underpinning processes.

The prerequisites pave the ground to the six-step approach of the preparation phase.

The Six-Step Approach

Designing trial, after having identified and contextualised the gaps, entails a non-linear six-step approach which follows a logical methodological process: from the trial objectives to the solutions selection. The preparation phase requires strong synergies, co-creation dynamics, and a certain degree of flexibility captured in its iterative nature: when decisions and or changes are made in, for example, step 4 (evaluation approaches and metrics), the data collection plan (step 3) may need some revisions (are we collecting all relevant data based on the metrics that we have in mind?).

The six-step approach is the "ladder" to go on stage: only if you climb that ladder can you get access to the stage, meaning to the execution phase. Each step is a step towards the trial execution, but it takes time to get there and crisis management organisations shall consider taking all the time needed to be prepared for execution. As for the entire TGM cycle, this phase is deeply rooted in co-creation. If, on the one hand, practitioners do have the last word, on the other, all trial's stakeholders have to be involved, based on roles and responsibilities, in key meetings so that a sense of ownership and engagement are ensured. Those aspects (ownership and engagement) are key to host a successful trial.

What do crisis organisations need to design a trial? What are the steps that they have to follow and how?

They have to climb the six-step ladder, as mentioned earlier. Practically, this entails the following:

1. Identify the trial objective. (What are your goals? See step 0.)
2. Determine the research questions to address the gaps. (See step 0.)
3. Define a data collection plan. (What do you want to collect during the trial)
4. Select the evaluation approaches and metrics based on the data collection plan. (What do you want to measure and how?)
5. Develop a realistic scenario in which the gaps and needs emerge (What is the scenario/plot that best fits with the needs and gaps you would like to address?)
6. Select the solutions to be trialled. (Which are the solutions to assess?)

The first step of the ladder is strongly gap(s)-dependent, meaning that the objective(s) or aim must refer to the gaps identified in step 0. To agree on the overall objective, the brain power of all four main roles would be needed. The trial owner and the three coordinators (practitioner, evaluation, and technical) have to sit around the same table and brainstorm about the main objective of the trial, based on the gaps identified. For example, if one of the gaps refers to shortcomings in the ability to exchange crisis related information among organisations (interoperability) in multi-event crises, the objective will be formulated as follows: improving cooperation and coordination among different countries and agencies using innovative solutions for large-scale and complex multi-event crises. Most likely, an objective or aim will not be able to capture the gaps and needs that a crisis management organisation is willing to explore in a trial. The identification of sub-objectives might be needed (e.g. improve and maintain in real-time a shared situational awareness among agencies and organisations). Each objective must be formulated in a specific, measurable, achievable, reasonable and time-bound (SMART) way. If the aim(s) is too broad, it could be challenging to draw meaningful conclusions and/or reflections on the potential innovation of solutions. Being as specific as possible requires having clear ideas on step 5 (scenario). It is very unlikely that, say, a trial owner does not have in mind a scenario even before climbing the ladder. Ideas on the scenario may help at this stage (e.g. multiple crises: a forest-fire with cascading effects on the critical infrastructures and human settlements), even if full development would require more time and additional work later. The objective(s) must also be measurable, meaning that it should be formulated in a way that allows measuring: What does improving cooperation and coordination entail? Does it mean being faster or more accurate? All these considerations should be carefully taken into account together with what would be realistically achievable and reasonable to explore within the timeframe of a trial. For instance, it would not be feasible to explore cooperation among all agencies and organisations, but it would be worth it to focus only on firefighters and medical services as this is a weak point. Additionally, it is also key to think in advance about the resources you can rely on in terms of people (how many CM stakeholders can realistically be involved?), technology (which are your legacy systems and which one will be actively used in the trial?), and time. The latter may be a challenge for first responders. In this regard, it is recommended to carry out trials in a short amount of time (one or two days maximum) so that you will have more opportunities to have the right people on board.

While identifying the trial objective is the right starting point, it is still too high-level. It is necessary to climb another step and define specific research questions to address in the trial. Like the objective(s), research questions are also gap(s)-dependent more than scenario-driven. Additionally, they should be aligned with the objectives identified in the previous step. For example, having in mind the objective mentioned earlier (cooperation and coordination among different countries and agencies using innovative solutions for large-scale and complex multi-event crises), the research question will be this: How does one improve and maintain in real-time a shared situational awareness by supporting the exchange of crisis-related information among agencies and organisations?

The following aspects turn a general objective into specific elements that will be measured in the trial:

1. improve and maintain (time dimension)
2. real-time (response phase)
3. shared situational awareness (response phase, specific multi-agent dimension)
4. agencies and organisations (main actors)

The effort which is required in step 2 is thinking about the gaps and the objectives in terms of measurability that will determine the overall evaluation approach. Research questions can be organised in a multi-level hierarchical structure: for example, main research questions and sub-research questions along the lines of the objectives. Both the trial owner and the evaluation coordinator are key here, but the questions have to be understood and approved by *all* trial stakeholders. It is crucial that all stakeholders are on the same page and have the same understanding of the core aims of the trial.

Research questions can be answered through evidence. The latter can be drawn from the data collected during the trial. To do this, the trial owners and the three coordinators have to decide what data will be collected, by whom, and how. Again, realism is important. The data collection plan is the third step of the preparation phase ladder and deals not only with what, whom, and how but also with why, meaning that it is essential to understand why a crisis management organisation, based on the trial objective(s) and on the research question(s) would need to collect a certain set of data. The key question is what would be needed to address the research questions. Collecting more data than necessary shall not be considered as an added value but only as an additional effort and potential "noise." For example, measuring shared situational awareness entails collecting data on, inter alia, the following aspects: quality and accuracy of the information shared (data: situational reports, shared logbooks, exact location of victims etc.) and time, meaning when a certain action occurs (data: time needed to locate victims or to dispatch ambulances). In this frame, collecting data on, say, the uses of social media is not relevant, unless you want to capture how shared situation awareness can be enhanced through social media and crowdsourcing. By "whom" depends on the resources available (both human and technical): data can be collected both through the testbed technical infrastructure (see Part 2) and also through observers during a specific session of the trial and in a given moment of the scenario ("how"). Data can also be collected through interviews and focus groups, as needed. The data collection plan should serve the purpose of mapping all the information you need (the data) to get to the final destination. Using a structured approach is necessary. A proper data collection plan calls for the use of a supporting tool (e.g. Excel) and for a comprehensive mapping: to achieve this objective, key performance measurement indicators in the three main performance measurement dimensions must be decided.

The three main performance measurement dimensions are key elements of the TGM as they ensure the robustness of the approach (see also Widera et al., 2018).

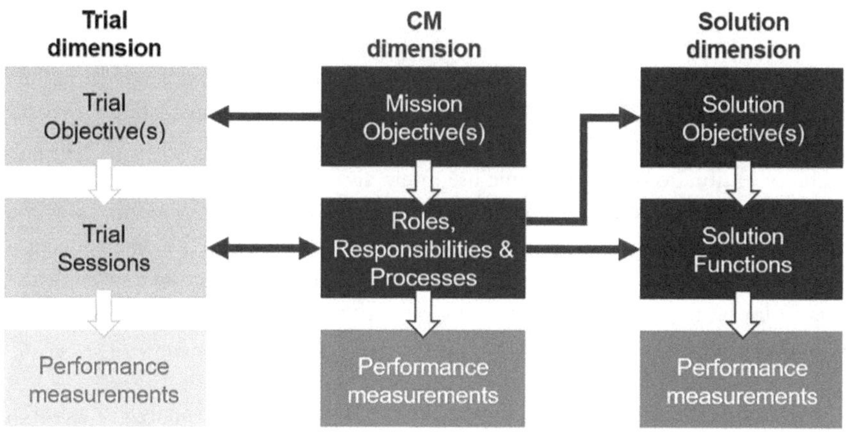

FIGURE 2.2 Three main performance measurement dimensions.

The three dimensions to measure during a trial refer to the trial as such (trial dimension), the crisis management dimension (e.g. the impact of a solution on the CM processes in which an operation is carried out), and the functionalities of the solutions. They do ensure the robustness of the approach as they provide a comprehensive picture, despite having different "weight." The most important is the crisis management dimension since measurement here focuses on how potentially innovative solutions address or do not address the gaps. The input to do this properly comes from step 0, specifically from the "as-is process" or baseline in which who, what, and how procedures take place should have been depicted. All crisis management operations and gaps are included in the baseline; hence, what organisations want to measure in relation to those operations is already captured there. To refer to the example mentioned earlier, measuring the accuracy of the information falls under the crisis management dimension as it goes beyond measuring the user-friendliness of a solution which enables writing and sharing situational reports. This dimension goes in-depth into the roles, responsibilities, and processes involved in an operation to measure the real added value of one or more solutions which can be inferred using a comparative approach. This is important to consider for the data collection: reference data from the legacy tools must be collected as well, unless valid data from past incidents or simulations are available. A mixed-method approach is strongly recommended since both quantitative and qualitative data may need to be collected.

The trial dimension deals with the overall trial organisation, the number of participants involved, the technical setup and potential failures, the training of the participants, the timeframe of the trial, potential delays, and so on. Capturing and measuring those aspects is important for the impact on the other two dimensions. If, instead of having 20 participants, only 10 with a very similar background show up "on stage," the impact on the crisis management dimension would be significant. The same holds true for technical failures, like a blackout during an important session of the scenario, which can cause issues in assessing the real potential of a solution. The use of questionnaires can be helpful for the trial dimension,

As far as the solutions are concerned, core functionalities as well as their usability must be measured. A structured way to do so is to use an international experts' standard (ISO), like ISO 9241-11, which provides a framework for understanding usability in situations where people use interactive systems.

The three dimensions outlined earlier bring us to step 4 of the ladder since the selection of the evaluation approaches and metrics depends on the data collection plan and how crisis management organisations are planning to make sense of the data. If listing how to ensure sensemaking is neither possible nor useful, we would highlight only a few key considerations:

- This step is concerned with sensemaking – basically how to analyse the data in a proper way.
- The evaluation coordinator had the lead on this, but the trial owner must be involved in all relevant discussions and decisions.
- The available techniques for sensemaking are data-dependent: a survey entails the use of a rating scale to measure opinions (quantitative method), while focus groups or interviews require qualitative methods and techniques, as needed (e.g. transcripts, coding, and qualitative data analysis software).
- A reality check should also be applied here: It is better to select approaches and metrics also on the basis of the skills of who is going to be involved in sensemaking. For instance, it is not recommended to use a data analysis software if there are no skilled people able to use the software in an appropriate manner, unless there is enough time to organise proper training.

We are now approaching the last two steps of the ladder which are essential before going "on stage" and executing the trial. Both steps (develop a realistic scenario, step 5, and select the solutions to be trialled, step 6) may come earlier into the picture, meaning that the idea of a potential scenario as well as solutions are most likely in the minds of those involved in the trial committee, as well as in the CM practitioners who have been consulted in the trial journey carried out so far. The difference between "having ideas" and climbing those steps lies in the outputs: a structured storyboard is the outcome of step 5, a list of selected solutions for the trial is the outcome of step 6.

Practitioners familiar with crisis management exercises are aware of the importance of scenarios which must capture problems, needs, and challenges a crisis management organisation wants to test in the frame of trials, scenarios must capture each of the gaps identified in step 0. The baseline is, again, key as it provides the foundations for writing a storyline. The scenario can be single to multi-hazard, depending on the gaps to explore. It must be written to trigger gaps-related conditions. For example, in relation to the gap shortcomings in the ability to exchange crisis related information among organisations (interoperability) in multi-event crises, the scenario would be cross-border and would involve different first responders, such as firefighters and medical services. To trigger this specific aspect, one or more sessions on, say, casualty management in parallel to fire attacks operations will be included. All roles and responsibilities must be decided and written down (who are CM practitioners/players involved?) together with injections (e.g. at this point in the timeline of the scenario, the fire reaches a chemical plant).

The formulation of a scenario must be as precise and realistic as possible. Crisis management organisations should aim for realism more than for complexity when writing scenarios. This is the reason that the trial owner and all three coordinators must be involved in this task together with crisis management practitioners who can assist in detailing scenarios based on their experience on the field. Ultimately, a scenario will look like a storyline with a departure point, several sessions, and a conclusion.

The solution selection is the very last step of the ladder. As mentioned earlier, most likely CM practitioners who have been consulted in the trial journey may have shared ideas about potential interesting solutions. Step 6 is when the ideas turn into a structured solution selection process and a list of solutions for the trial.

First and foremost, solutions can vary from, say, a modelling tool to a training course. Hence, the process of selecting them depends on the objectives and the gaps that should be addressed in the trial. Moreover, to select one or more solutions (there is no need to trial an endless list of solutions as it would only add layers of complexity – see additional data), solutions providers must be involved. It is highly recommended to organise *ad hoc* meetings with them to discuss at least the following issues:

1. How does the solution contribute to crisis management? Meaning, in frame of a specific mission (rescuing people from a flooded area), what the solution would contribute to and how? This would lead to the second important point
2. How can be integrated into the existing crisis management operations?
3. What is the maturity level?
4. Does the solution address the gaps and the challenges raised by the practitioners?
5. What references on solution application exist?

If, on the one hand, solution providers should be involved, on the other, CM practitioners should also have a say and provide their feedback after the ad hoc meetings.

At this point, the trial committee can take a more informed decision by considering the following:

1. If the solution(s) can be used to address the gaps and to answer the research questions.
2. If appropriate training can be provided.
3. If the solution(s) requires a specific technical setup.
4. If the solution providers are willing to participate and contribute to the trial-related tasks and meetings.

Execution Phase

Although the name of the second TGM phase appears to cover the actual execution of the trial only, it also refers to three important meetings which can be understood as milestones assessing the maturity of the trial development. Thus, when thinking of the execution phase, one has to concentrate on *preparatory* tasks first. A good comparison might be the performance in a theatre: while the theatre script, casting, costs, and location were elaborated during the preparation phase, the main part of the execution phase aims at bringing all participants (such as stage managers, set crew, electricians, and costume designers) together to achieve alignment on what has to be done by whom, when and how and to train the actual performance.

The main goal of the trial integration meeting refers to the alignment between all participants of the trial on what will happen when, how, and by whom. The first and second dry runs serve training and quality assurance purposes to avoid frictions, unexpected disruptions, or failures during the execution. Once the trial committee recognises a sufficient maturity of the preparations, the final trial run can be initiated: in other words, the number of dry runs mainly depend on the quality of the work carried out before (i.e. during the preparation phase); it might happen that one dry run allows for the trial execution, while also a third or fourth dry run might be required in case the set objectives are not met.

The Trial Integration Meeting

The main objective of the first preparatory meeting in the execution phase, the trial integration meeting (TIM), aims at connecting and operationalising the six-step approach results between the trial committee, further known trial staff and the (pre-)selected solution providers. The scratched trial context, information about the (pre-)selected solutions, and finally both the baseline and draft innovation line are of key importance to present, discuss, and critically reflect. The expected outcomes cover not only a clearly defined list of needs from both the practitioners (such as the implementation of innovative solutions into their operating procedures) and solution providers (such as historical data to feed the solutions) but also an elaborated and agreed innovation line. The latter further allows a clear data integration plan and its impact to the scenario design.

It must be stressed that the TIM should be planned carefully in terms of time and accommodation. It might happen that interviews with CM practitioners are needed to concretise initial thoughts and ideas. Thus, various collaborative tools should be applied to support the co-creative participation during the meeting (such as white boards, flow diagrams or joint documents). The structure might follow the presentation and discussion of the trial setting. The ultimate goal is to reflect the practitioner's needs in the light of established processes and workarounds in the baseline (the as-is situation of the involved CM organisation[s]). Once all participants agree on both the current situation and the gaps, an innovation line should be co-developed by the practitioners in dialogue with the solution providers describing the potential changes or innovations stemming from their suggested solutions. While the practitioner coordinator and trial owner concentrate on the development of the innovation line, the technical coordinator must think of specific solution use-cases associated with the generated to-be setup of the innovation line. Accordingly, the developed innovation line supports a first revision of the data collection plan and the final evaluation plan. On the other side, the technical coordinator collaborates with the involved solution providers on a particular solution interaction plan and a respective data integration plan. By doing so, a proper alignment between the conceptual and technical artefacts is ensured. The according results must be then transferred to a trial script, which serves as the screenplay of the trial.

The difficulty of the TIM definitely lies in the interconnectedness of each decision and the outcomes. A small change or adjustment (e.g. withdrawal of a preselected solution) has, by design, an impact on the previous decisions (e.g. the identified gaps). Patience and joint decision-making can reduce the complexity of the desired goals. Nevertheless, it must be stressed to leave enough space to discuss critically the main decisions and also to document the most likely happening disagreements or arguments: The main mitigation tool must be seen in the dedicated roles and responsibilities, ensuring a proper backing within the trial committee; for example, the evaluation coordinator should have the final word when looking at the data collection plan, while the data integration should be finally confirmed by the technical coordinator. Last but not least, every participant in the TIM must be aware that the further process might impact the decisions made during previous steps. This openness should be ensured at least until the final dry run. In any case, after the TIM, all participants should be aware about what each of them should do, how, and when from a conceptual perspective of the trial. In other words, the theatre scene is set and ready to be trained during the dry runs.

Dry Run 1

The initial dry run (DR1) aims mainly at testing the technical setup, the data collection plan, and the "handling" of solutions to be assessed during the actual trial. It is important that next to the trial committee and solution providers also respective CM practitioners contribute so that unbiased feedback can be given towards the trial scenario (being depicted along the baseline and innovation line), the developed data collection plan and the intended solutions training. A proper reference to structure the DR can be the trial scenario defined by the solution use-cases. Thus, not only the solution interaction and data integration plans can be tested, but also a first check of the collection and measurability of the data collection plan can be evaluated. For this purpose, a role-play with the participating CM practitioners and solution operators (at this stage represented by the solution providers itself). Depending on the required degree of the testbed technical infrastructure (to connect the trialled solution and the legacy systems) and potentially common simulation space (to imitate scenario-specific developments), the technical coordinator of the trial is the best candidate to lead the walkthrough to identify what is working well, which further adjustments are necessary and which elements of the trial should be revised conceptually. To achieve such a state, it is highly recommended to execute the DR1 in the same environment in which the actual trial is envisioned to happen.

Given the fact that the overall aim is to get a first (technical and data collection) stress test, the trial committee should follow a clearly structured agenda and meticulously document the progress and comments from the participants. It must be clear that this stress test seeks to identify inconsistencies in the story line, technical issues, and other kinds of frictions related to the trial and its underlying goals. Therefore, an error-friendly atmosphere can support and facilitate a proper trial environment. Main results can be seen in proof of concepts of the data collection and evaluation plans as well as solution interaction and data integration plans. In accordance with the error-friendly environment, a to-do list covering all identified improvement potentials can be seen as another key result of the DR1.

Dry Run 2

Once all the identified to-dos in the DR1 are fixed, the trial committee is able to proceed with the second (and hopefully last) dry run (DR2). While DR1 mainly covered the conceptual and technical feasibility, the DR2 serves as a full rehearsal. Thus, it fully follows the trial scenario and respective trial script. Its structure might be identical with the final trial run. The overarching objective is to ensure that the data needed to answer the underlying trial or research question can be collected as planned. Using the initial comparison of a trial as a theatre performance, it is the filled questionnaires by the "playing" CM practitioners and observers, the recorded data of the connected legacy systems and solutions, and the deployed common simulation space (i.e. the information about the executed CM operation) replacing the standing ovations by the audience.

One might ask what the difference to the actual trial execution might be. Perhaps the overall agenda might be denser and tighter (e.g. no formal introductions or networking activities in between the sessions); perhaps the participating CM practitioners do not fully match the requested actor profiles (e.g. early stage practitioners instead of experienced responders); perhaps the collected data will be only roughly processed or aggregated to prepare the evaluation phase (e.g. exemplary quality assurance of the applied data collection tools). However, if the trial committee has the required resources, a perfectly planned and executed DR2 could theoretically replace the final trial run. This scenario is not very likely because of the limited resources within the CM domain, though it might happen under certain circumstances. However, a DR2 can also fail in terms of the minimum goals: it might happen that the practitioners give negative feedback towards the degree of realism of the scenario and/or the depicted innovation line; unforeseen technical hiccups might occur because of the deployment in a new environment (e.g. strict

IT security settings of the host organisation or limited bandwidth; the collected data from the technical infrastructure or the practitioner questionnaires are not filled out properly). In this case, adequate countermeasures can be executed. The observed open issues might be few and not too complex; here, quick ex post fixes could help to go ahead with the final trial execution. If, however, the identified issues are more complex and require a wider revision of the trial plan, the postponement of the trial run and the addition of another dry run might be the only way out to still ensure the trial goals.

The ideal result of DR2 should allow a freezing of the key trial decision, in particular the fixed trial script, technical setup, and the data collection plan. It must be kept in mind that the actual trial run comes along with many other tasks, such as logistical arrangements or preparation of dissemination and communication activities, which will keep the trial committee busy.

Trial Run

After investing a lot of time and energy into the design and preparations of the trial, its actual execution can be described as the "theatre premiere." Given the TIM, DR1 and DR2 experiences and stress-tested results, the trial committee can trust the efforts taken to avoid any kind of disruption of a smooth trial execution. Most active participants might be thrilled and excited to see the trial be started. Nevertheless, there will certainly be several minor unexpected situations, which must be met flexibly and highly effectively. Therefore, the trial team should meet early enough to check whether everything is up and running. It might be useful to again go through the overall setup and document well with clear checklists. The advantage of the initially assigned roles becomes visible at this stage: every coordinator within the trial committee has clear responsibilities and tasks, which eases the execution through sharing the required tasks. Well-aligned individual efforts are merged to a smooth teamwork.

Nevertheless, the trial committee coordinators will be quite busy not only with checking the setup but also monitoring the progress. A special emphasis must be given to a proper introduction and training of the solutions to the "actors" being the CM practitioners as well as the data collection. The latter should be checked on the fly but also after every single attribute. A close collaboration between the technical and evaluation coordinator should be ensured regarding backups of each solution deployed during the trial. And just in case a serious disruption occurs (e.g. lost connection between the solution and the testbed or loss of collected data), the trial owner must have the right to intervene by pausing the execution until the issue has been analysed and fixed. Last but not least, the whole team should also celebrate the trial run because a significant amount of work needed to reach the trial goal has been reached. What the audience contributes to the ensemble in a "theatre premier" might be compensated by the feedback from the involved practitioners, observers, and visitors gathered through interviews or hot-wash-up sessions. The biggest applause and glory of the "trial ensemble" must be seen in the data that has been collected during the execution. In both cases, a proper celebration should be planned in (e.g. social event) to also motivate and move from the execution to the evaluation phase.

Evaluation Phase

If a "theatre performance" can serve as a proper comparison to the trial execution phase, then the TGM evaluation phase might be compared to the tasks of a "committee inquiry." While each and every set of collected data (such as time measurements or quotes from practitioners) must be analysed adequately (e.g. comparisons of time stamps during the executed baseline and innovation line runs or open codings of the received qualitative answers to the questionnaires), the findings for all three performance measurement dimensions must be synthesised to draw and disseminate conclusions within the respective peer groups. Thus, the evaluation process according to the TGM does not ask the evaluation coordinator to be locked in an office and work on the data but to organise a collaborative process within the trial committee (trial staff) and wider trial network (trial participants and relevant stakeholders). In other words, it is not just analytical tasks to be carried out as desk research, but the evaluation process must be understood as a participative process of sensemaking and communication.

Data Quality Check

Once the whole trial committee receives their well-deserved rest, the evaluation coordinator is in charge of initiating the evaluation process by verifying and preparing the collected data. The main contributor to the evaluation coordinator is the technical coordinator, who is aware of where data from the various sources (such as the filled-out questionnaires or recorded data of the applied tools) has been stored and whether any frictions occurred during its collection. The main objective is to ensure an adequate quality of data before it is processed for the sake of evaluation (i.e. answering the trial questions). The objects of interest are mainly the collected raw data, which might contain numerical (e.g. usability-related indications on a Likert scale or time stamps), textual (e.g. observer comments or qualitative feedback of practitioners), or graphical elements (e.g. snapshots of GIS tools or automatically generated diagrams through the trialled solutions).

To assess the quality, the raw datasets must be checked for completeness, formatting, and correctness. More specifically, the evaluation coordinator must check whether all planned data was collected appropriately: every participant and observer who was tasked (regardless of the format) to give feedback should be available at the foreseen directory. Here, not only the number of feedback forms must be verified but also its content. In case of missing (mandatory or important) answers, it still might be possible to double-check with the participants, if the missing data can be shared ex post, or ask for the reasons for its absence. The same applies for all the various data that was collected by either parts of the technical testbed infrastructure (e.g. time stamps of decisions made by the practitioners throughout the scenario in the after-action review tool) or the applied solutions (e.g. received field unit messages to create a common operational picture).

Once the availability of the planned data to be collected is checked, the evaluation coordinator must also ensure whether each of the files, physical (paper-based questionnaires) or digital (different file formats), is not broken. In case there are files which cannot be accessed or processed (e.g. bad handwriting or corrupted electronic files), the evaluation coordinator might first need to estimate its importance to the overall assessment. According to the importance, the respective data should get fixed if possible (e.g. contacting the technical coordinator whether alternative formats can be made available) or to be dismissed. If the latter is the case, it is of utmost importance to document and describe deviations from the initially planned data collection approach.

Finally, as a transition to the data analysis step, a preliminary structuring according to the performance measurement dimensions should be executed. By doing so, the later evaluation tasks can be also shared within the trial committee (e.g. the trial director could take over the analysis of the trial dimension, while the technical coordinator would navigate best through the solution dimension-related data). The main outcome of the data quality check can be seen as a complete, clean, and pre-structured dataset.

Data Analysis

As for the data quality check, it is the evaluation coordinator being responsible for coordinating and executing the actual data analysis. The main counterpart within the trial committee is the practitioner coordinator, who best knows the backgrounds, contacts, and first impressions of the CM practitioners who participated in the trial. Since the data quality has been assessed carefully in the previous step, the main task within the data analysis step covers an itemisation, aggregation, and visualisation of the preliminary structured dataset.

To have a starting point, the raw data was already assigned to the three performance measurement dimensions: the trial, CM, and solution dimension. The trial dimension has further a predefined sub-structure according to the quality criteria set in the preparation phase. Here, one can first have a look whether all thresholds were reached by aggregating the raw data: How was the degree of realism perceived between all participating actors (e.g. practitioners perspective vs. solution providers)? Did the involved practitioners meet the required profiles (e.g. in terms of experience or functional match)? The aggregated answers give a first indication about the intended quality requirements of the trial. For the CM dimension, the time-wise flow of the trial is a promising approach to get a first idea on the results of the simulated operation. Depending on the predefined performance metrics, the evaluation coordinator might observe some characteristics or patterns of the executed operation. For example, if the task was related to achieve a better situational awareness, a direct comparison to the set ground truth known by the trial committee gives some first insights whether the innovation line led to better or worse results than the baseline run (or other comparable reference data such as historical data or simulated outcomes). In other words, the operational tasks and goals per episode of the trial scenario serve as a structure guide to prepare the collected data for a robust assessment. Finally, the raw data in the solution dimension must be structured according to each trialled solution and each solution is differentiated according to (1) the results on usability and (2) the feature-based assessments.

Depending on how the data has been collected (e.g. paper-based questionnaires or digital data collection tools), all inputs must be aggregated according to the described order and ideally stored at a central place. It must be noted that some solutions might be producing data itself (e.g. reports by the CM practitioners in different formats or simulation run outputs). In this case, the evaluation coordinator might need some support from the solution provider; for example, if applying a common operational picture (COP) tool, it might be important to save screenshots of the final situation assessment episodes but also access to the individual COP inputs (e.g. common alerting protocol messages between the applied devices by the field staff and the central COP database).

Once fine-structured and aggregated data is available, the evaluation coordinator and/or the respective responsible staff can start the actual analysis. Theoretically, all appropriate data analysis approaches, regardless of their formats, can be applied here (e.g. open coding of long qualitative feedback or sensitivity analysis of quantitative results). Practically, the final set of the applied specific analytical methods very much depends on the selected evaluation approach and the nature of the collected data, but also according to the available analytical capabilities within the trial committee and the available resources. While the trial dimension might be simply checked by comparing the target values or intended thresholds with the achieved results (e.g. perceived degree of realism of the trial scenario in relation to the intended degree of realism), the CM and solution dimensions require some analytical work. However, for all three dimensions, it is recommended to visualise as much as possible. For example, pie charts give intuitive overviews describing the backgrounds of the trial participants (e.g. countries of origin or operational experience in years). The processing approaches of the CM dimension highly depend on how the perceived CM performance was described and defined by the involved practitioners. It might contain only a few KPIs, such as the response time of an operation (i.e. the time between the emergency occurrence, like an emergency call, and the arrival on the scene or the delivery of a rescue service, like securing the danger spot) or the number of saved persons in need (e.g. number of evacuees in relation to number of persons in need of an evacuation). Here, a direct comparison to the reference data visualises numerically whether and to which extent the implemented innovative solution impacted the observed performance. However, there might be more complex definitions of CM performance, such as an achieved and perceived situation awareness. In these cases, the evaluation coordinator will rely on what was agreed with the respective CM practitioners during the preparation phase. For example, in the case of situational awareness, one possible way to identify the impact of the innovative solutions by not only comparing the fed and visualised field reports on a COP-tool but also to incorporate qualitative feedback by the decision-makers towards the visibility or understanding of the COP. Here, the actual analysis would then consist of a direct comparison between the ground truth (defined by the trial scenario), the innovative geospatial information (innovation line) and the traditional COP (baseline). Finally, within the solution dimension the application of spider diagram has proven as a well understandable overview on the aggregated usability assessments on both the innovative solutions and the compared legacy systems. Besides, the qualitative comments shared by the applicants must be considered to better understand the given rankings. The spider charts allow a direct comparison between the legacy systems and the innovative solutions. Besides, for each trialled solution an aggregated view on the (perceived) feature- and maturity-based assessments can be visualised. Theoretically the same questionnaires might be executed for the legacy systems, which can be though understood as an optional choice since comparisons to the current systems can be directly investigated through dedicated questions.

Many of the resulting overviews (in terms of numbers, charts, or comments) give the evaluator already valid information about the underlying trial questions. These insights are quite important, which is why it is recommended to note down several patterns, inconsistencies or aha moments. Data does not lie, but it can only make sense in a context (e.g. the combination of a clear trial question, a specific pattern and the background of the reader). Therefore, the notes of insights are a great source for the moderation and collaborative sensemaking of the results in the next step, being the data synthesis.

Data Synthesis

The key objective of synthesising data covers the ability to draw valid conclusions from the trial and to agree on the main takeaways of the assessed solutions. Again, the evaluation coordinator should moderate this process in a close collaboration with the trial owner but also the broader trial team. The starting point for the data synthesis step builds the analysed data in combination with the observed notes from the previous step. Although the evaluation coordinator has the underlying gaps and trial objectives in mind when analysing the data, it is now time to present and discuss the results within the team. Therefore, it is recommended to block enough time and a proper setting, ideally a physical meeting.

The best approach is to follow the already well-known structure of the three performance measurement dimensions. Here, the evaluation coordinator might present, dimension by dimension, the main findings by using tables and visualisations prepared in the previous steps. The trial committee should together go through each of them and discuss patterns, surprising results, inconsistencies, or even confirmations of their own beliefs. Once every observation is discussed and noted for each dimension, it might be useful to visualise the findings on a flipchart or a PowerPoint slide. Here, the participants should ask themselves if some of the observations between the performance measurement dimensions are interrelated and how. For example, if a particular solution received a rather bad usability assessment, it should be related to the respective evaluation results of the executed training: Here, a rather badly experienced training might be the cause of an according usability assessment; if, however, the training has been assessed positively, it might be worth to again look into the comments provided by the applicants of the solutions. In other words, the trial committee first must ensure a shared understanding of the evaluation results in its overall setup. Once all expected or unexpected results are checked (session per session, solution per solution), the trial evaluator can recall the initial gaps, trial objectives, and questions trying to conclude the learnings in the group. Here, everyone should be heard, meaning that also solution providers might be contacted to share their feedback, so it can be incorporated to the overall evaluation results.

Assuming that the trial committee consists of one or more CM practitioner organisations, it is further recommended to present and discuss the results with their closer networks. By doing so, the novelty, potential impact and transferability of the findings can be further investigated. Here, especially the reference of the baseline and the specific deduction of an innovation line can be also stressed so that the broader CM community can benefit from the results. Pointless to mention that all these tasks must be properly documented to share the feedback with all stakeholders and, if of relevance, external partners and peers. Why and how to share this knowledge is explained in the next and last step.

Dissemination of Results

The main objective of this task is meant to ensure the sustainability of the results. Here, the trial owner and the practitioner coordinator can proudly look back together at a quite extensive process leading to highly relevant knowledge. The evaluation report prepared in the previous steps contains all relevant information and direct conclusions from the trial, but also indirect feedback from domain experts. Since time is a scarce resource in CM, it might be good to prepare some easy to digest information sources, such as articles, website posts, or brochures.

However, there are already some well-accepted formats to trials in particular. The trial owner should not miss them because the whole trial team gained from them already:

- The solution dimension results might be uploaded to the portfolio of solutions.
- The trial can be briefly described at the trial guidance tool.
- The specific trial setup (including trial question, evaluation approach, etc.) can be added to the knowledge base of the trial guidance tool.

There will be a lot to share, which also allows for smaller inhouse events, conference or summit participations, or even the organisation of an own trial-related event. Here, the participants get the chance to not present their findings but also to learn further from CM practitioners who reflect on the trial evaluation or who had similar assessment formats. It might happen that the discussion about the trial leads to thoughts on broader topics (e.g. response-specific thoughts), and these findings can be also implemented into the lessons learnt library which go even beyond the scope of the initial trial question. In any case, the end of the trial smoothly transits into the next innovation cycle improving the response capabilities.

Research Ethics and Societal Impact Assessment

As a final note, it must be stressed that throughout the whole TGM cycle research ethics must be considered, while considering the potential societal impact assessment for the trialled solutions is highly recommended.

Given the rather context-bound consideration of research ethics, each TGM step has to properly take into account that the data gathering, processing, and analysis for the three performance measurement dimensions might affect data protection rules and requirements. Here, especially the general data protection regulation of the EU (GDPR) sets a number of guidelines, procedures, and rules to protect privacy principles of the trial participants. Besides, also, national laws and discipline-perspective research ethics (e.g. medical trials) have to be taken into account. From a methodological perspective, it is therefore hardly possible to provide generic questionnaires or specific requirements a priori. However, because of the high importance of research ethics the TGM provides specific research-ethics-related questions, which are made precise through checklists for all phases and steps. Additionally, a dedicated section describing the respective key GDPR principles (e.g. transparency, data minimisation, or data storage limitation) in the TGM handbook (Fonio & Widera, 2020, pp. 90–93).

Besides, it must be noted that CM-related innovations might have a strong impact on the societies, in which they are deployed. This impact can be intended and positive but also unintended consequences must be investigated before the implementation of new solutions. Therefore, it is of key importance to ensure that the implementation of CM solutions maximises the added value while reducing societal costs and burdens. Here, a specific tool – a societal impact assessment (SIA) tool – was developed during the DRIVER+ project and became an integral part of the TGM. Similar to research ethics, SIA can only be executed in the specific trial context, the respective participants, and selected solutions. The process consists of the following five steps:

- Identify stakeholder groups/communities.
- Collect background information.
- Get an overview of legislation and policies.
- Identify and predict impacts.
- Describe mitigating measures and follow-up.

The tool itself is summarised and exemplary applied in the TGM handbook (Fonio & Widera, 2020, pp. 86–89). For its application, it is recommended to consider the entire SIA framework (Bergersen & Reuge, 2017).

Outlook: So What?

The trial does not end "on stage." The execution is, in fact, a potential new beginning. Crisis management organisations who have been involved in organising a trial will have plenty of lessons learnt to share with the wider crisis management community. Lessons learnt may refer to different aspects: the trial dimension and how some constraints can impact on trials (or the other way round), the co-creative approach and work processes put in place with solution providers, the innovative ways in which socio-technical solutions can help addressing gaps, the added value of carrying out trials, ethical dimensions to take into considerations, how to best use resources, and so on. It is recommended to disseminate the results not only internally with those who were directly involved in evaluating the solutions and / or organising the event, but also externally so that the experience gained can be of relevance to others in the same domain. Organising an internal meeting first to share and discuss with all the stakeholders what has been "learnt" and then presenting key takeaways to the external CM world would be the right approach to go beyond the needs and challenges of just one organisation. This should not be considered as an "extra mile"; rather, dissemination is inherently part of the TGM and it is necessary to reflect upon what a crisis management organisation learnt and what other organisations can benefit from. Learning entails implementing changes: such as purchasing new solutions, using in more effective ways existing legacy tools, rethinking work processes, and so on. The TGM stops here. The journey for a methodological approach stops when the methods have been applied. It will be up to crisis management organisations to take decisions and implement changes, as needed. New beginnings are in their hands!

References

Bergersen, S., & Reuge, E. (2017). *Societal Impact Assessment Framework (D840.11), Driving Innovation for Crisis Management for European Resilience (DRIVER+)*. https://driver-project.eu/wp-content/uploads/2017/11/Societal-Impact-Assessment-Framework-Version-1.pdf.

Fonio, C., & Widera, A. (Eds.). (2020). *Trial Guidance Methodology Handbook. DRIVER+ (Driving Innovation in Crisis Management for European Resilience)*, Brussels, 2020. https://tgm.ercis.org/.

Grunnan, T., & Fridheim, H. (2017). Planning and conducting crisis management exercises for decision-making: The do's and don'ts. *EURO Journal on Decision Processes*, 5(1–4), 79–95. https://doi.org/10.1007/s40070-017-0065-0.

iso.org. (2018). *ISO 9241-11:2018(en) Ergonomics of Human-system Interaction – Part 11: Usability: Definitions and Concepts* [online]. www.iso.org/obp/ui/#iso:std:iso:9241:-11:ed-2:v1:en. Accessed 7 July 2020.

Kahneman, D., Sibony, O., & Sunstein, C. R. (2021). *Noise: A Flaw in Human Judgement.* London: William Collins.

Mazzuccato, M. (2015). *The Entrepreneurial State.* Great Britain: Penguin Books.

Mazzuccato, M. (2018). *The Value of Everything.* Great Britain: Penguin Books.

Robbins, S. P., Campbell, T., & Judge, T. (2010). *Organizational Behavior.* Harlow: Financial Times Prentice Hall.

Rogers, E. M. (1962). *Diffusion of Innovations.* New York: Free Press of Glencoe.

Widera, A., Fonio, C., Lechtenberg, S., & Hellingrath, B. (2018). Measuring Innovations in Crisis Management. *Proceedings of the 15th International Conference on Information Systems for Crisis Response and Management.* https://www.researchgate.net/publication/325283498_Measuring_Innovations_in_Crisis_Management.

PART 2

Technical Infrastructures to Assess Potentially Innovative Solutions

Part 2 deals with all relevant technical aspects which are key to carry out trials. It describes in detail the essential components of the open-source DRIVER+ testbed, developed to enable the trialling of solutions within realistic and controllable environments (Chapter 3). The explanations of the technical infrastructure are coupled with use-cases, as well as boxes for developers who may be interested in getting an understanding of, for example, the software stack advised by DRIVER+. Support tools and services are also outlined. In addition, the knowledge management and inference toolset for the CM domain hosted on the so-called portfolio of solutions (PoS) is outlined (Chapter 4). The focus is on the repository of solutions, the gaps explored, and the trial guidance tool that were developed in the course of the project and are still available. Part 2 can be of interest to a wide variety of stakeholders: from developers to practitioners, from solution providers to future trial owners. It offers practical considerations that cannot be overlooked when planning trials. It is a technical snapshot into what is needed and what is available.

DOI: 10.4324/9781003256977-5

3

TESTBED TECHNICAL INFRASTRUCTURE

Erik Vullings, Martijn Hendriks, and Steven van Campen

Introduction

The importance of crisis management (CM) is growing, as societies are getting ever more interconnected and natural and man-made incidents have an enormous impact on human lives and well-being. The main challenge is the scale and unpredictability of these crises. In order to manage them better, there are several options available, such as improving procedures and methods, training of personnel, new tooling, or adapting the environment and infrastructure. Whichever option is chosen, it is a best practice to have the new procedure, training, or tool tested and evaluated before implementing and introducing it into an existing crisis management operation, especially considering the huge impact a new procedure or tool can have on existing crisis management operations.

To support this common need for assessing CM solutions, a realistic, controllable, and observable context is needed. Realistic, so it is more likely that the innovative solution will also work during a real crisis. This implies that a crisis needs to be staged in such a way that the solution and the CM staff act as if they are dealing with a real crisis. Controllable, so the solution can be tested against the majority of use-cases which it should support. And finally, observable, so a proper assessment of the solution's operation can be done at the end of the trial, as described in chapter 2 about the trial guidance methodology (TGM).[3]

In addition, after selecting a solution, this same environment that was used for the assessment can also be used to train the staff in operating it – crisis-induced learning.

To summarise, for a successful assessment,

- a fictive crisis must be staged, offering a sufficiently realistic experience for the CM practitioners and a realistic environment for the assessed solution,

DOI: 10.4324/9781003256977-6

- which can be controlled such that all applicable use-cases for assessment occur,
- in which data and observations can be gathered for a proper evaluation at the end.

To facilitate all of this, the DRIVER+ project developed the testbed technical infrastructure (in short, testbed). The testbed is a set of open-source software applications, offering CM organisations a way to create a realistic, controllable, and observable environment to trial solutions – both software-based and physical tools. The testbed consists of the following:

- The *Common Information Space (CIS)*: An integration layer to easily share information between existing software (i.e. legacy) systems and new solutions. For example, to share information between an existing crisis management application and a new resource management tool. The CIS thus assures that a solution can function in a realistic setting among other systems and data sources. Details are provided in the section on the Common Information Space.
- The *Common Simulation Space (CSS)*: An integration layer to easily share simulated incident information among several simulators and with the applications connected to the CIS. Together, those simulators create a realistic and immersive crisis scenario – for example, by sharing a simulated flood progression map with a 3D simulator for practitioners and first responders, and with an operational crisis management application. The CSS creates a simulated reality consisting of as much information on the crisis as needed to let the solutions function, to immerse the participants in the fictive crisis, and to create a ground truth for the trial staff in which realistic use-cases can be triggered. The CSS is described in detail in the respective section.
- *Tools* to set up, control, and observe this environment before, during, and after a test, dry run, and trial session: This set of tools consists of software application to set up and test the entire collection of software needed for a trial, control the execution of the trial scenario containing the use-cases to trigger, and collect digital data from the CSS and CIS and collect human observations. This is explained in more detail in the section on the support tools and services.

The rest of the chapter explains the ins and outs of the testbed. It gives a complete overview of all components of the testbed, it explains in detail what the CIS and CSS are and what can be done with the other tools, and it provides illustrative use-cases. Furthermore, it describes the steps involved in setting up the environment and (dry-)running a first trial. It concludes with a perspective on the future of the testbed, considering that the DRIVER+ project has ended. All DRIVER+ software is released as open-source and can be found on GitHub as part of the DRIVER+ organisation: https://github.com/DRIVER-EU.

Take-Home Message

The DRIVER+ testbed is still being actively developed and continues to be used in several projects, trials, and exercises. It offers a flexible way to connect new and existing crisis management systems and feed them with simulated information or scripted data from a fictive incident. By choosing only the testbed components that you need, you can create a lean, adequate, and powerful environment for assessing new solutions or for training.

Overview of the Testbed Technical Infrastructure

This section starts with an overview of all possible components of the testbed and all its users. However, not all components or users are always needed. The testbed is a facilitative infrastructure, scalable to the needs of a specific trial. Technical staff is therefore advised to select and configure only those components needed to run a valid assessment of the solution(s).

Following the components overview, the CIS is explained in more detail, and a use-case describes how the CIS can integrate an innovative and an operational crisis management information system (i.e. linking a solution and an operational system). Then the CSS is described, integrating multiple simulators, and a use-case is presented, which integrates four off-the-shelf simulators to stage a fictive flood scenario. Next, the other tools to set up, control, monitor, observe, and evaluate a trial are explained, together with a use-case of these. Technical details are provided in a separate break-out section, aimed at software developers.

The Complete Testbed and Its Users

Figures 3.1, 3.2, and 3.3 show a small-scale, a medium-sized, and a full-scale testbed. A minimum use of this infrastructure would be to trial one solution in combination with an operational system, in which the testbed functions to link

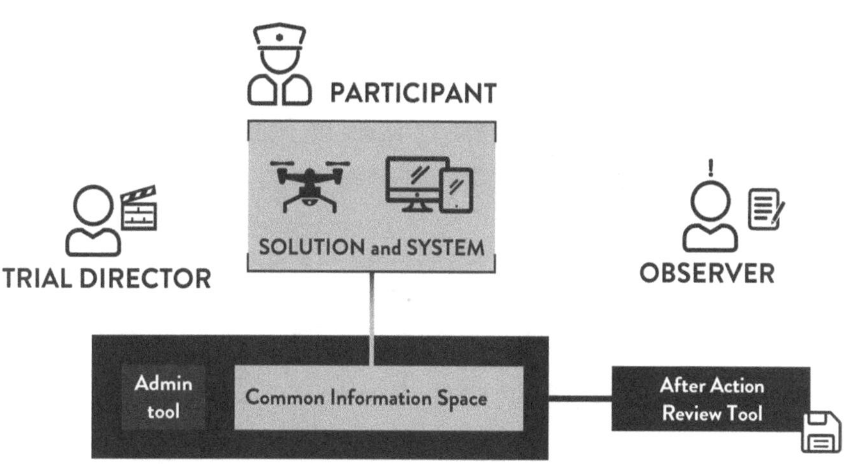

FIGURE 3.1 A simple setup of the testbed and its users.

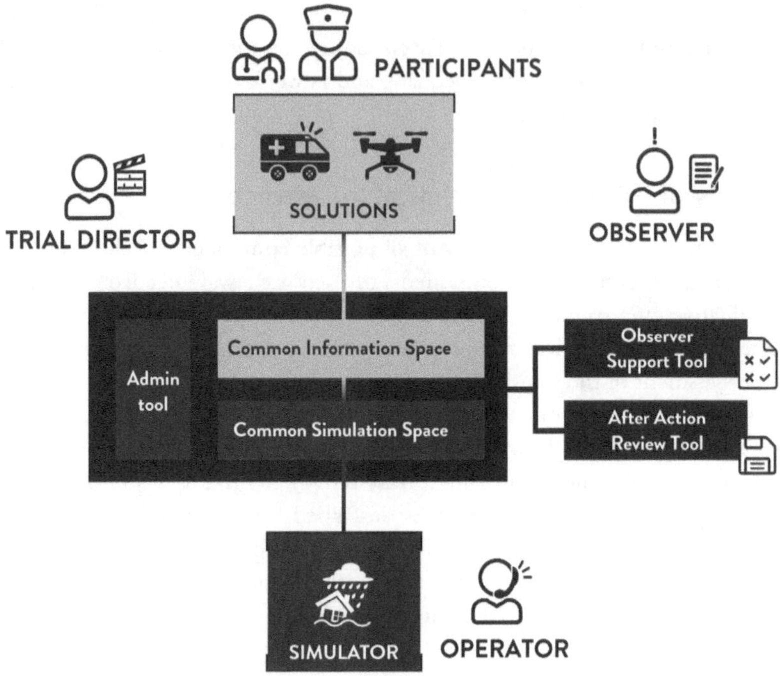

FIGURE 3.2 A medium-sized setup of the testbed and its users.

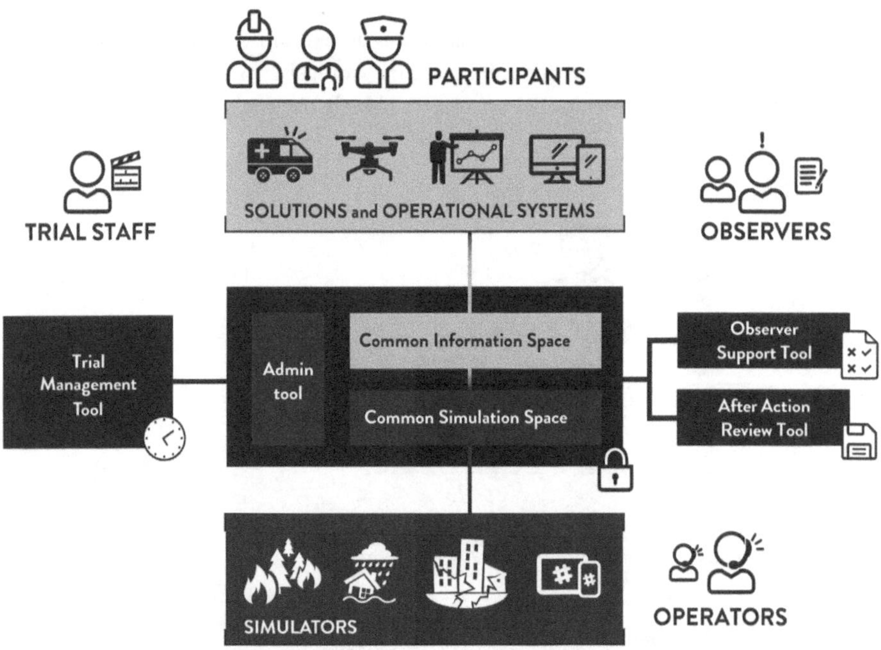

FIGURE 3.3 All components of the testbed and their users.

these two via the CIS and to log the digital data on how this setup was used by the CM practitioner (i.e. participant). The trial staff in this case consists of one trial director, who leads the trial run and operates the testbed, and one observer using pen and paper to meticulously note down (qualitative) observations.

An example of a medium-sized setup to assess two solutions is shown in Figure 3.2. Compared to the minimum use in Figure 3.1, a simulator is added, which is used by an operator. Furthermore, the observer is given the digital observer support tool to collect and store the observations made.

Figure 3.3 illustrates all components of the testbed, needed to conduct a large-scale trial, similar to those run in the DRIVER+ project. This setup consists of the following *users* and *components*:

- Participants (e.g. experienced CM practitioners) react to the fictive crisis at hand. To manage this crisis, they use the solutions and the currently operational systems.
- These solutions and operational systems are linked to the CIS, such that these can exchange data with each other, can be fed with simulated data, and that data can be logged for evaluation.
- Operators (being part of the trial staff) use the simulators to provide information to the Participants via the solutions and/or operational systems.
- Participants can also use one or more of the simulators to acquire sufficient and realistic immersion in the fictive crisis.
- The simulators are linked to the CSS, so they can exchange data with each other, to feed the CIS and thereby the solutions and operational systems, and such that simulated data can be logged and stored for evaluation.
- The trial director, role-players, and other staff use the trial management tool to play out and control the scenario and the admin tool to monitor the technical infrastructure.
- Observers use the observer support tool to log what the participants do. This information is fed back to the trial director and other staff.
- When the trial has ended, the trial director and operators use the after-action review tool to gather all data and observations from the CIS, CSS, and observer support tool, such that this can be analysed and synthesised in a report containing a timeline of all events.

Using the Testbed in Exercises

The testbed is actually suitable to facilitate both trials and exercises (also known as digital tabletops or simulation-based training). This dual use became apparent during the development of the testbed's components in the DRIVER+ project. The designs are based to a large extent on experiences and action research by the founding developers, who facilitated military and civil CM exercises (Hendriks et al., 2019).

Where trials are aimed at assessing socio-technical innovations (i.e. solutions or "new things"), training and exercises are aimed at teaching, improving, and testing the behaviour of practitioners in an existing situation (i.e. with operational systems and procedures, or "as it should be done"). When looking at trials and exercises from a facilitative and technical perspective, both share a large overlap: both require an immersive play through of a crisis scenario, both require thorough methods for observation and evaluation, and both require the same set of people to participate. The following sections could therefore be read from the point of view to facilitate a trial but as well to run an exercise supported by off-the-shelf technology.

Note, that using the testbed to optimally facilitate an exercise requires an initial investment for software developers to set up the infrastructure. But once done, trialling or exercising is easy. The testbed could even be used operationally – for example, to link up two operational CM systems, thereby creating a new, innovative set of tools. However, this operational use is outside the scope of this chapter.

Common Information Space

The Common Information Space (CIS) connects any combination of solutions and operational systems (i.e. software tools) so information can be shared and logged. The CIS should be used in case

- solutions and/or operational systems need to exchange data, and/or
- one or more solutions and/or operational systems need to be fed with fictive crisis data, and/or
- the exchanged information must be logged.

The CIS is used to facilitate data exchange between solutions in a transparent and reliable way, in order to enhance the collaboration and the effectiveness of crisis management while using these solutions. Having a set of components connected via the CIS is more likely than using only a standalone solution, because a solution is (almost) always used in combination with other systems and external data sources. In the total socio-technical context, it will for sure be used in combination with other already-operational systems and experience shows that solutions often need to be fed with data from the real world or from the (simulated) crisis.

Exchanging messages via the CIS is done by using current emergency management data exchange standards, like Common Alerting Protocol (CAP) or Emergency Data Exchange Language (EDXL). Other data exchange standards from outside the emergency management field, like, for instance, GeoJSON, are also supported. This facilitates exchange of understandable information between different organisations, even if they use different data formats (syntactic interoperability) and different languages and/or taxonomies (semantic interoperability). Main benefit is that the systems connected to the CIS do not have to adapt to all the different data formats of other systems yet can still exchange information with them. If a solution or operational system is not yet using such data exchange standards, their data inputs or outputs first need to be transformed into common standard formats.

To connect the solutions and operational systems with simulators, the CIS can be connected to the Common Simulation Space (CSS) via the so-called CIS-CSS gateways. Data from the simulators is filtered and translated into data that can be understood by the solutions connected to the CIS and requests from the solutions can be relayed back to the simulators.

For a valid assessment of a solution, it is crucial to know how it is used and in which context. The testbed provides tools to configure the CIS and log all data that is produced by a solution, operational system, simulator, or testbed component, thereby making it available for monitoring the CIS status in real-time and for evaluation afterwards.

One major aspect of the developed CIS concept is data protection and security, which is considered necessary in order to create trust among the integrated organisations and their systems. This is achieved by a trusted registration process for all organisations and an encapsulation of all messages exchanged via the CIS.

Use-Case 1 – A Operational System Exchanging Data with a Solution

Within the DRIVER+ trial in The Netherlands (trial NL), a flood scenario of an urban area in The Hague was played out. Its focus was on planning and communication regarding large-scale evacuation of parts of the city. This requires both governmental agencies, institutions, and critical infrastructure companies to communicate clearly about this emergency. At the time of the trial, the operational governmental crisis management system, called LCMS, was restricted to be used within the main emergency management branches of government. This makes crucial crisis information inaccessible via this system for agencies and companies that are also involved or affected by it (e.g. electrical or public transport companies). The innovative solution CrisisSuite from Merlin Software was included in this trial to serve as the crisis management system used by non-LCMS users. The infrastructure also included a data communication bridge between LCMS and CrisisSuite to enable the information exchange between all organisations participating in the trial.

The testbed's CIS facilitated this data communication bridge. Figure 3.4 depicts part of the data exchange diagram created for this trial, focusing only on the communication between LCMS and CrisisSuite via the CIS. Both LCMS and CrisisSuite include the functionality for a user to create geographical plots and situation reports. For this trial, it was requested for those two information sources to be synchronous across both systems: whenever a user of LCMS would publish changes to a report or plot, it would need to be automatically visible in the CrisisSuite system, and vice versa.

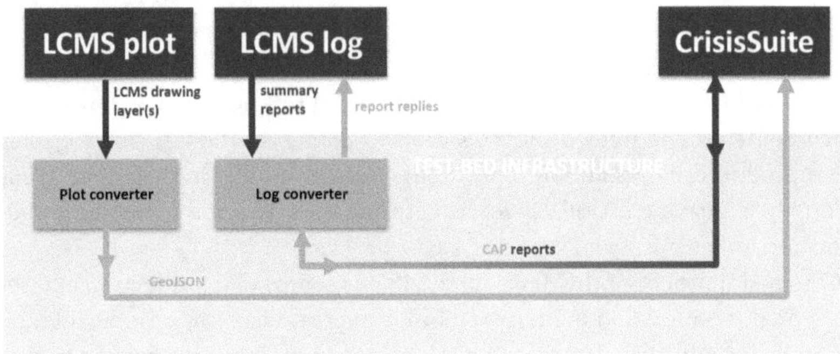

FIGURE 3.4 Part of the data exchange diagram depicting LCMS and CrisisSuite connected to the CIS, including the data exchanges between these systems.

By technically connecting both systems to the CIS with the CIS adapters and using the CAP data exchange standard for the situation reports and GeoJSON for the plots, the CIS provided the message infrastructure for LCMS to send plots to CrisisSuite (see Figure 3.5 for an example) and for both systems to send incident reports to each other. During the trial, this implementation proved very stable in synchronising these data and provided a good example of how the solution could improve the overall crisis management information exchange. By having the data exchanges going

FIGURE 3.5 Screenshot of Merlin Software's CrisisSuite: (note no. 3) that was synchronised with LCMS to display plot information concerning the flooded area and suggested evacuation routes. In addition to this synchronised information, organisation-specific information (like locations of interest) is also shown.

through the CIS, the after-action review tool allowed trial assessment to further analyse what and at which frequency information was exchanged between both systems. This, including the comparison with a test run without the solution, proved to be very fruitful in understanding to which extent the solution was bridging the crisis management gap of information mismatch between governmental LCMS users and CM organisations not able to access LCMS.

Common Simulation Space

The CSS connects standalone simulators together, so they can exchange data with each other to present a fictive yet realistic and immersive crisis. The CSS should be used in case

- one or more digital simulators are needed to feed a solution and/or operational system, and/or
- participants need to be immersed in the staged crisis in a more realistic manner than by providing them a spoken or written description of the crisis situation, and/or
- the trial scenario is complex and trial staff need a "ground truth" providing them at every point in time a realistic overview of the entire crisis, and/or
- it is needed to digitally log the evolution of the crisis scenario over time for detailed evaluation.

Participants, solutions, and operational systems connected to the CIS typically require information from a fictitious crisis (e.g. number of resources present at a certain dispatch location or at the incident scene). The CSS is the component within the testbed that provides a framework for simulators to jointly generate and maintain a simulation world needed for the solutions and operational systems and for the participants to get a sufficiently realistic impression of the fictitious crisis to resolve. Immersion is essential to create a trial context realistic enough to conduct valid solution assessments.

Dependent on the trial scenario, simulators are to be selected, based on the following:

- Whether solutions or operational systems need data from the simulated crisis, which they cannot get from each other. Examples of simulated data can be sensor data, the positions of (simulated) crisis management units, sending messages (e.g. fictive social media posts and emails), and providing realistic visualisations (e.g. 3D incident views for first responders).
- Whether participants need extra information about and immersion in the simulated crisis such that they can complete their current standard operating

procedures and CM processes (e.g. eye-level view of the crisis, simulated by a virtual reality application or by staging this in the form of physical items on a live-exercise terrain).
- Whether information in the scenario needs to be pre-calculated/pre-simulated for realism (e.g. a realistic wildfire progression).

To satisfy these needs and create a complete enough simulated reality, often multiple (off-the-shelf) simulators are needed because each has its specific purpose and limitations (e.g. a wildfire calculation model, a social media simulator, or a 3D incident scene visualiser). The CSS allows multiple simulators to focus on their part of maintaining the current state of the simulated world (i.e. the simulated "ground truth"). The CSS in itself is thus not a simulator, and the open-source version of it available online does not come with simulator functionalities. Software developers have to connect each simulator to the CSS, the same as each solution needs to be connected to the CIS. Details are provided in the technical break-out section later in this chapter.

In order to communicate state changes with other simulators linked to the CSS, standardised message formats are preferred, but self-created communication messages are also allowed. This differs from the messages being sent over the CIS, because the CIS is aligned to and should adhere to current emergency management data exchange standards (see also the information on CIS presented earlier). Simulators all have their own data model of how their simulated world is represented. The CSS allows these simulators to agree on a communication form that the applicable simulators understand to create and maintain a jointly simulated world.

To direct the simulated world towards a desired scenario relevant for the trial, the CSS is connected to the trial management tool, which can send out messages to change the simulated world (i.e. injects directly processed by simulators) – for example, to initiate a dike breach, to let a container explode, or to drive ten ambulances to the incident scene.

Via the CIS-CSS gateway, solutions and operational systems can be fed with simulated crisis data. Although technically using the same underlying implementation, the split between CIS and CSS is made to ensure that the two spaces of simulated truth inside the CSS and perceived/communicated truth inside the CIS are kept separate. The CIS and CSS are linked by direct data exchanges, or gateways can be used. Gateways can filter which data is completely, partially, or temporarily fed from CSS to CIS. These filters restrict participants from getting (all) the correct information out of the simulators (e.g. instead of seeing the actual flood map as it is simulated, only the part of the flood map that is measured/observed by sensors is shown, so the participants have to deal with incomplete information). So whereas trial/exercise staff can see all (correct) information and the ground truth provided by the simulators, participants may only be able to see part of that information or may deliberately receive incorrect information.

Use-Case 2 – Four Simulators to Stage a Flooding Crisis

Say, a trial is to be organised with the following characteristics:[1]

- A gap has been identified for large-scale evacuations – for example, when facing a natural disaster – that off-site command teams (i.e. silver level) have difficulties optimising the evacuation strategy, as they rely on on-site observations (i.e. bronze command level) of the state and evolution of the affected area, resource positioning, and resource availability.
- The common operational picture (COP) tool currently in use by the emergency services functions well for communications between on-site and off-site teams and shows on a map the currently affected area – plotted by on-site users – and the GPS coordinates of all the region's resources and their availability. It, however, does not show how the affected area and resource availability will or could change over time (e.g. a flood or wildfire front is not stationary, and resources become unavailable for a new task when they are deployed or due to the disaster itself). Therefore, the current map plot cannot be used to optimise evacuation strategies.
- An innovative evacuation and resource planning software package is selected as a solution to be assessed. It shows great potential, enabling CM practitioners to calculate an optimal evacuation strategy, based on the geographical characteristics of the entire region (e.g. road map and evacuation centres), the evolution of the affected area, the number of evacuees in this area, and the resources available for evacuation.
- By diving into when this gap occurs and what would be a challenging use-case for the solution and the participants, it is decided to stage a flood, affecting a densely populated area in The Netherlands. In this area many resources are stationed by default, some potentially affected by the flood. The flood spreads quickly through the urban area, as this area contains many streets, canals, and low dikes but few nature areas able to absorb flood water.

To facilitate this setup, the trial staff faces the following challenges:

- To have a realistic flood progression, a calculation model is needed. The characteristics of the urban area and the level of detail, needed to accurately see how the flood spreads through this area, are very complex. Even for senior flood experts, it is too complex to draw out a realistic flood map over time when they have to base it only on their experience.
- The operational COP-tool needs to be fed with a large amount of resource positions and their availability based on whether they are inside

or outside the flooded area. As there is no real flood and it would be an enormous amount of work to drive around and track many real response units, a resource simulator is needed that can send GPS coordinates of simulated units and their availability to the COP-tool.

- The evacuation strategy solution also needs the positions and availability of all resources. It also needs the flood progression over time, to determine where inhabitants need to be evacuated and to determine which roads are available at what time.
- To immerse the commanders in the scenario – especially for those on-site – 3D visualisations are to be provided of the positions they are stationed. These visualisations need to be dynamic over time, such that they show an up-to-date flood level based on the time in the scenario.
- During this staged flood, tactical and strategic command teams (i.e. silver and gold levels) need to communicate with many more organisations than only blue-light emergency services. For security reasons, only blue-light services have access to the COP-tool. Hence, tactical and strategic commanders also need a means to communicate with other organisations (i.e. role-players). In reality, email is mostly used for these communications. But the real, operational email system should not be used during the trial, because then fictive trial emails would circulate among real-world emails, greatly risking a mix-up of real incidents with the staged flood. Moreover, real-world email conversations cannot be controlled by the trial staff, while monitoring, control, and logging is essential for evaluation.

Software technicians solve these challenges by connecting four simulators to the CSS and thereby indirectly to the CIS:

- *Flood calculation model* (e.g. 3Di) calculating a realistic flood progression over time occurring in the region in which the participants are working in reality. A fictive dike breach is triggered, and the simulator creates a detailed overview of water heights based on the time after breach and the region's height and land-use data. This flood model is fed to the 3D and resource simulators and to the evacuation solution.
- *Resource simulator* (e.g. XVR Resource Management) to simulate emergency response units stationed in the region and potentially being affected by the flood. The resource availability and their positions are fed to both the COP-tool and the evacuation solution. The flood model is used to determine resource availability (i.e. those in flooded areas are unavailable) and to calculate their routes (i.e. no driving through flooded areas).
- *3D incident scene simulator* (e.g. XVR On Scene) to provide immersive visualisations for emergency response units at the flooded areas and to

create news broadcasts to immerse the higher command level participants. The water heights visualised are based on the flood model (i.e. the position of visualisation and scenario time).

- *Email generator* (e.g. TNO's email simulator) to provide a standalone email system for the participants and role-players. It has roughly the same functionality as the email system used in real life. The simulated system is, however, only available during the trial and only to the participants and staff, to ensure it is safe and excluded from real-world CM operations. This prevents mails on the simulated flood from being picked up by crisis management actors not participating in the trial.[2] By using the trial management tool, staff can easily trigger prepared emails to be sent out. All emails created are logged, such that these can be used afterwards in trial evaluation.

The operational COP-tool and the evacuation solution are linked to the CIS, such that these receive the flood map and the resource positions and availability. The CIS and CSS also ensure all data exchanged by the simulators, the COP-tool, and the solution are logged. Figure 3.6 displays how all this data is exchanged.

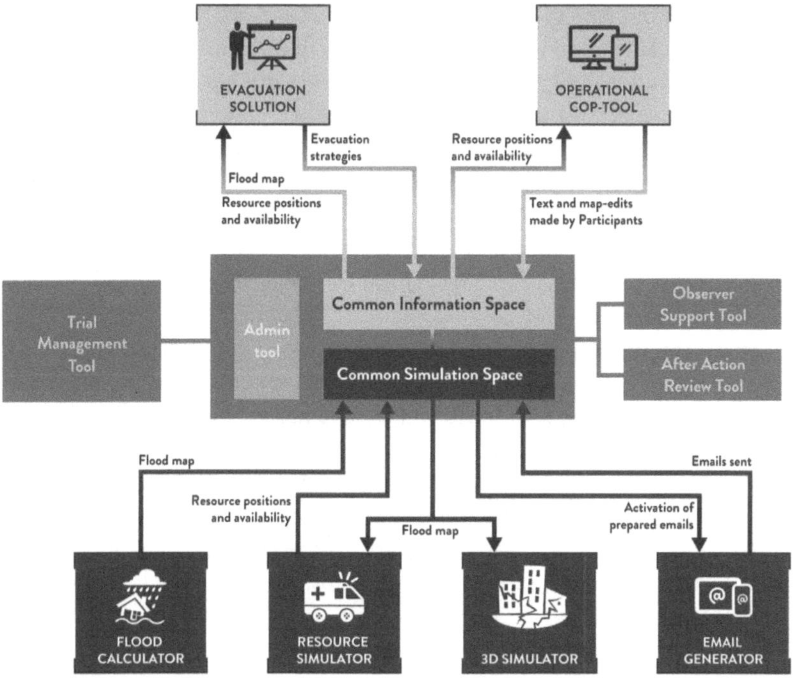

FIGURE 3.6 Data exchange diagram for this use-case of one operational COP-tool, an evacuation solution and four simulators.

The CIS and CSS Technology Stack for Developers

The core of the testbed consists of the CIS and CSS. Both are based on the open-source version of Apache Kafka. All connected solutions, operational systems, and simulators publish their messages to one or more topics. They subscribe to one or more topics to receive messages. Every data exchange flow designed for a specific trial (e.g. a CAP message from one operational system to solutions that need to receive it) should have its dedicated Kafka topic and message type, such that data exchange between all solutions, operational systems, and simulators can be easily configured and managed.

Since there are no direct dependencies between simulators and solutions, it is very easy to add or remove one or to replace it with another that exhibits similar functionality. The core of Apache Kafka is the broker. In an enterprise setting, there may be multiple brokers to improve performance and uptime. In all of the DRIVER+ trials, however, a single broker running on a laptop or on an online server proved to be sufficient.

Implementation wise, the CIS and CSS are identical. The distinction made in this chapter is purely conceptual and does not affect the technical deployment. When connecting solutions and operational systems to the CIS, standardised message formats should be used (e.g. CAP) since they are readily available within the crisis management domain. Also refer to "CEN Workshop Agreement CWA 17513:2020 en – Crisis and Disaster Management – Semantic and Syntactic Interoperability" for additional information. For the CSS, standard message formats are not required due to the different data formats used by simulators. Software technicians are, however, strongly advised to adhere to the design principles as described in the "CEN Workshop Agreement CWA 17515:2020 en – Building a Common Simulation Space," which also provides typical message schemas for the information that was exchanged.

Even though Kafka accepts any kind of message, early on the choice was made to use the Kafka schema registry: It enforces that the messages in a topic adhere to a so-called AVRO schema. In this way, AVRO validation prohibits that syntactically "wrong" messages are published, as that could induce parsing errors for the receiver of that message. Also, in many programming languages, the AVRO schema can be converted to a class, improving the integration development time. Examples of the AVRO schemas that were used can be found in https://github.com/driver-eu/avro-schemas.

To connect your solution, operational system or simulator to Kafka, there are many (open-source) adapters available. However, not all of them offer AVRO validation out-of-the-box, which is why the DRIVER+ project created several adapters that encapsulate the common Kafka functionality and added AVRO support, as well as heartbeat functionality and simulation time support. The heartbeat functionality allows trial staff to check whether all services are up and running. The simulation time offers a synchronised clock, which is very useful if the trial does not take place in real-time. The DRIVER+ project adapters are freely available at https://github.com/DRIVER-EU/ in the programming languages Java, C#, JavaScript/TypeScript, and Python. A REST end point is also implemented to be used in connecting to either CIS or CSS.

When running the core of the testbed, normally some additional tools were included to support development during the preparation for a trial. The basic software stack advised by DRIVER+ consists of the following:

- *Apache Kafka*, the broker that supports the exchange of messages between solutions and simulators.
- *Apache ZooKeeper*, internally used by Kafka to support state management of connected services.
- *Confluent Schema Registry*, to register all AVRO schemas that will be used in a trial.
- *Kafka UI* https://github.com/provectus/kafka-ui, to inspect the schemas that are registered, to inspect the messages that are exchanged, and to send test messages.

All these systems can be easily deployed using a single Docker Compose file. When starting the testbed, the AVRO key and value schemas that you want to use must be registered with the schema registry. Although you could do that manually, this is very cumbersome. There are two alternatives:

- A *bootstrapper docker service* created by DRIVER+ (a.k.a. the silent producer in the docker hub) that uploads a folder of schemas to the schema registry, creates the required topics and quits again.
- The *admin tool* (see its description in the next section).

Especially regarding the admin tool, there are some extra technical details to be aware of. As part of the testbed configuration, this tool can create PKI certificates for each connected service, so only trusted services are allowed to connect to the Kafka topics. Furthermore, it invites services to connect to a topic, so a service can only connect to topics it has been invited to. This feature, although not often required, can be useful when sharing sensitive data, as it will prevent other services from accessing this sensitive data.

All DRIVER+ software is released as open-source and can be found on GitHub as part of the DRIVER+ organisation: https://github.com/DRIVER-EU.

Support Tools and Services

This section discusses the optional software that enhances the testbed to do the following:

1. Control the progression of a trial, assuring all use-cases are played out.
2. Collect observations, which is often a crucial necessity for innovation assessment (and also for training assessments).
3. Monitor whether the entire environment is still technically and human-wise running as it should to conduct a valid assessment.
4. Collect and analyse all data produced during the trial, such that the synthesis step of the evaluation phase can be conducted.

Since all components are optional, only choose the ones you really need and expand when required.

Trial Management Tool (TMT)

When the testbed is running, one of the most useful additions to control your trial (or exercise) is the trial management tool. The TMT is used to create and run a digital, time-based scenario, consisting of injects that kick-start simulator action (e.g. start progressing a simulated flood), messages that are directly fed into the CIS solutions, and messages to role-players – for example, to call participants. It also provides an overview to trial staff of the current status, what is about to happen, and what should be done to make the trial progress to touch upon all assessment criteria. Furthermore, it controls the time service (explained in the following section) to facilitate one joint simulation time clock during the trial for all solutions, operational systems, and simulators to adhere to (if they technically can). The TMT allows for easier control of the trial scenario compared to a paper-based or spreadsheet-based script.

During the preparation phase, a scenario is created in the TMT. A scenario consists of one or more storylines, and each storyline consists of one or more injects. Injects are messages that are sent to the testbed and are intended for a solution or simulator. For example, a flood map intended for a command and control (map) application, a first incident report for an incident management application (see Figure 3.7), or a simple command for a 3D simulator to initiate a bomb explosion. Injects may also be messages intended to instruct role-players, for example, to start a conversation as mayor with the incident commander, demanding information about the incident, or to send a simple email, in case the email service is installed too.

In the simplest of cases, these injects can be prepared in advance and fired (published) manually when needed. Alternatively, the injects can be arranged in time, similar to a project's Gantt chart planning, and then fired automatically. The scenario specification, as implemented by the TMT, is currently being standardized into a CEN Workshop Agreement by the STRATEGY project (https://strategy-project.eu).

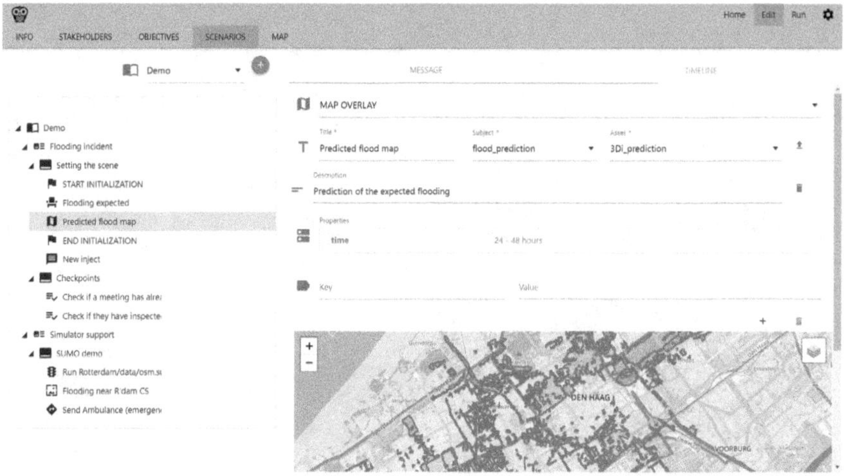

FIGURE 3.7 Preparing an inject in the TMT, in this case a map overlay indicating the incident area.

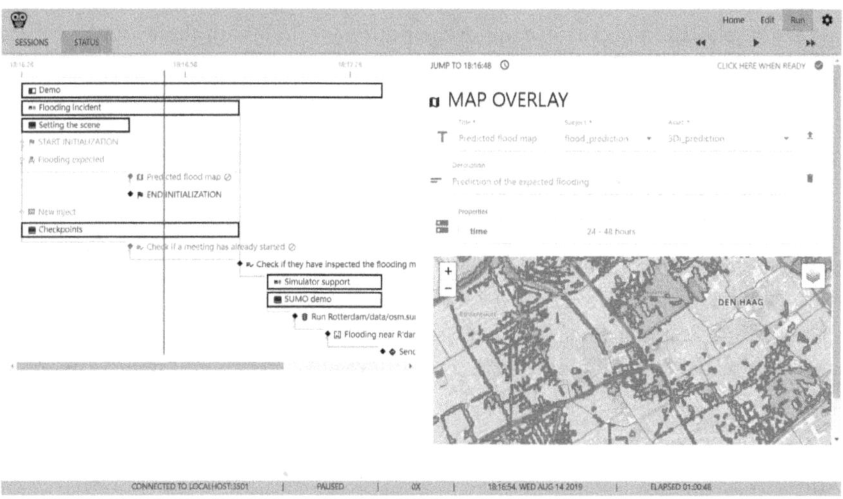

FIGURE 3.8 During the execution of a trial, the moving vertical line indicates the current time.

Time Service

When the trial is not running in real-time, the time service comes in handy. It is a simple application that publishes the current simulation time. This simulation time is a fictional time, fitting the trial scenario (e.g. a night-time incident, which is played out while, in reality, it is day-time). Solutions can listen to this time service,

FIGURE 3.9 The time service features its own clock and calendar, displaying the simulation time.

so they display the simulation time in their interface and mark their own messages with the current simulation time.

The time service also features a wall clock display, showing the trial's current fictive time and date (see Figure 3.9). Optionally, messages from the TMT can be displayed here, too – for example, to announce a break or a general message for all participants.

In the top left corner of the time service's interface, there is a hidden menu button to set the time, such that the application can be used standalone (i.e. without the TMT). However, normally, it is considered part of the TMT, which controls the start time and which can pause it or speed it up. These should be tested well in advance since not every solution and simulator may be able to deal with an external application controlling their clock.

Admin Tool

Software technicians use the admin tool to configure the complete testbed setup (e.g. which solutions, systems, and simulators exchange data over the CIS and CSS and in which format) and to monitor the health of each testbed component during trial runs (i.e. whether all components are still connected and are still sending and receiving all data they need). Whenever the CIS and CSS are used, the admin tool is recommended too.

A simple trial with only a few solutions or simulators is easy to oversee. However, when it scales up, the admin tool can be used to check the current configuration, configure the testbed, and catch and display errors emitted by the solutions, systems, or simulators (all testbed adapters provide a logger to log errors via the

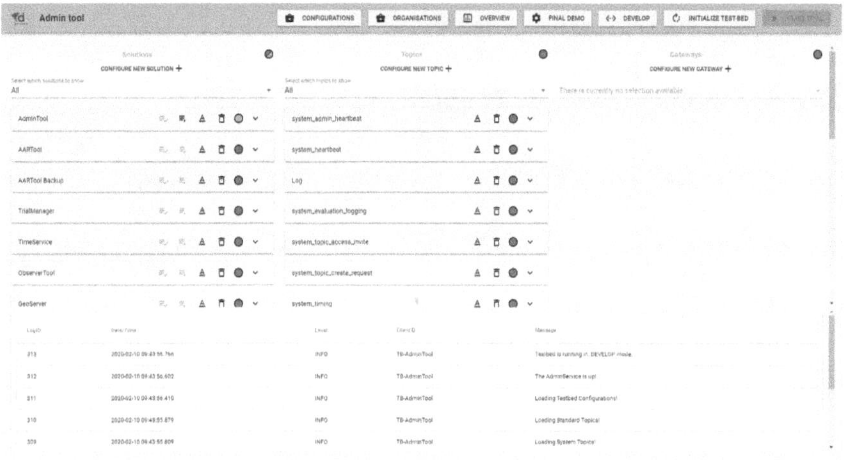

FIGURE 3.10 The admin tool displays the active solutions and simulators, as well as the error log (at the bottom).

CIS/CSS, which become visible in the admin tool). The tool can also provide extra security when sensitive data is to be shared among only a few systems connected to the CIS/CSS.

After-Action Review Tool (AAR Tool)

To support a proper evaluation, the after-action review tool collects all messages sent to the testbed. This includes data exchanged within the CIS and CSS, messages from the TMT, emails exchanged using the email service, and observations made via the observer support tool. Besides capturing all messages, it also displays a real-time timeline showing the most recent messages (see Figure 3.11), it can be used to inspect a message, and it generates a report with a UML timeline (i.e. sequence diagram; see Figure 3.12). So apart from using it for after-action evaluation, this tool is also used by trial staff during a trial run to monitor the state of the technical infrastructure.

Observer Support Tool (OST)

The OST is an open-source alternative for other off-the-shelf observation and form tools, with the primary advantage that it logs all observations directly to the testbed. It can operate both online and offline, and it can be controlled via the testbed (e.g. to [de]activate observation templates relevant in specific timeframes of the scenario, or linked to use-cases being played out sequentially). Trial organisers are always advised to use a well-thought-over tool for collecting and storing observations, as these are essential for evaluation.

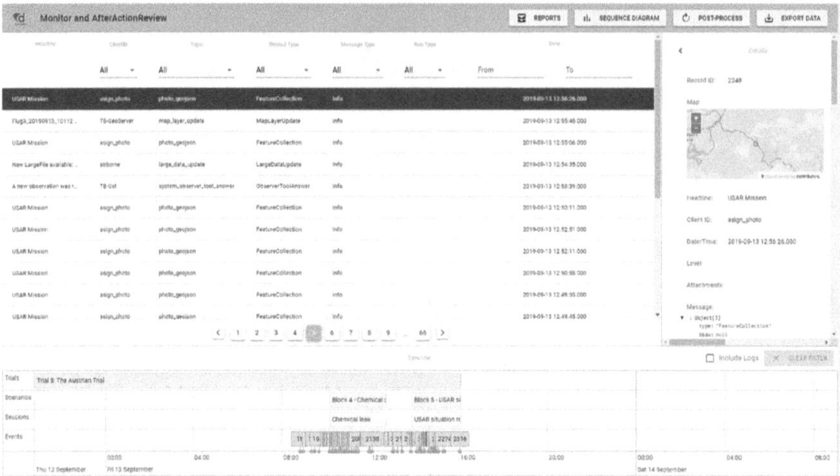

FIGURE 3.11 Main view of the AAR: it displays messages on a timeline and in a list, and you can inspect the selected message.

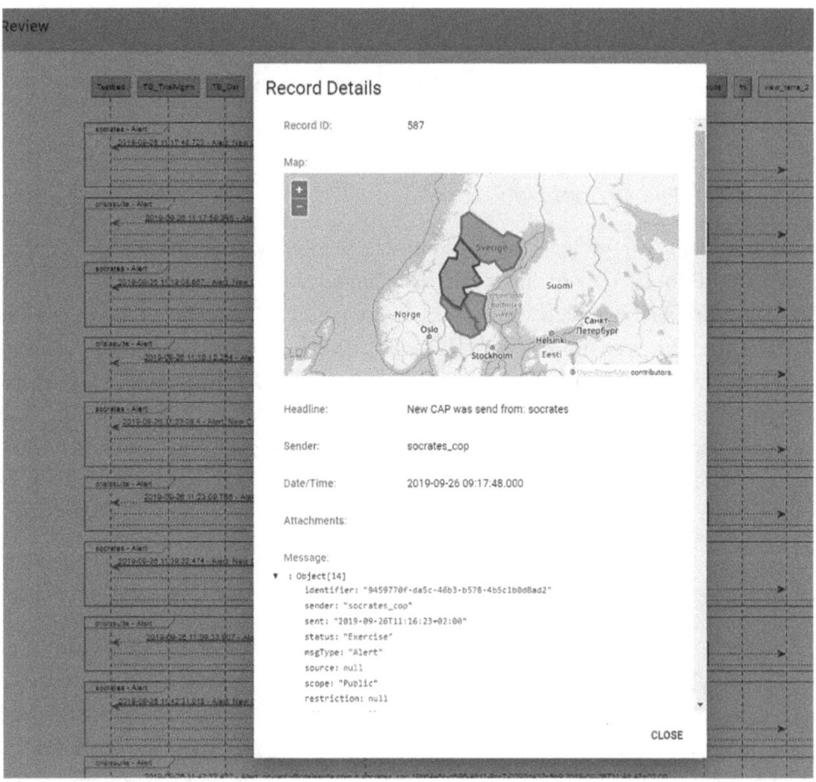

FIGURE 3.12 Reporting view in the AAR: the exchange of messages can be reviewed in detail. In the background, the UML sequence diagram is shown that details the message exchange between applications.

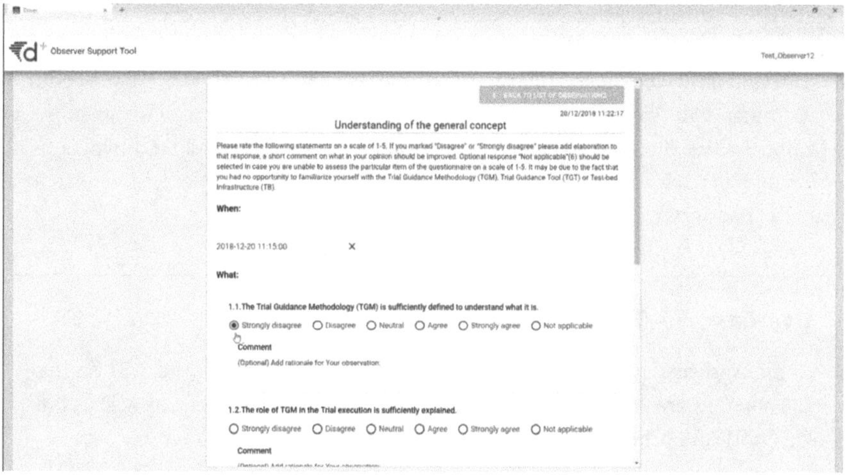

FIGURE 3.13 Observer interface where the observer can enter his or her observations.

With the observer support tool (OST), observations can be collected during the trial, using observation templates (i.e. prepared questions that an observer should try to answer) based on the chosen data collection strategy. By feeding these back to the TMT, staff can monitor the progression of the trial.

The OST is an advanced questionnaire tool: observers log into a web page and answer specific questions about what they observe, which can then be used during the evaluation. The questions, which are prepared in advance in the tool's admin interface, can also be updated during a trial. Answers are immediately shared with the testbed, so the trial staff can get a real-time impression of what is going on during the trial. This allows them to monitor what the participants are working on, track the evolution of the scenario, and control its direction when needed (e.g. via the TMT).

The OST can also be installed locally on a smartphone, tablet, or PC (as a progressive web application). When there is no Internet connection, observations are stored until the Internet connection is restored.

FTP Service

The testbed is capable of sending and receiving quite big messages. But when sending very large files above tens of Mb, it is better to upload them to the FTP service and send a link instead. The uploaded files can be shared "publicly" within the trial, so everyone can see them, or hidden using an obfuscated link. The testbed adapters developed in DRIVER+ offer support for this functionality.

Email Service

Although there are numerous open-source email services available – and even a regular email service can be used – there are several reasons to consider the

DRIVER+ email service: all messages are logged to the testbed, so they are included in the AAR reports; it listens to the simulation time, so the displayed time in your email service's web client matches the simulation time; and email accounts are automatically created (and deleted) for the purpose of a trial. Furthermore, it will not pollute the operational email with fake messages generated during a trial, nor will a participant or role-player inadvertently email an incident report to a real email address.

Use-Case 3 – TMT + Time + OST + AAR

As an example, assume the complete testbed is setup (see Figure 3.3), so the CIS and CSS are working, the TMT, and time service are running, as well as the OST and AAR. Chronologically, the following is done:

1. At the start of the trial, the trial director signals the start using the TMT.
2. The TMT kick-starts the simulation time and the time service sends time messages to all connected solutions, systems, and simulators. The simulation date and time are also displayed on the time service's clock interface in the rooms in which the participants and staff are located.
3. Next, the TMT sends a message to the OST, which loads and displays a list of scenario specific observation templates applicable at the start of the trial scenario.
4. Observations, made by the observers in the OST, are recorded in the AAR and are displayed to the trial director and other staff members in real-time (i.e. using either the OST admin or AAR interface). The observations logged are also directly included in the generated AAR report.
5. Data exchanged over the CIS and CSS are recorded as well and will be part of the generated AAR report.
6. The trial director monitors the state of the scenario, reads observations submitted, thereby understands what participants are doing, and hears from operators about what they do in the simulators. The director also hears about the full infrastructure health. He or she decides which injects to fire next (i.e. manually in the TMT or automatically by the TMT).
7. Injects fired cause data to be made available in the CIS/CSS and can cause actions to take place in the simulators. Injects also instruct role-players to contact participants. Furthermore, the TMT instructs the OST to make other observation templates available.
8. This, in turn, leads to new observations being made and sent to the trial staff, repeating steps 4–7.
9. In case technicians spot an error message while monitoring the infrastructure's health via the AAR tool or another staff member reports a serious issue, they first analyse whether this error/issue has a negative effect on the progression of the trial and on the validity of a solution's

assessment. This is discussed with the trial director to determine any of the following:

a. The error is logged, but no action is taken on it.
b. The error is deemed to seriously affect the assessment's validity, and therefore, the use-case currently being played out is prematurely ended (see also steps 12–14 about ending the trial run).
c. The current trial run is paused or even stopped, for technicians to resolve the issue, before restarting the run (see also step 11 about time jumps).

10. When a break is scheduled or improvised, the trial director announces this and pauses the simulation time via the TMT in the time service, and this service's interface indicates the paused time in the rooms.
11. When the scenario calls for a time jump (e.g. to move from the day before the flood occurs to the hours after which a dike fails; see also chapter 5), the simulation time is fast-forwarded or rewound by the TMT and time service, the service's clock indicates the new simulation time, and participants are briefed about the new situation. It should be noted that technically, a rewind might require the start of a new trial run.
12. The trial director stops the trial run when all use-cases and scenario elements (of that run) needed to achieve the trial objectives are played out. He or she then clicks the stop button in the TMT, resulting in the simulation time to stop. All staff and participants are informed the trial run has ended.
13. A final set of observation templates and questionnaires is made available in the OST. Observers, participants, and staff hold an after-action-review session and fill in these forms. This crucial data is again logged by the AAR tool, like step 4.
14. Finally, using the AAR tool, all data logged in the testbed is saved and stored (in a primary storage and in a backup) such that it is available for detailed evaluation.

Technical Implementation and Development Steps

As with any experiment or exercise, setting up a trial takes a serious amount of preparation and planning before actually conducting the trial and understanding the results from it. This also applies for small-scale trials or experiments. Chapter 2 already explained that success lies in meticulous and well thought out preparation. This section focuses on the technical preparation steps. Although all technical work might seem overwhelming, it is – like the testbed's components – scalable: a small trial requires less work, yet all steps need to be touched upon.

All expertise involved from the start converges towards a common goal: executing the trial. These forms of expertise can be roughly categorised into different yet very cohesive aspects:

- *Scenario* – tasked with designing a scenario that highlights the CM gaps and fits the solutions to be trialled.
- *Technical* – consisting of developers and facilitators of solutions, systems, simulators, and the CM organisation needed to integrate all systems required for the trial.
- *Evaluation* – focusing on defining, collecting, and analysing concrete performance indicators and methods to get better insights into whether or not a (set of) solution(s) is an improvement to the defined gap.
- *Facilitating/event* – concerned with practitioner and staff scheduling, security, logistics, catering, communication, visitors, and media, at either physical and/or virtual locations being used before, during, and after the trial run(s).

The technical aspects of preparing and running a trial follow the TGM's phases and steps. The main goal of this work is to support solutions, systems and simulators to participate in the trial, connect these to the testbed, set up all other tools needed, and design and configure the data exchanges.

Like the TGM, technical preparations follow an iterative approach. Each iteration converges all aspects towards the actual trial run. It is a continuous process of refining, implementing and testing use-cases, dealing with technical and practical limitations (e.g. effort, hardware availability), and adjusting the setup. This iteration requires clear communication between all stakeholders involved, especially since the mentioned aspects can influence each other quite drastically. Several milestones, preferably in the form of face-to-face meetings, are created to help structure that process.

Trial Integration Meeting

After the trial owner has discovered and selected solutions that fill a gap in its specific CM situation, the trial integration meeting (TIM) is the first (face-to-face) meeting where all the different aspects come together. Each aspect or expertise should already familiarise themselves with the envisioned objectives, setup, (im)possibilities, and conditions for the trial. The TIM should bring everyone up to speed on each of those plans and allow possible conflicts to be discussed and resolved before technical work has already begun.

Pre-TIM Technical Work

Technical TIM preparations entail getting the selected solutions acquainted with the testbed. This includes the following:

- Reading into the workings of the testbed.
- Choosing a specific CIS adapter to connect with.
- Running a local deployment of the testbed to start a first test implementation.

Understanding the Scenario

The scenario experts already have a concept in mind and/or documented of a trial scenario that highlights the selected gaps and contains use-cases to assess how well solutions bridge those gaps. This focuses the technical aspect of the trial by prioritising functionalities a solution offers. A solution might have a large range of use-cases that can be played out. A good understanding of the scenario should give insight into which use-cases are relevant.

- How does the CM organisation operate? What are its command structures and standard operating procedures?
- What is the first draft of the scenario?
- Which level of command is this scenario designed for? Who are the actual participants in the trial and which levels and external organisations have to be simulated?
- How do the solutions fit into this scenario? What do they offer? What innovation is to be assessed related to the gaps?
- Which currently operational systems and simulators are required for this trial? Do they need to interact with any other technical system?

Understanding the Evaluation

Like with the scenario aspect, the evaluation experts have an idea on how to evaluate whether a solution supports bridging a gap. This might require additional technical configuration, such that the data needed for evaluation is actually present and stored correctly.

- What is being evaluated?
- What evaluation method is going to be used (e.g. is there a baseline run without the solutions that will be compared with a trial run including the solutions)? How are simulators, operational systems, and scenario progression integrated into this method?
- Which data is required for this?

Understanding the Solutions

Every solution selected might have a different context in which it operates. It is good to know upfront how such a solution works and to which extent it fits into the envisioned trial. Understanding these solutions and their requirements to operate should give a better insight into the integration work that needs to be done for this trial.

- How is the solution normally used? Does it require a trained operator, or can a trial participant directly use it?
- What are the hardware requirements for the solution to function properly, if any? This might also include a stable Internet connection or a plane for aerial photography.

- Which data is required as input for the solutions to function, if any? Which information is normally shared with other systems?

Understanding the Simulators

In order to actually trial solutions instead of using them directly in the operational crisis management system, some form of simulation is required. Like with training the incident command in dealing with a forest fire, it is not desired to actually start a real forest fire.

- Which organisation departments, command levels, and/or external organisations are required but not participating in the trial and thus are required to be simulated?
- Do solutions require simulated data (e.g. GPS positions of emergency vehicles to be visualised)?
- Is simulation needed to get a better immersion for the participants in the trial scenario (e.g. the progress of a forest fire)?
- Does the trial require person-to-person simulation like re-enactment of emergency calls or incident reports?

Practical Considerations

Apart from these four aspects, trial organisers will likely face some extra practical limitations. Budgets for technical developments, practitioner participation, and IT hardware are limited. The location at which the trial will be executed has limited capacity. And the interests of all stakeholders of a trial should be balanced. These considerations should be made clear during the TIM, such that all involved can optimise their work within what is possible.

Designing a Data Exchange Diagram

After having discussed these topics, the technical aspect can focus on creating a first version of the data exchange diagram (DED) based on the decisions that have been made during the TIM. This DED provides a clear overview of the integration work for the trial, describing the following:

- Which and how solutions, simulators, and operational systems are being used.
- If and how they have an interaction with the participants and/or the testbed.
- Which information is being transferred from one system to another.

The DED serves as a blueprint for all technicians involved in this trial. An example DED can be seen in the next break-out section.

Drawing the DED is closely linked to detailing the use-cases on which the solutions are to be assessed. A use-case highlights one single interaction of a participant and the solution, thereby exposing all data exchanges between the systems that are part of that interaction. An example of a use-case would be a participant sending a report via one of the solutions, which in turn converts that report into a defined data exchange standard to be sent over the testbed, being received by another solution displaying that report to its users (e.g. see "Use-Case 1 – A Operational System Exchanging Data with a Solution," shown earlier).

Example of a Data Exchange Diagram

Throughout the implementation phases of a trial, a clear design overview of the integrated components as a whole is very useful for understanding the desired technical implementation. The data exchange diagram (DED) provides such an overview that has proved to be crucial in communicating about scenario and evaluation requirements, user interaction flows, and technical concerns.

An example of a DED can be seen in Figure 3.14, which was made for the trial NL. This trial included six solutions (ZKI, Keep Operational, HumLog, 3Di, SIM-CI, and CrisisSuite) and one operational system (LCMS) connected to the testbed (outlined in the middle). Three simulators (flood calculator, traffic simulator, and XVR Crisis Media) were also linked to provide simulated data of a staged flood.

From top to bottom, this diagram shows (by means of the directed arrows) the following:

- The interactions participants have with all solutions, the operational system and simulators.
- Which data these systems send to and receive from the testbed.
- How simulators are designed to provide their data to the testbed and participants.

One of these flows is the forecast flood maps from 3Di. This solution provides these forecasts to a subset of the trial participants. Whenever the participants (i.e. 3Di users) publish those forecasts, 3Di sends them to the testbed so that these are received by HumLog, SIM-CI, CrisisSuite, and LCMS. These solutions will in turn use these forecasts for further analysis and/or visualisation.

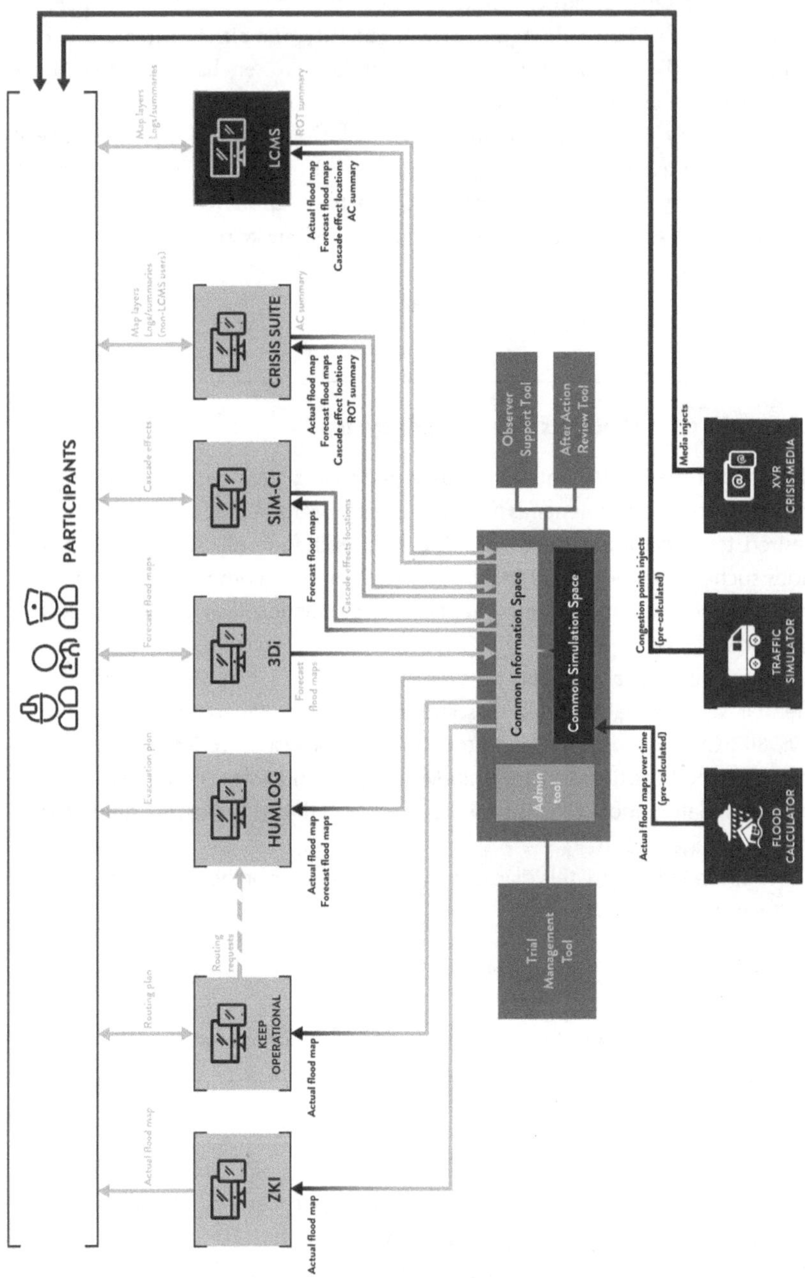

FIGURE 3.14 Example of a data exchange diagram created for DRIVER+ Trial The Netherlands.

Dry Run 1

While the TIM is an essential step in the design of the trial, dry runs focus on testing the implementation created based on those designs. Particularly, dry run 1 is focused on the technical aspect. Technicians of solution providers, operational systems, and/or simulators have focused on their set of use-cases to be implemented. In dry run 1, all aspects come together and it is checked whether these use-cases are functioning correctly. Whatever the scale of the trial, it is highly advised to always organise a dry run 1.

Technical Testing

The main goal of dry run 1 is to test whether the defined use-cases are correctly implemented. This requires a one-by-one checking of those use-cases (e.g. roughly following the chronological flow of the trial). It allows all involved to get a better idea of how participants and staff will interact with the solutions, systems, simulators, and tools, and how all components interact with each other in the context of the trial. During dry run 1, it is also wise to discuss progress and possible changes to the technical implementation, scenario, and evaluation method.

Platform Infrastructure

For a valid test, dry run 1 should take place under similar infrastructure conditions as the actual trial run. This means that all technical systems should be located at the (physical or virtual) locations, using the hardware and infrastructure as much as possible like those used during real execution. Any problems or risks that might be found during dry run 1 should be discussed and mitigated before dry run 2.

After dry run 1, all involved should have a better understanding of how the trial will technically run. It is to be expected that insights gained during this dry run will lead to changes to parts of the data exchange diagram and use-cases. However, these changes should not diverge the scope of the trial or its technical facilitation to a large extent.

Dry Run 2 and Trial Run

Having tested the technical implementation of the trial in dry run 1 and having resolved issues and risks afterwards, dry run 2 serves as the dress rehearsal for the trial. From a technical standpoint, dry run 2 and the actual trial run(s) should not be any different. Whatever the scale of the trial, it is paramount to always perform dry run 2, assuring all aspects are checked and ready for the actual trial execution.

Technical Testing of the Complete Infrastructure

Dry run 2 starts with individual tests for each use-case. These tests should also be aimed at testing all hardware being used in the trial. Where dry run 1 tests can be performed on one machine per solution, during dry run 2 all equipment that can be used during the trial should be tested. This also includes testing simultaneous interactions that could occur during the trial run. Any unwanted system or user behaviour should be discussed and mitigated.

Since this is the first time all components are connected, the technicians can see if the current hardware and infrastructure is capable of handling the load required by the integrated systems and their data exchanges as a whole. In order to test this, load/stress tests can be performed. These tests increase the volume of interactions and data load to a level higher than expected during the trial run. Finding out which component (and at which load) experiences defects or latency allows technicians to apply optimisations if required.

Monitoring

Dry run 2 focuses on testing, which in turn requires monitoring of the infrastructure to spot any defects. This monitoring is also required during the actual trial run. Unwanted behaviour at user, component and infrastructure level might occur. Monitoring might detect these defects before it is affecting the critical execution of the trial, enabling technicians to resolve it pre-emptively.

The testbed has certain monitoring tools already implemented, mostly residing in the admin tool and the AAR tool. While resolving the issue at hand might seem like a high priority, discussing and documenting any anomalies and their resolutions with both scenario and evaluation aspects is crucial for a proper execution and evaluation of the trial. It sometimes might even be better for evaluation purposes to leave a detected defect as is and log it instead of pausing or delaying the trial run to resolve it.

Technical Support for Evaluation

Right after the execution of the trial, logged data from solutions, simulators, and operational systems need to be stored. Monitoring logs, residing in the testbed's support tools, also need to be stored for evaluation. It is strongly advised to save and store these data and to make backups whenever possible – for instance, during breaks of the trial run. This to make sure possible future anomalies will not affect already recorded data.

After hardware, system, and infrastructure clean-up has taken place, the technical aspect in principle has finished its tasks. Note that technicians are often consulted during the evaluation phase to understand what exactly happened during the trial and how this affected the use of the solutions. Good technical documentation is very supportive in this.

Future Perspective

The DRIVER+ project ended, but the testbed still lives on. For example, it is actively used in a number of projects, maintenance is still continuing by all partners, and the software is still evolving. A few of these are listed in the following paragraphs.

In the STRATEGY project (https://strategy-project.eu/), funded by the EU as part of the Horizon 2020 program, the testbed is used to support several tabletop sessions that are used to validate new crisis management standards. The testbed will also be used as part of the final (live) exercise.

In the ASSISTANCE project (https://assistance-project.eu/), also funded by the EU as part of the Horizon 2020 program, the testbed is used for integrating a hazard cloud service, a command-and-control application, an early warning system, and a Traccar-based GPS tracking service (www.traccar.org).

The Dutch National Institute of Safety (www.nipv.nl) is deploying a similar, Apache Kafka–based infrastructure to connect operational crisis management systems and share information.

Note that the testbed is not even limited to the CM domain. For example, for the Lithuanian government, the testbed is used to support an OSINT (open-source intelligence) research environment. And in a TNO (www.tno.nl) programme for the Dutch Ministry of Defence (Gendarmerie) and National Police, the testbed connects a command-and-control application with a 3D platform, XVR On Scene (www.xvrsim.com), which is used to evaluate possible solutions against future threats.

Take-Home Message

The main point to take home is that the testbed, consisting of many loosely coupled components, and its setup can be a bit overwhelming. So before rolling out your own testbed, it is important to consider which components are essential for your needs. Start with these, and expand when needed.

Notes

1 This setup of four simulators is a fictive example, based on a combination of simulator setups as used in trial France, trial The Netherlands, and the final demonstration.

FIGURE 3.15 The testbed receives information from the 3D platform (e.g. entity positions and state), and it can send messages back too, like simulating an explosion and creating victims.

2 Although seemingly unlikely, it has happened during a few CM exercises that non-participating organisations picked up exercise communication or were contacted by exercise participants about the staged crisis, mistakenly thinking the exercise's crisis was really happening. This causes unwanted overhead at exercise/trial staff to resolve such communication issues.
3 https://www.merlincrisis.com/

References

CEN Workshop Agreement - Crisis and disaster management - Semantic and syntactic interoperability (May 2020) https://www.cencenelec.eu/media/CEN-CENELEC/CWAs/RI/cwa17513_2020.pdf

CEN Workshop Agreement - Building a Common Simulation Space (May 2020) https://www.cencenelec.eu/media/CEN-CENELEC/CWAs/RI/cwa17515_2020.pdf

Hendriks, M., Vullings, E., van Campen, S., & Hameete, P. (2019). An interoperability framework for trials and exercises. *Proceedings of the 30th International Forum for the Military and Civil Simulation, Training and Education Community (ITEC)*. https://www.itec.co.uk/__media/libraries/technologies-and-architectures/35---Martijn-Hendriks-Paper.pdf.

4

DRIVER+ ONLINE KNOWLEDGE MANAGEMENT AND INFERENCE TOOLSET

Dražen Ignjatović, Georg Neubauer, Denis Havlik, and Todor Tagarev

Introduction

Crisis and disaster management is a complex and changing profession, with a large number of stakeholders, tasks and processes that need to be adequately addressed at all stages of the crisis management cycle, from mitigation and preparedness to response and recovery. Such processes can be supported by various crisis management (CM) solutions, with each solution addressing one or more of the stakeholder's needs. In this context, a CM solution is defined as a combination of tools and processes that is designed to help stakeholders perform their tasks or execute a process. Moreover, a concise description of specific stakeholders' needs is called a CM gap.

Many solutions are already (off-the-shelf) available or are being developed by different solution providers from industry, small- and medium-sized enterprises (SMEs), or research organisations. Unfortunately, discovering relevant solutions is a challenge for practitioners and decision-makers, and assessing which of these solutions optimally addresses "their" gaps even more so. Here are a few reasons: (1) lack of agreement on common CM gaps, (2) lack of conventions for describing and semantically annotating the solutions in a structured way, (3) conflict of interests between solution users and providers, (4) high effort required for independent testing of the solutions, (5) sheer number of solutions available, and (6) lack of adequate search, matching, and discovery mechanisms addressing these issues.

To help overcome these problems, a dedicated online knowledge management and inference toolset for the crisis management domain was developed within the DRIVER+ project (Ignjatović et al., 2019). The main functions of this toolset include a repository of (CM) solutions, an inventory of CM gaps (gaps explorer), and a tool to support development and documentation of trials (trial guidance tool). PoS supports the crisis management stakeholders in discovering the answers to questions of high interest to them, such as "Which CM solutions are currently

DOI: 10.4324/9781003256977-7

available, envisaged or in development?" "Which of these solutions are addressing the issues (gaps) we have?" "Do other organisations have similar needs/gaps as we do?" "How do we assess which of the solutions we discovered can best meet our needs?" and "What did other stakeholders already learn?"

In the crisis and disaster management domain, the whole toolset is usually referred to as the portfolio of solutions, or PoS, and this naming convention will be also used in this chapter.

Need for the Toolset and How It Was Made

Why DRIVER+ Invested in Making the Portfolio of Solutions

The main reason for developing the toolset was that, up to that point, no such web-based knowledge management system existed to help the various stakeholders in disaster and pandemic management. The PoS enabled practitioners to identify solutions that had the potential to fill the gaps they faced, and the owners of innovative solutions to describe their solutions and clarify what needs those solutions targeted and how. In addition, the PoS helped the crisis management community understand what needs (strategic gaps) practitioners were truly interested in, and it helped practitioners not only find innovative solutions but also prepare trials to assess the performance of these solutions. In addition, the PoS was developed to help all stakeholders improve mutual understanding by introducing a common vocabulary and taxonomy of crisis management functions.

Development Process and Technical Details

Two key concepts were used to assist in the discovery of innovative solutions:

- The development of a repository to store information about innovative solutions.
- The development of a taxonomy of crisis management functions used to describe these innovative solutions.

The PoS is designed as a web-based knowledge management system that stores information about CM gaps, solutions, and experiences in deploying these solutions and makes this information available to end-users through a web interface. This ensures that all information about the innovative crisis and disaster management solutions is easily accessible to anyone with access to the Internet, bridging the gap between communities of interest geographically dispersed throughout Europe.

For the development of the PoS platform, a mixture of two software development approaches was used:

1. Agile approach.[1]
2. Prototyping approach (Alavi, 1984).

The first approach was chosen because it was a good fit for the needs of the project and the focus was on collaborative development rather than a predefined structured development process. It encouraged collaboration and teamwork among partners while maintaining contact with end-users, which allowed for quick responses to changing specifications. A monthly sprint was chosen, and new features were developed over the course of a month in the development environment. At the beginning of each month, which in this case was also each sprint, these newly developed functions were migrated to the production environment so that they could be used immediately by the relevant stakeholders. The second approach was taken to improve end-user engagement. This involved presenting and demonstrating smaller models of the desired features to ensure they would meet user needs.

The implementation of the PoS is based on Drupal 8[2] – an open-source content management framework, which offers a variety of contributed modules to further extend the core functionality but also allows the development of custom modules to adapt to the specific needs of the project. The implementation steps can be divided into two groups, those required for the visual elements and those required for the functional elements of the PoS. The first group was formed by the development of custom visual elements. The project required the development of a custom theme for the website and all other visual elements to represent the visual identity of the project. To make the website more attractive and user-friendly, the presentation of the content was also customised using the state-of-the-art graphical user interface technologies (cascading style sheets, JavaScript). The second group was implemented based on the functional requirements for the PoS, using a mix of features provided by the Drupal framework along with the community-developed modules and features developed individually during the project. Figure 4.1 shows a possible core structure for sorting crisis and disaster management solutions.

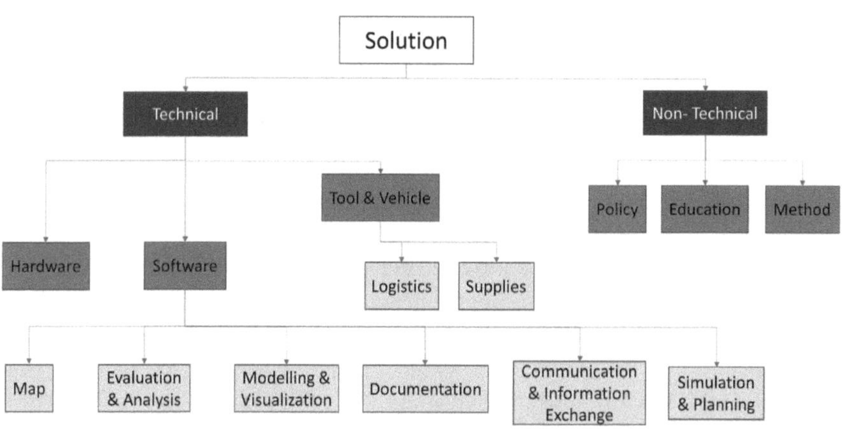

FIGURE 4.1 Classification of crisis management solutions. A tree structure of crisis management solutions illustrates the proposed structure for classifying solutions.

To enable easy discovery of solutions, PoS has implemented advanced search algorithms, mainly based on the core Drupal Search module and Solr,[3] but also modified and fine-tuned according to the specific requirements of the project. In addition, a custom-matching feature has been developed to further improve solution discovery by matching them based on the CM features they cover. To assist solution providers in describing their solutions in a systematic and standardised way, two custom functions have been developed and implemented – namely, contextual help and validation. The first aims to provide contextual help to the user in any part of the website by identifying the user's location on the website based on the URL and displaying the corresponding help information. The second aims to automatically validate input and warn the user if the solution description does not meet the defined criteria.

Using the Portfolio of Solutions as a Practitioner

Overview

As mentioned, the PoS toolset consists of several knowledge management applications – namely, the *gaps explorer*, *solutions explorer*, and the *trial guidance tool.*

- *Gaps explorer* allows the users to search for crisis management gaps, as well as to add new gaps. It automatically discovers the solutions that are addressing the gap, thus providing an alternative way to discover innovative solutions addressing specific gap(s).
- *Solutions explorer* allows the users to directly search for solutions, using a combination of full-text and faceted search (allowing users to narrow down search results by applying multiple filters based on faceted classification of the items), as well as to add their solutions to the PoS. Application also automatically lists "similar solutions" to the ones found by search, thus providing a possibility for more organic solution discovery. It allows documenting and presenting the innovative crisis management solutions that are already used by practitioners, as well as those that are actively developed and tested (e.g. by start-up providers or in other research projects).
- *Trial guidance tool* aims to support the correct implementation of the trial guidance methodology (TGM). It allows users to explore the results of previous trials and to plan their own trials and trial-like events. The main focus of the TGT is on facilitating the trial preparation by supporting the users in defining the trial context, managing the trial team, defining the relation of the trial to strategic crisis management gaps, and implementing the iterative six-step approach to design the trial, as well as provide support in other phases and steps of the TGM.

In addition, the site also hosts the following:

- The taxonomy of CM functions and other taxonomies that are used to tag and discover information.
- An extensive crisis management terminology.

- The knowledge database: Annotated references to scientific papers which are related to the crisis management domain, collected during the systematic literature review.[4]
- Country profiles: Information on crisis management organisations in different countries.

Exploring Gaps

The role of the gaps explorer is to systematically compare the needs in crisis management with the characteristics of existing solutions and to select potentially suitable solutions. To enable this functionality, the gaps explorer relies on two main elements: the taxonomy of CM functions and the implemented matching algorithms. Using the terms from the taxonomy, both gaps and solutions were tagged, and matching algorithms were developed to detect overlaps between the terms and calculate the relevance of a given solution to a given gap – the higher the overlap, the higher the score and also the possibility that the automatically suggested solution can be used to bridge the gap.

The DRIVER+ project initially identified a list of 21 strategic gaps that were relevant to the project. The sustainability of the "strategic CM gaps" was one of the issues encountered by the PoS development team. During the development of the service, it was clear that the DRIVER+ project would end in 2020 and that the PoS platform had no mandate to define strategic CM gaps for Europe. As a result, the initial list of gaps was used as a basis, and the gaps explorer was developed as a tool that enables a systematic description of gaps.

In addition, two organisations started publishing their own lists of strategic gaps:

- Common global capability gaps were identified by the International Forum to Advance First Responder Innovation (IFAFRI).[5]
- Strategic gaps and the challenges, policies, and recommendations were elaborated by the EU DRMKC's (Disaster Risk Management Knowledge Centre) "Gaps Explorer" in a pilot project on forest fires.[6]

Since these organisations have the mandate to define strategic CM gaps, PoS was redesigned to make the gaps more visible, and the existing DRMKC and IFAFRI[7] gaps were integrated into the gaps explorer (Figure 4.2). The result is that PoS automatically finds the solutions that could fill these gaps, which is a win-win for both solution providers and practitioners who need such solutions.

Finding Solutions

Since PoS is intended to be a repository for available solutions in the area of crisis and pandemic management, it was very important to make searching for this information easy and user-friendly. To achieve this, several search functions were implemented and designed to be used in combination with each other. The first of these functions is the so-called full-text search, which was implemented in such

FIGURE 4.2 The gaps explorer user interface.

FIGURE 4.3 PoS solutions overview.

a way that search algorithms query all available texts for the term(s) searched for, not only in the name of the solution or in its description but in all sub-entities belonging to the solution entity, as described in section "Why DRIVER+ invested in making the Portfolio of Solutions." This feature is based on Solr,[8] a reliable, scalable, and fault-tolerant search engine that provides distributed indexing, replication, and load-balancing queries, as well as automatic failover and recovery and centralised configuration. Using Solr also made it possible to fine-tune the search algorithm so that the results of a search query are as accurate and relevant as possible. Other search engines such as ElasticSearch were investigated, but Solr provided the best results for the project's needs. The second search function implemented is faceted search, allowing users to narrow down search results by applying multiple filters (e.g. innovation stage, crisis cycle phase). As mentioned earlier, several taxonomies were developed and used to describe the solutions and gaps within the PoS. First, the desired area of interest could be defined by applying a filter (e.g. displaying only results on solutions used in the preparation phase of the crisis management cycle). Since the search functionalities can be combined, a list of solutions narrowed down in this way can then be further queried for specific terms using the full-text search. The final search function offered by PoS is automated search. By leveraging the common terms from the taxonomies used to tag entities in PoS, automatic matching of gaps with potential solutions occurs as soon as the information is entered into the database, making it available for the user in real-time. In addition, similar solutions can be automatically identified by calculating the overlap of taxonomy terms used to label different solutions. In this way, it is possible to determine which other solutions could be used to bridge the same gap. The combination of these search capabilities aims to provide users with the most relevant information based on their individual search criteria, ultimately achieving the goal of supplying end-users with information on available innovative solutions as well as gaps in the CM domain.

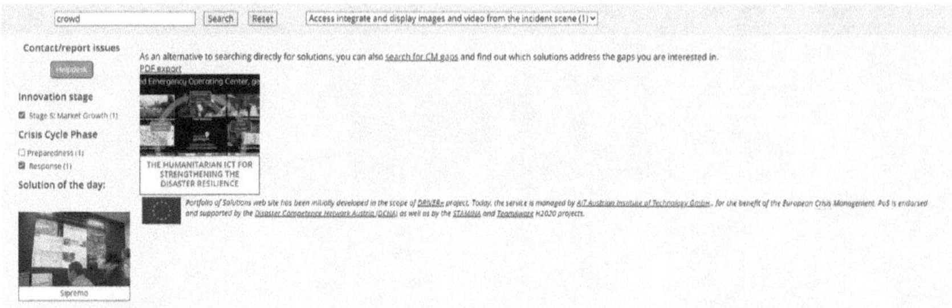

FIGURE 4.4 PoS search functions example.

Seeing Other Trials and Setting Up Your Own

The TGT is the final part of the toolset. It is a web-based software tool developed to assist trial managers and high-level crisis managers in implementing the trial guidance methodology (TGM) at the different steps of the trial phases. It derives directly from the TGM and can be considered a software guide of the methodology.

It ensures that practitioner needs are met along with trial objectives by following the six-step approach defined in the preparation phase, as well as by providing assistance in all steps of the execution and evaluation phases. The TGT also allows for validation of the results of each step to ensure that they are performed as intended. Considering that the TGM is multi-faceted and elaborate methodology, that can be experienced at first as complex or as an innovative way of working, effective and successful implementation requires systematic guidance, which the tool provides.

The TGT is also a knowledge base that contains the results of the DRIVER+ systematic literature review (SLR), as well as lessons learnt from previous trials, and serves as a reference for the future. The TGT provides examples of trial objectives and helps users define their own. The tool also provides examples of best practices based on past experience and helps formulate structured and pragmatic data collection plans for assessing trial results by providing ad hoc templates.

The TGT also allows users to formulate trial scenarios and store them in the tool for later use (e.g. in another trial). As part of the toolset, it is directly linked to the solutions explorer and gaps explorer, which provides the ability to identify potential solutions to the gap described in TGT. In addition, the tool introduces test cases that can be defined and shared across trials to help CM practitioners meet trial objectives and answer research questions. Trial managers and their teams can use the tool simultaneously to improve their collaboration. One of the most important features of the tool is the PDF export, which allows data from the TGT to be imported directly into the trial action plan, which is a document used by trial organisers to document all relevant information about a trial. An integrated help system accompanies the user every step of the way, providing support and examples

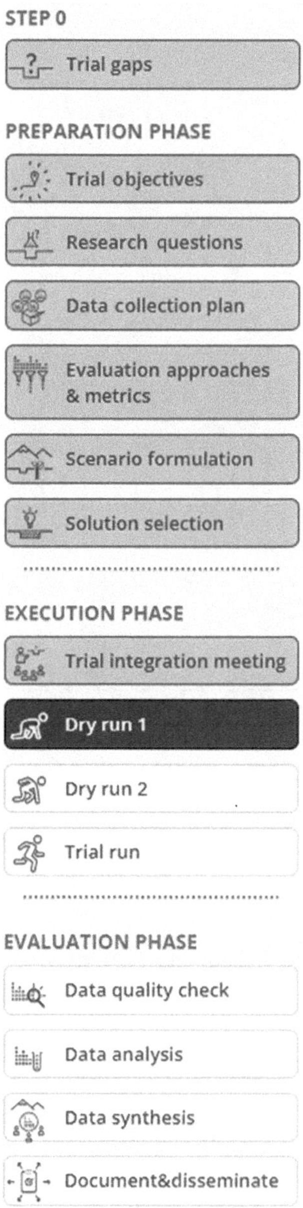

FIGURE 4.5 Trial phases with corresponding steps as depicted in the TGT.

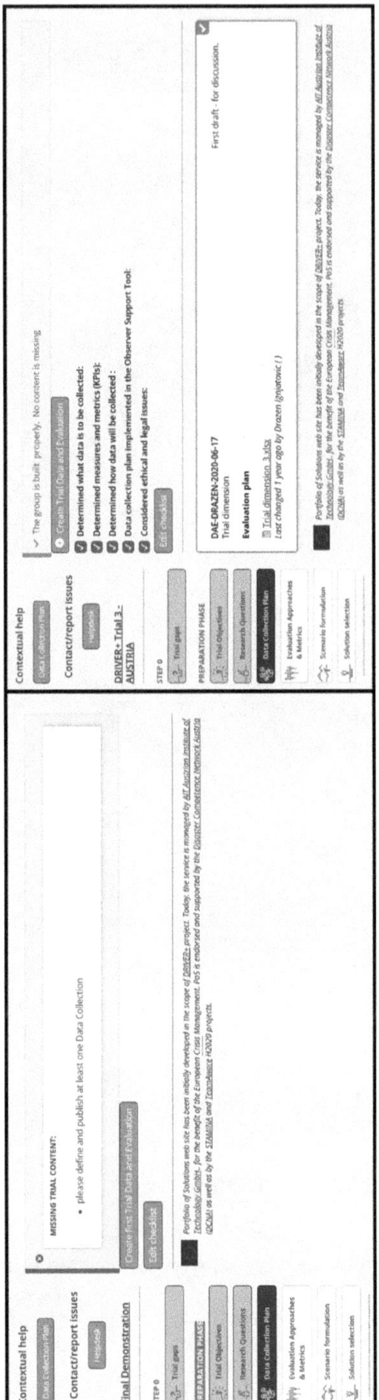

FIGURE 4.6 Validation in TGT – missing input (left), correct input (right).

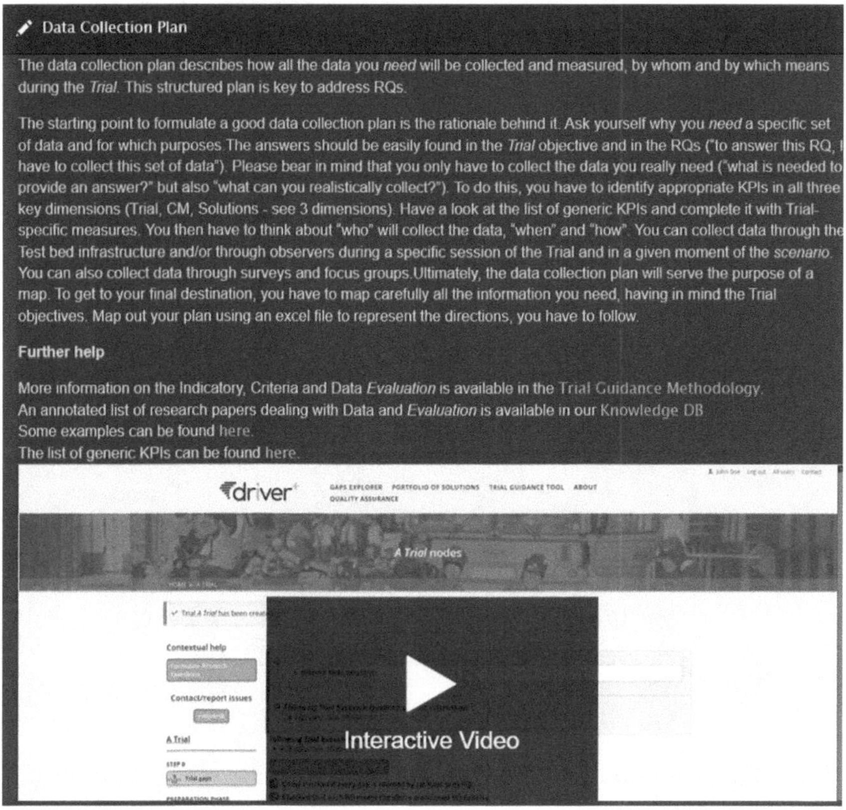

FIGURE 4.7 Contextual help.

of the actions required. By using the TGT, the TGM is followed, which assures a systematic and guided approach to evaluating potentially innovative solutions.

Using the Portfolio of Solutions as Solution Provider

The main functionality of PoS, aimed at solution providers, is that they can add their own solution(s) to the repository. Of course, a provider can also use it like a practitioner to search for solutions, get inspired, compare the existing offer, or even organise a trial.

Advertising Your Solution

To ease up entering a new solution to the PoS, the website offers a step-by-step interface that guides the user through this process. To support this, a workflow has been developed that consists of a few steps. The first step is to add a basic

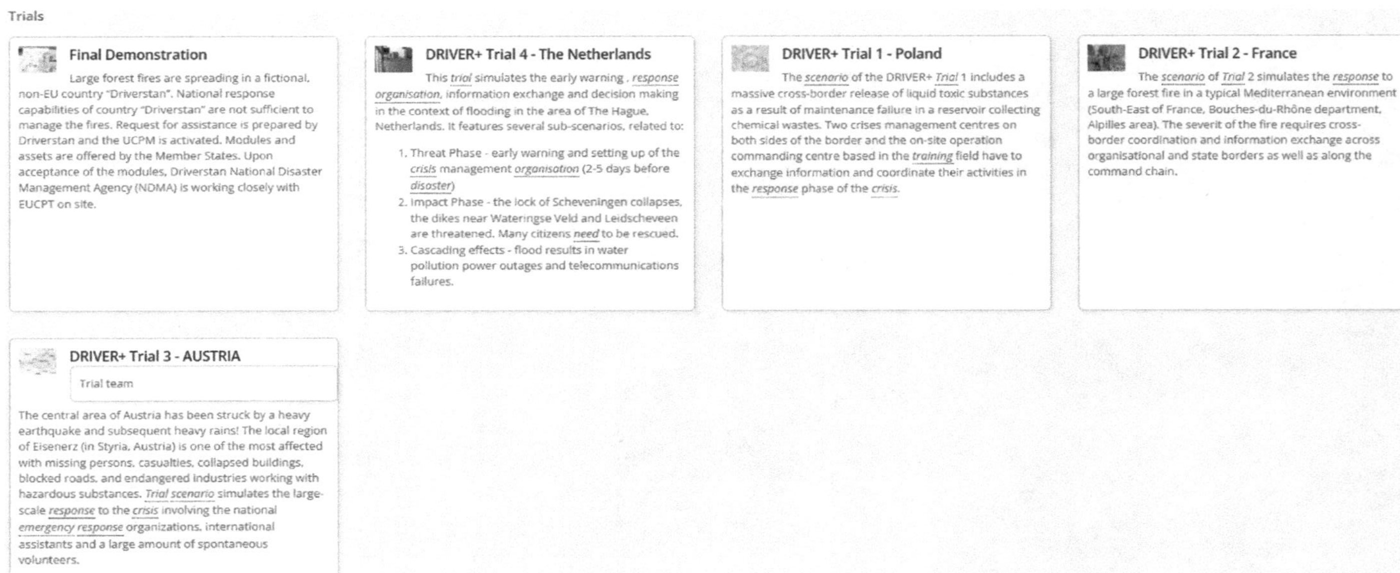

FIGURE 4.8 DRIVER+ trials in TGT.

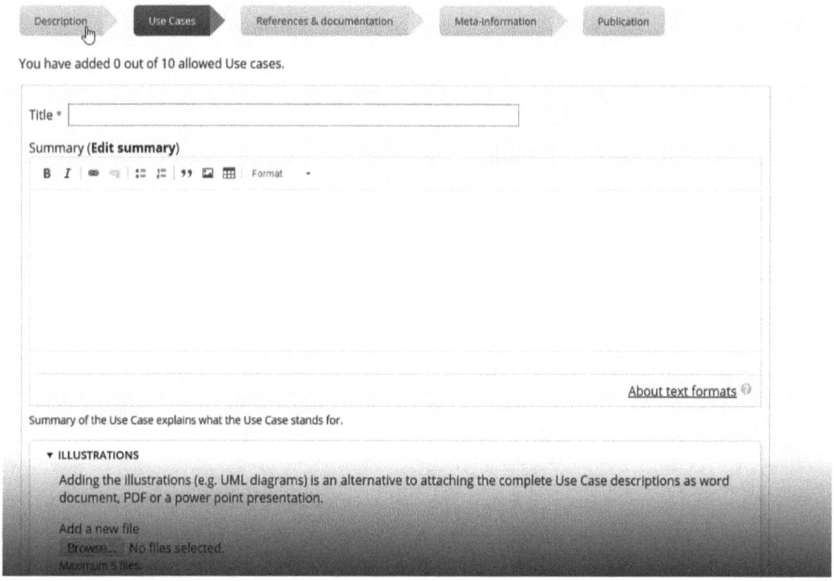

FIGURE 4.9 Solution description workflow.

description of the solution with some illustrations. The following step supports the definition of use-cases in which the solutions can be used. In this step, CM function terms are selected and added to the description of a use-case. In the next step, any references (e.g. participation in a trial) or solution-related documentation (e.g. user manual) can be added. Furthermore, meta-information about the solution, such as TRL level, crisis cycle phase, and so on, is added in the next step. Finally, a last review of the description is performed, and the user decides whether it is complete and can be openly published on the PoS.

To meet the requirements for different permissions where each solution provider together with their team can add, edit, and remove solution descriptions and perform the same operations on related entities on the website, a contributed "Group" module was implemented and configured according to the requirements. In the PoS implementation, a solution description was designed as the main entity with three additional entities: "Use cases," "Used/tested in," and "Documentation," as shown in Figure 4.10. A taxonomy vocabulary for "CM functions" was added to the "Use case" entity, which was then used to implement the search and match function on the one hand and to achieve one of the main goals of the project on the other hand – to ensure that all innovative solutions that exist in PoS are classified and found based on a common taxonomy across the field.

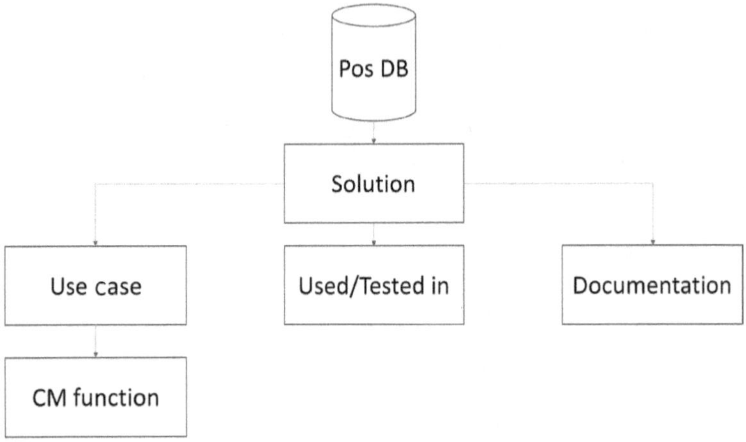

FIGURE 4.10 The portfolio of solutions structure diagram. A tree structure to illustrate the relationships between entities in the PoS.

Quality Assurance

A unique feature of the PoS platform (for the crisis management domain) is that it is open to contributions of all interested parties but requires that all of the users are employees of an organisation with legitimate interest in crisis management. Anyone requesting a PoS user account must disclose their real name and affiliation. These requests are screened by site editors and legitimate accounts are enabled. In combination with standard automated spam detection mechanisms, this effectively assures that most of the spam bots and malicious users are blocked from posting or changing the site content. Once the user's identity has been proven, they are free to publish anything they see fit on a site – under their real name and with a clear indication which organisation they work for. Quality of the information on the site thus heavily relies on the assumption that PoS users have a real-world reputation at stake and therefore are not going to publish anything lightly. To help these users, PoS also offers interactive tutorials and automated quality controls and suggestions for content authors. This includes the completeness check (automatic check if all fields are filled with entries) and a periodic check of the "last changed" date as a way to differentiate between maintained and abandoned contents. Any issues found are sent to the user per email, with a suggestion how to improve it in monthly QA emails that are automatically generated by the site. To further improve the content quality, a simple automated mechanism for reporting any mistakes or omissions found is available on every page of the site. Finally, special quality assurance views to the underlying data have been implemented that allow the site editors to easily detect potentially low-quality content and either hide it from the site or contact the authors and suggest improvements. In the worst-case scenario, the authors can even be banned by the administrator from the site and their contents removed. Finally, all the contents that are considered incomplete or "abandoned" (not updated for more than two years) are automatically sorted below presumably higher-quality contents in all PoS views. Combination of all these measures results in a reasonably high content-to-noise ratio for the site users, while giving a lot of freedom to content authors and keeping the need for editorial work low.

Additional and Subsequent Functionalities

CM Taxonomy

One of the tasks in the DRIVER+ project was to create a taxonomy that would facilitate the classification of solutions in the PoS database. Considering the broadening spectrum of crisis management tasks (e.g. in counterterrorism and response to chemical and biological threats) and new concepts such as resilience and agility, the project team was tasked to go beyond current practice and design a functional taxonomy: a taxonomy of crisis management functions. Initially, the PoS database was expected to include information on solutions developed or demonstrated within the project. These solutions were selected in order to address specific gaps in the available crisis management capabilities identified in the project proposal or in the initial phases of project research. Yet the working assumption was that the PoS will continue to be used years beyond the end of the project and will be continuously enriched by adding newly identified gaps and diverse solutions provided by numerous companies and research organisations. Hence, the taxonomy was expected to serve the navigation of the PoS in the long term, without the need for frequent amendments, and by solution providers and crisis management practitioners that are not necessarily familiar with advanced theoretical concepts and developments beyond their area of professional expertise. Therefore, the designers of the taxonomy aimed to make it comprehensive and, to the extent possible, easy to use. These two considerations are examined in some detail in this section.

Comprehensive Coverage of the CM Taxonomy

The comprehensiveness of the taxonomy was considered from several perspectives. First, the functional orientation of the taxonomy allows to capture not only solutions based on new technological advances and respective equipment but all capability components. In fact, the term "capability" designates the potential for performing a particular function or task under specified conditions. This potential is provided by the combination of organisation, doctrine and procedures, material and infrastructural components, and personnel with their skills, leadership capacity, and training levels. Thus, the CM taxonomy allows to capture solutions aiming to improve any of the capability components and not just equipment. Second, the taxonomy accounts for the role of all contributors – professional responders, other governmental agencies and public services, volunteers, non-governmental organisations, private companies, and local communities – through the functions they perform in crisis management. Third, the design of the taxonomy takes into account the full scale of potential consequences of a natural or man-made event – such as an incident, a disaster or catastrophe, or a crisis – where each level up the ladder of consequences necessitates the involvement of additional players and resources. Respectively, the taxonomy reflects all levels of command and management. It accounts for responsibilities of various players in policy- and strategy-making, operational and tactical levels of crisis management, planning and organising, development, deployment, and the use of capabilities. Fifth, it incorporates possible jurisdictions and geographic scope of crisis management, including local, regional, and national, and in particular cross-border cooperation and European levels of preparedness and response. Sixth, it is an "all hazards" taxonomy, taking into account both traditional hazards and recent or newly emerging vulnerabilities and risks related to, for example, terrorism; the transfer of radioactive and nuclear materials; chemical, biological, and explosive threats; and threats from cyberspace. Finally, the taxonomy covers most advanced crisis management concepts at the time of its design, including concepts for addressing uncertainty, such as adaptiveness, foresight, organisational agility, resilience of communities, systems, and networks. In fact, the only constraint followed in the development of the taxonomy was its designation to support interactions in the disaster management community. Hence, no attempt has been made to reflect requirements specific to crises of a financial, social, political, or military nature.

Commonality of Implemented Constructs and Terminology

The attempt to capture all these considerations inevitably led to a somewhat complex design of the taxonomy. The taxonomy is structured at four levels and, in its most recent version (version 2.1, May 2020), contains respectively 10 functional areas, 54 functions, 262 sub-functions, and 103 tasks. Where available, the design incorporated widely accepted delineations of crisis management functions. At its first level, the taxonomy includes three operational, three preparatory, and four common functional areas. The functional areas "Protection," "Response," and "Recovery" correspond fully to the high-level functions in the Universal Task List used by the US Department of Homeland Security (Department of Homeland Security, 2005) and combine the constructs underlying the UN Sendai Framework for Action (United Nations, 2015), UN Office for the Coordination of Humanitarian Assistance (UN OCHA, 2021), and the European Union Civil Protection Mechanism (European Parliament and the Council, 2013). Likewise, the three preparatory functional areas of "Mitigation," "Capability development," and "Strategic adaptiveness" build on the constructs of the UN Sendai framework (United Nations, 2015) and the EU ResiStand project (Stolk et al., 2017). Finally, the four common functional areas of "Crisis communications and information management," "Command, control, and coordination (C3)," "Logistics," and "Security management" resemble and amend the functional constructs used by the UN Office for the Coordination of Humanitarian Assistance (UN OCHA, 2021), the US Department of Homeland Security (Department of Homeland Security, 2005), and the EU ResiStand project (Stolk et al., 2017). For a detailed mapping to these and other relevant and widely used documents the reader may refer to Figure 4.6 in Tagarev and Ratchev (2020). The taxonomy also adheres to the terminology (Kähler et al., 2020) adopted in these documents. At the second level of the taxonomy, the arrangement of functions follows core management elements of planning, organising, command, coordination, and control, as established by Henri Fayol (Fayol, 1949). At the third and fourth levels – of sub-functions and tasks, respectively – the organisation implements the so-called OODA (observe, orient, decide, act) loop, proposed by the US fighter pilot John Boyd and well known in the security community (McKay & McKay, 2021).

Lessons from the Implementation of the Taxonomy

Version 2.1 of the crisis management taxonomy is available as supplementary material to an article presenting the logic underlying its design (Tagarev & Ratchev, 2020). Although considerable efforts were invested to make it easier to comprehend and use by practitioners and solution providers, they still found it challenging. Therefore, in the PoS, its use was complemented by semantic search and other, more traditional taxonomies oriented to the type of hazard, the intensity of the consequences of an event, the most relevant phase of a typical crisis management cycle, and the stage of innovation of a solution. The primary purpose of the taxonomy was to facilitate the organisation of information and knowledge on disaster management and the search for solutions by navigating the PoS. Yet it soon became clear that it brings benefits in examining other related issues, such as risk management planning (Sliuzas et al., 2020) and elaborating research-based general crisis management policies (Group of Chief Scientific Advisors, 2021).

Opening Up PoS Information in Other Systems

As mentioned earlier, the PoS was developed as a web-based service which allows its data to be shared in various forms. In addition to providing information to a human user who browses the PoS content directly, it also provides the ability to export this information for use by other applications and services. There are two main ways of sharing data: The first is exporting all solution-related information as a PDF file, which facilitates analogue transfer of information when desired (e.g. information about some interesting solutions could be printed out to review in a meeting). The second is a machine-readable export, which is provided via a RESTful interface in JSON format that can be accessed by any third-party application. An example of such a use-case is CMINE,[9] an online hub for crisis management experts in the EU and beyond. Its goal is to foster innovation and research in crisis management through cross-sector dialogues between different stakeholders on capability gaps and potential solutions. Through the provided RESTful interface, CMINE collects the information about published solutions in PoS and presents it to a wider audience.

Conclusions and Ways Forward

The PoS toolset consisting of the gaps explorer, the PoS, and the trial guidance tool is operated and maintained beyond the lifetime of the DRIVER+ project. One example of application is the use of the PoS toolset in the frame of the H2020 project STAMINA. In order to support the trialling of 21 solutions for pandemic management developed in the frame of STAMINA, it was first necessary to validate the suitability of the taxonomy of CM functions for the purpose of pandemic management. It turned out that only small extensions were necessary in order to support matching between pandemic management solutions and gaps specified by end-users acting in the management of COVID-19. An analysis of the pandemic gaps identified in STAMINA can be represented by 27 pandemic management functions. Cross-examinations with the DRIVER+ CM functions revealed that most of these functions are already included in the DRIVER+ taxonomy, albeit not always in the context of health or pandemic management. The only function that was added to the taxonomy is "contact tracing." This was the basis to specify both gaps and solutions in the gaps explorer and the PoS, respectively. An added value of the combination of solutions for crisis management and pandemic management in the same repository is the possibility to identify solutions a priori designed for crisis management and not specifically pandemic management that have the potential to support pandemic managers and vice versa. The extended and adapted PoS toolset offers not only the possibility to support the execution of the pandemic trials, but it also makes it possible to make the STAMINA solutions visible to members of multiple security domains encompassing pandemic management.

As described earlier, the trial guidance tool offers the possibility to support the planning, execution and evaluation of trials based on the trial guidance methodology. Trials focus on the evaluation of socio-technical parameters of solutions. A very demanding challenge is the systematic evaluation of large-scale civil protection exercises. In the frame of the Union Civil Protection Mechanism (UCPM) project INEGMA-E2, a versatile evaluation methodology is developed, addressing the different types of existing exercises. The concept of the trial guidance tool will be adapted in order to explore the possibilities of automated evaluation of such exercises in the future.

Acknowledgements

Appreciation to David Lund for the English grammatical correctness check of the manuscript.

Notes

1 https://www.scrum.org/.
2 https://www.drupal.org/.
3 https://solr.apache.org.
4 Note: the knowledge database and the country profiles are not being actively updated and only contain information collected during the course of the project.

5 www.internationalresponderforum.org.
6 https://drmkc.jrc.ec.europa.eu/knowledge/Gaps-Explorer/forest-Fires.
7 www.internationalresponderforum.org.
8 https://solr.apache.org.
9 www.cmine.eu.

References

Alavi, M. (1984). An assessment of the prototyping approach to information systems development. *Communications of the ACM*, 27, 556–563. https://doi.org/10.1145/358080.358095.

Department of Homeland Security. (2005). *Universal Task List: Version 2.1*. United States. Department of Homeland Security. Office of State and Local Government Coordination & Preparedness. www.hsdl.org/?abstract&did=458805.

European Parliament and the Council. (2013). Decision No 1313/2013/EU of the European Parliament and of the council on a union civil protection mechanism. *Official Journal of the European Union*, 347, 924–947. https://data.europa.eu/eli/dec/2013/1313/oj.

Fayol, H. (1949). *General and Industrial Management*. London: Pitman.

Group of Chief Scientific Advisors. (2021). *Strategic Crisis Management in the EU: Improving EU Crisis Prevention, Preparedness, Response and Resilience*. European Commission, Scientific Advice Mechanism. https://ec.europa.eu/info/sites/default/files/research_and_innovation/groups/sam/scoping-paper_crisis-management-in-the-eu_june_2021.pdf.

Ignjatović, D., Havlik, D., Neubauer, G., Turptil, S., Gonzales, F., & Regeczi, D. (2019). The portfolio of solutions. *Proceedings of the 27th Interdisciplinary Information Management Talks* (pp. 199–206). https://idimt.org/wp-content/uploads/2019/08/IDIMT-2019-proceedings.pdf.

Kähler, E., Neubauer, G., Van Berlo, M., & Stelkens-Kobsch, T. H. (2020). The relevance of tailored terminologies for large scale international projects. *Proceedings of the IDIMT – Interdisciplinary Information Management Talks Conference* (pp. 181–186). https://idimt.org/wp-content/uploads/2020/07/IDIMT_proceedings_2020.pdf.

McKay, B., & McKay, K. (2021). *The Tao of Boyd: How to Master the OODA Loop*. Art of Manliness. www.artofmanliness.com/articles/ooda-loop/.

Sliuzas, R., Jackovics, P., Thorvaldsdóttir, S., Kalinowska, K., Tyrologou, P., Resch, C., Castellari, S., & Greiving, S. (2020). Risk management planning. In A. Casajus Valles, M. Marin Ferrer, K. Poljanšek, & I. Clark (Eds.), *Science for Disaster Risk Management 2020: Acting Today, Protecting Tomorrow* (pp. 66–77). Luxembourg: Publications Office of the European Union.

Stolk, D., van der Lee, M., Petiet, P., Liedtke, C., Lindner, R., van Zetten, J., Karppinen, A., Anson, S., & Kroen, I. (2017). *Assessment Framework for Standardisation Activities; Deliverable 1.3*. Rijswijk, the Netherlands: ResiStand Project; TNO.

Tagarev, T., & Ratchev, V. (2020). A taxonomy of crisis management functions. *Sustainability*, 12(12), 1–32. https://doi.org/10.3390/su12125147.

United Nations. (2015). Sendai Framework for Disaster Risk Reduction 2015–2030. In *Proceedings of the Third UN World Conference on Disaster Risk Reduction* (pp. 14–18). www.undrr.org/publication/sendai-framework-disaster-risk-reduction-2015-2030.

UN OCHA (United Nations Office for Coordination of Humanitarian Affairs). (2021). *Themes*. www.unocha.org/themes.

PART 3

Implementation and Evaluation of Innovative Solutions in Crisis Management

Part 3 focuses on the beating heart of the DRIVER+ journey – namely, how new solutions and approaches were trialled and assessed by crisis management organisations. This was the moment in which the trial guidance methodology, the testbed technical infrastructure with all the additional tools and services, and the knowledge management and inference toolset were applied and implemented in different contexts. It was also the moment in which lessons were learnt and the outputs were refined according to the main takeaways. This is the reason that this part is so rich with experiences, considerations, lessons learnt, and both technical and less technical perspectives that can be of an added value for future trial owners and for the wider crisis management community. The "trials perspective" offers a snapshot into three trials – the trial in the NL, the trial in AU, and one carried out at the same time in two locations: NL and PL. "Simulation as Decision Support" goes deep into modelling and simulation, always having as a background knowledge the data collected in the DRIVER+ trials. "Situation Awareness Approaches" looks at models and technologies that can improve decision-making.

DOI: 10.4324/9781003256977-8

SUB-PART 3.1

Trials Perspective

5

THE TRIAL IN THE NETHERLANDS

Konstanze Lechner and Carsten Dalaff

Background / Scenario

From May 21 to 23, 2019, the fourth trial within the DRIVER+ project (trial 4) took place in The Hague, the Netherlands, at the Safety Region Haaglanden (SRH). The German Aerospace Center (DLR), a research organisation, teamed up with the trial host organisation SRH to design the trial in an end-user-driven approach, addressing real capability needs of crisis management (CM) practitioners. The event was attended by 140 people from 13 countries. The majority of the organisational staff was of course from the host country. As the scenario was conducted at all three levels of the Dutch national emergency system, including an international request for assistance, there was a broad representation of Dutch experts representing ten different emergency centres/crisis teams of the Netherlands. However, there were also significant contributions from other countries, such as Germany and Poland, reflecting the trans-European network and cooperation within the DRIVER+ project.

One of the greatest natural hazards in the Netherlands is flooding, as a large part of the country is below sea level. Extreme weather conditions in combination with a possible dam break or lock failure could lead to severe flooding of large areas and thus devastating damage. For this reason, the authorities and crisis managers, such as SRH, are focusing on good preparedness and safety measures in case of a disaster. The systems and processes are to be permanently improved and, where possible, supported with innovative solutions.

The overall objective of trial NL was, therefore, to improve cooperation and coordination between authorities and organisations during severe floods by using innovative solutions to help manage large-scale and long-term crises.

An extreme coastal flood was chosen as the scenario for this trial, in combination with an expected storm. In addition, a medium probability of a technical

DOI: 10.4324/9781003256977-10

failure of the ship lock in Scheveningen was assumed. A possible breach of the coastal protection at Scheveningen may lead to the flooding of large areas in The Hague (with water depths of up to two metres). The Hague city centre may be threatened. If the event occurs, thousands of people are at risk of being trapped, with loss of life also expected. The predicted flooding will therefore require decisions to evacuate residents of the threatened area. The influx of water will further affect vital infrastructure and lead to loss of electricity, drinking water, and heating (the event takes place in winter) in the area. Traffic, whether cars, buses, or trains, will be hampered by flooded roads, debris, and missing manhole covers. To minimise casualties, attention should be paid to emergency supplies so that the population can be evacuated efficiently before, during, and after the disaster.

The trial was prepared and executed as a tabletop (in-door) event based on a scenario run in a simulated environment created in the testbed technical infrastructure. Actions were taken by the participants in a realistic information environment, based on currently available legacy tools and means, crisis management plans, rescue procedures, and good practices of the trial practitioners. Focus was on coordinating the flood event by SRH only.

The tabletop exercise scenario was divided in two phases:

- Threat phase: There is a serious threat of flooding due to the severe meteorological circumstances.
- Impact phase: The attack has been committed and flooding occurs. An appeal is made for additional (international) emergency response.

Each of the phases consisted of two blocks (see Table 5.1). On day 1, the blocks were played in actual time. The start of block 2 was three hours after the start of block 1 (two hours duration of block 1).

Each block lasted 2 hours and was divided similarly:

- Plenary session to describe the general situation.
- Regional operational team (ROT) to start the session to assign tasks to all action centres and crisis teams.
- Working session to complete the tasks based on the solutions.
- Final meeting of the ROT to agree on the advice for the mayor of The Hague.

The practitioners were asked to make an assessment of the flood crisis situation and to design and recommend a course of action to address the crisis. They focused exclusively on these tactical issues. In each block, the practitioners were given a realistic task to get the situational picture (the common operational picture) and to advise the strategic level (i.e. the mayor of The Hague) about the crisis. In a real flood threat, practitioners also have operational tasks, such as coordinating first responders on the ground, crisis communication, and so on. These operational tasks were not part of the tabletop exercise.

TABLE 5.1 Trial Setup

Day	*Phase*	*Block*	*Trial Time at Start of the Block*	*Focus*	*Objective*
1	Threat	1	31 hours *before* the expected dike breach	Situational awareness and determining the cascading effects of possible flooding	Assess The Hague city centre
		2	28 hours *before* the expected dike breach	Formulation of mitigating measures, with focus on evacuation	Formulate two evacuation strategies; define actions/measures to mitigate effects of possible flooding
2	Impact	3	16 hours *after* the dike has breached	Damage assessment	Assess damage in the flooded area (The Hague city centre) and mitigation measures
		4	16 hours *after* the dike has breached	Damage control and recovery	Answer questions of international organisations and planning police personnel and about mitigating measures

The emphasis was mainly on the use of the solutions in order to assess their added value in executing the given assignments by practitioners (the TGM "innovation line"). The time available for the tasks was limited to the length of the block (two hours). The baseline for the scenario in which the same tasks were performed without using the solutions was set before the trial. In the baseline, practitioners still had to get an overview of the situation and advise the strategic level by using the available information.

Crisis Management Perspective

The general purpose of the trial in respect to crisis management tasks was to improve cooperation and coordination among agencies and organisations, use innovative solutions, and provide support in handling large-scale and long-term crises.

Actors Involved

The trial required the engagement of stakeholders (end-users and decision-makers) from all levels of crisis management (local, regional, national, and international), such as representatives of the Ministry of Infrastructure, the national police, regional fire and rescue services, crisis management experts, representatives of the water sector, representatives of non-governmental organisations (volunteers), representatives of decontamination institutes, and representatives of other institutions responsible for disaster relief in similar situations.

The practitioners represented ten different emergency response cells, each comprising two or more individual practitioners (such as the action centres police, fire brigade, or municipality). Each of them was tasked to execute different emergency response assignments with the use of the trialled solutions.

Objectives

Three high-priority crisis management capability gaps were selected by practitioners in ad hoc workshops:

- Gap 1: Limitations in the planning of resources (qualified personnel and equipment) for response during large-scale and long-term crises.
- Gap 2: Shortcomings in the ability to exchange crisis-related information among (emergency management) agencies and (consulted) organisations (also related to as interoperability).
- Gap 3: Shortcomings in planning and managing the side effects of large-scale evacuation of population in urban areas.

All these gaps have been discussed and validated during the DRIVER+ gaps assessment workshop 2 in January 2018 and subsequently prioritised by the trial committee.

The resulting main research questions were the following:

1. How can simulation tools improve resource planning activities in large-scale and long-term disaster operations?
2. How can net-centric data exchange improve information sharing between relevant parties and thus improve the shared understanding of the current situation?
3. How can simulation tools support the planning and management of a large-scale evacuation under consideration of real-time traffic information?

Innovative Solutions

In response to a call for applications, 25 applications were initially received (13 external and 12 internal to the project), and 11 solutions were shortlisted (5 external and 6 internal) after the double-blind review, the meta-review process, and a final teleconference. The results of the review, practitioner comments and the suitability of the solution for trial 4 then resulted in a further shortlist. Through further face-to-face interviews and rehearsals, five innovative crisis management solutions were ultimately selected for their ability to address a number of gaps identified by practitioners at the start of the project. The main features of the selected solutions are captured here:

3Di (Nelen & Schuurmans):

- Provision of a flood prediction in the threat phase and update of this prediction based on actual water level in the impact phase.
- Calculating the effects of mitigation measures (pumps, barriers).

Airborne and Terrestrial Situational Awareness (components KeepOperational and ZKI, DLR):

- Overview of actual flood state based on aerial images.
- Route calculations that avoid the flooded area.
- Provision of damage assessment maps in 2D and 3D based on the derived inundated area.

CrisisSuite (Merlin Software BV):

- Establishing information exchange to provide a common operational picture to supporting crisis teams without access to the legacy system of crisis management professionals.
- Log situation reports and actions.

HumLog (University of Münster):

- Creating an evacuation plan for neighbourhoods, hospitals, and so on.
- Calculating organisational logistics (esp. planning of personnel).

Sim Safe (SIM-CI):

- Predicting cascading effects on critical infrastructure (power, telecommunication, and public transport).
- Update of this prediction based on actual water level.

The functions of the individual solutions are described in detail in the following sections.

3DI (Nelen & Schuurmans)

3Di is an interactive water simulation model that enables crisis managers to construct a common operational picture of the dynamics of floods and allows a quick calculation of the effects of mitigation measures.

Within 3Di, a tool gives users the possibility to edit the digital elevation model (DEM-edit function) during a simulation. Using this tool, it is very easy to calculate the effects of adding a mobile flood wall or breaching a levee in any chosen location to prevent floods in another location. Combined with the already present option to add positive and negative discharge points to the simulation, which can simulate pumps, this gives crisis managers the possibility to calculate the effects of several mitigation measures in a very short time span (minutes) and use those to support them in the decision-making process. Since crisis managers only have limited resources available, knowing the consequences of different choices can be very helpful for them in the hectic process of crisis management.

3Di has short computation times with a high spatial resolution, an accurate prediction of floods, and a realistic visualisation of model outputs. Furthermore, it has a very easy interface, comparable with Google Maps, and can be operated via a touch table. This interface is tested with people that had no model experience, was

FIGURE 5.1 Screenshot of solution 3Di.

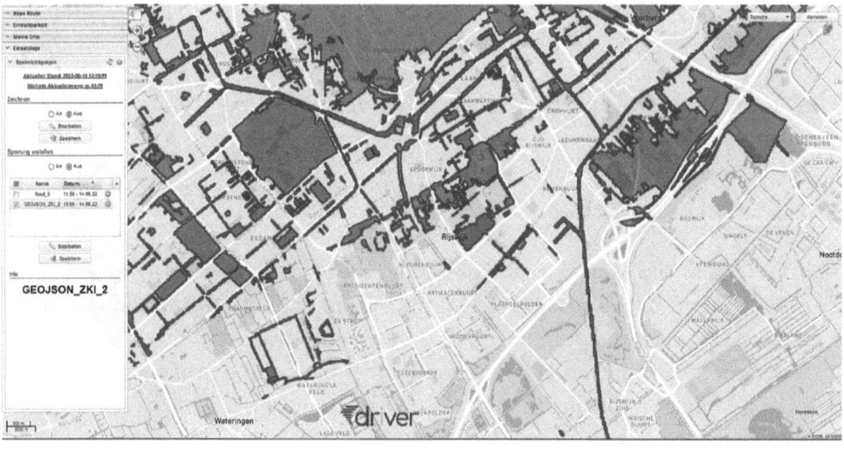

FIGURE 5.2 Screenshot of module *KeepOperational* (solution *Airborne and Terrestrial Situational Awareness*).

supervised by experts, and has been proven to be usable for this user group. 3Di was successfully tested in various disaster exercises in the Netherlands. Outputs from the model (predicted flood waves and river flow) can be exported to other systems. All the input and output data are in common formats (GIS).

Solution 3Di is mature (TRL9). It is in use since 2014 and has been applied in numerous flood analysis and calamity exercises in the Netherlands, the UK, Taiwan, Vietnam, Australia, South Africa, Kenya, Grenada, and Saint Lucia.

Airborne and Terrestrial Situational Awareness (DLR)

The solution relates to the impact of the transport system on the efficiency of professional responders during crisis management and contributes to overall situational awareness.

The solution is composed of two modules, discussed in the following sections.

KeepOperational

To ensure efficient planning and managing of large-scale evacuation as well as planning of resources in order to provide the affected population or logistics with goods and services, this module provides reliable transport and traffic information – for instance, web-based visualisation of current traffic situation based on various traffic data sources, routing advice and visualisation of the timely reachability in dependency of the current traffic situation, and traffic prediction. A special feature of this module is its capability to optimise the routing and to provide routing options depending on the actual availability of the traffic infrastructures in the affected area. For this purpose, the routable network can be adapted by considering either automatically generated vector-based input datasets or manually created network adaptations (i.e. the solution provides an interactive functionality that allows the user to manually adapt single edges or wider areas of the network).

Center for Satellite-Based Crisis Information (ZKI)

Regarding the scenario of flooding in The Hague, this module derived crisis-related impact information, such as surface water extent or affected infrastructures and buildings at different time steps from aerial and satellite imagery. On the one hand, the flood extent information was one of the input datasets for the previously mentioned transport and traffic management related support in order to update the routable traffic network to optimise the existing routing functionalities in the hazard area. On the other hand, 2D maps and interactive 3D visualisations were provided for professional responders in order to improve situational awareness, to support damage and needs assessment and to facilitate decision-making processes.

From a maturity point of view, the solution relates to stage 2 (research and development) of the stages of innovation. However, the solution has an overall

FIGURE 5.3 Screenshot of module *ZKI* (solution *Airborne and Terrestrial Situational Awareness*).

technology readiness level of TRL 6, which means it has already been demonstrated and tested in a relevant environment.

CrisisSuite (Merlin Software BV)

CrisisSuite is an online crisis management software that enables organisations to successfully manage information during a crisis. CrisisSuite supports the net-centric working methods of crisis teams by creating a universal picture of the crisis and share it horizontally and vertically with the other teams in the crisis organisation. It also assists in maintaining an effective crisis meeting structure, and it decreases the administrative workload for the people managing the crisis.

CrisisSuite is a solution that different organisations can use to manage their information on an operational, tactical, and strategic level during a crisis. The solution is made up of the following components:

1. Organisation: Define people with their contact details in different groups (roles, teams).
2. Plans: Store existing crisis management plans, and link them to the relevant people, roles, or teams.
3. Alarming: Alarm people about a crisis situation and request a response.
4. Maps: Draw a geographical overview of the impacted area.

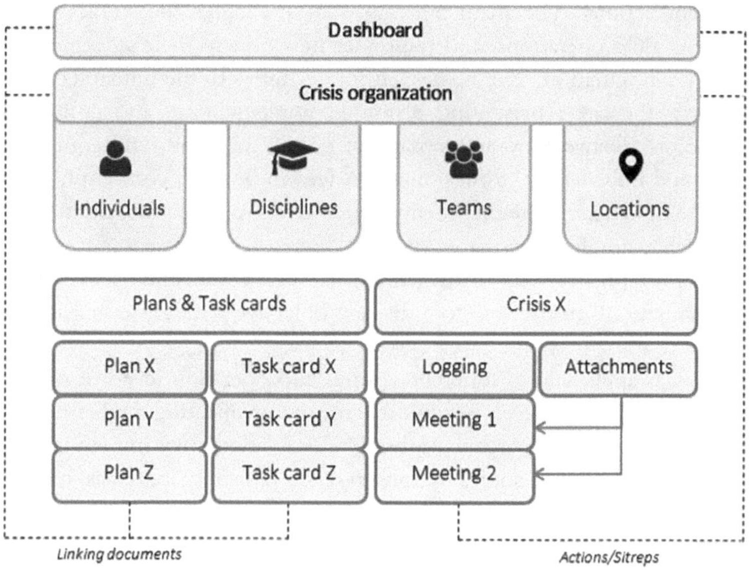

FIGURE 5.4 CrisisSuite structure enabling organisations to successfully manage information during a crisis.

5. Logtool: Create a logbook of all the processed crisis information, decisions, and actions.
6. Action: Keep an overview of all actions and their status.
7. Attachments: Store any documents (e.g. pictures) that are relevant to the crisis.
8. SitReps: Create situation reports (SitReps) regarding certain topics in the crisis.
9. CrisisSuite app: This app for mobile devices enables the user to receive notifications and view information on the go.

The technology readiness level of *CrisisSuite* is TRL 9: actual system proven through successful mission operations. The system is currently in use by organisations in different sectors, such as the energy sector, healthcare sector, and government sector. It was used in the trial as a crisis information system for the crisis teams Stedin, HTM, and IO. Both LCMS and CrisisSuite were linked in the frame of the trial: Information (maps and reports) from CrisisSuite was also available in LCMS; the situational report from LCMS was available in CrisisSuite.

HumLog (University of Münster)

HumLog is a performance assessment platform that serves logistic processes in crisis management. It can operate on both current operational logistics network and fictional (planned) network configurations. The functionality comprises strategic planning support and tactical and operational decision support by assessing and comparing the network performance under given situations and realistic crisis management actions. The main target groups are higher-level crisis managers, coordinating field operations, and responder networks in their strategic network design and configuration. The name "HumLog" refers to the humanitarian logistics context, thereby representing planning, implementing, and controlling of efficient, cost-effective flow and storage of goods, materials, and equipment, as well as related information, from point of origin to point of consumption aiming to meet the beneficiaries' requirements under given resource capacities. HumLog is an adaptable simulation environment for discrete event-based and agent-based simulations. It represents crisis management activities within and between humanitarian organisations on the way to a defined objective whilst assessing the overall performance.

HumLog is applicable to different settings and operation levels of crisis management. On a strategic level, it supports creating or updating a relief network of one or multiple responder organisations. The simulation environment can analyse different possible network configurations in terms of facility locations, relief goods stock values, warehouse capacities, human resources, and others. By executing so-called "what if" scenarios, the performance of a planned network can be simulated under a fictive crisis event, which can make use of historical crisis event data to represent a realistic scenario.

On a tactical level, these capabilities can also be utilised to compare possible execution plans and to estimate their expected outcomes. Given a relief network of one or multiple responder organisations in an actual crisis event, crisis management commanders can test different strategies in addressing the crisis without any influence on the real-world resources. The simulation capabilities can thereby be freely adapted to the needs of an individual organisation, like the construction of sandbag barriers, evacuation of the population, or allocation and transportation of human resources. The simulation will take all available resources and capacity restrictions

FIGURE 5.5 Screenshot of solution HumLog.

into account to provide a complete and sophisticated execution protocol, which can then be used by command staff to compare execution strategies.

Sim Safe (SIM-CI)

Sim Safe visualised the event of flooding and its cascading effects on critical infrastructures in The Hague by means of a digital twin city. With its simulation, crisis managers could see how water spreads through the area. They saw the digital twin city of The Hague, including buildings and critical infrastructures, such as roads and the electricity and telecoms networks. Some of the cascading effects calculated and visualised by the tool were electricity failure due to flooding and the subsequent loss of mobile reception. In addition, impact on society was calculated as well. Crisis managers experienced the direct and indirect effects of flooding on the critical infrastructures and eventually on people, residents, and businesses in the area.

Based on this simulation, the crisis managers could align, prioritise, and communicate their crisis response plan. This crisis response plan was based on accurate predictions of floods and on the effects calculated and visualised, such as the availability of roads (important for evacuation routes), mobile reception and electricity in each area in the city.

In general, Sim Safe enables integration of different types of information and can help answering questions like "When does the water reach particular areas?" "How much time is there until the first power failure?" and "Until what time can we communicate with residents in certain areas and are roads still available?" This provides operational crisis managers and decision-makers with insights into the integral character of a flooding or disruption in general.

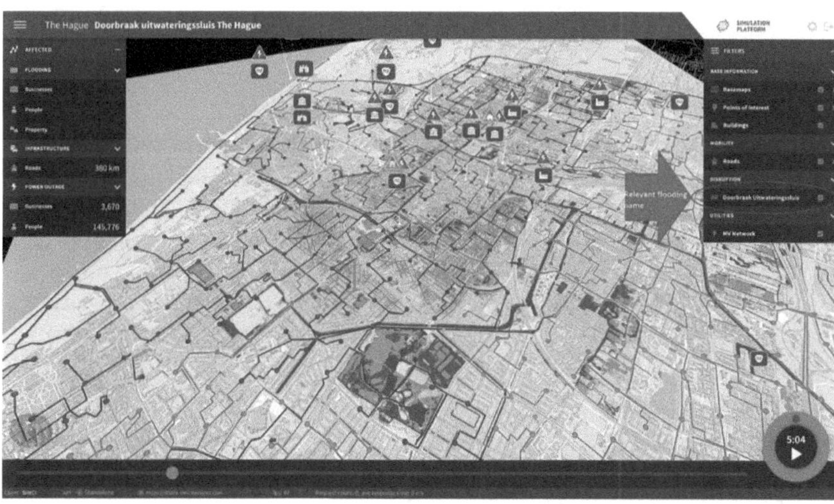

FIGURE 5.6 Screenshot of solution Sim Safe.

The sole Sim Safe model is mature (TRL 9). The simulation engine integrates the Sim Safe model with different network models into a spatial environment and visualises this into a 3D simulation. The combination of technologies is used in a number of projects in the Netherlands.

Planning, Execution, and Evaluation of the Trial

The planning execution and evaluation of this trial was done according to the TGM and is based on five trial event stages, which reflect the different stages of trial preparation:

1. Step zero: The initial stage, which ends with Workshop Zero.
2. Preparation meetings: The main preparation stage, which ends with dry run 1.
3. Dry run 2: The maturation stage, which ends with dry run 2.
4. Trial run: The final preparation stage, which ends with the trial itself.
5. Evaluation: The recapitulation stage, which ends with the trial evaluation report.

The overall planning process started officially in March 2018 with Workshop Zero and the execution was conducted in May 2019 in The Hague, the Netherlands. Table 5.2 shows the schedule of the different trial stages and the key task.

A peculiarity of this trial was the split responsibility in form of the trial owner (DLR, a German research institution) and trial host (SRH, a Dutch practitioner organisation) which was not the case in other DRIVER+ trials. Therefore, this also enabled the testing of the TGM in the context of a shared responsibility between two organisations owning the role of trial host and trial owner. In addition to the overall teamwork of the whole involved team, the detailed definition and identification of the different roles with their responsibilities can be mentioned as one important success criteria. The key functions of trial 4 managed the interfaces and the multiple interconnections within the so-called trial committee.

In addition to the key roles, more responsibilities were required and defined. As there were the solution training coordinator, safety officer, methodology validation conductor, evaluation coordinator, review leader, media officer, and an external cooperation manager. In addition to the content and technical responsibilities with the support teams, the deployment of a trial furthermore required support from the area of event managing, like a catering crew, cleaning crew, PR team, VIP host team, and a medical team to be prepared providing medical assistance in case of real accident during the trial.

Trial 4 was conducted on May 22–23, 2019, during a two-day tabletop exercise. After the official welcome, a briefing required a duration of more or less half a day, including special briefing before the block, via evaluation interviews after the block and the final block debriefing. The usage of innovative new solutions in a crisis situation requires training to be familiar with the solution. The

TABLE 5.2 Trial Schedule

	Dates	Description	Needed For
Preparation	**Step zero**		
	March–April 2018	• General onformation on the trial.	• Trial baseline, gap selection.
	Step zero – initial stage, ended with Workshop Zero.		
	May–October 2018	• Trial committee and further key functions set up. • Baseline system description. • Gap analysis. • Trial objectives (specific). • Research questions. • Timeline of preparatory activities.	• Trial management, planning. • Scenario, innovation line – implementing solutions in trial.
	Workshop Zero		
	November 9, 2018	• Research questions – confirmation.	• Baseline for section 6 trial scenario development.
	After Workshop Zero		
	December 2018–February 2019	• Solutions – technology availability list. • Key performance indicators of solutions – selection. • Detailing timeline of preparatory activities. • Identification of external participant profiles. • Detailing scenario scope and story. • Information / data exchange / evaluation plan. • Participants – practitioners and solution providers.	• Confirmation of participation. • Solution assessment. • Scenario development. • Trial screenplay. • Trial management. • External stakeholders' involvement.
	Trial integration meetings and dry run 1		
	Three months before trial to February 22, 2019	• Solution integration procedure. • Preparation of material for external stakeholders and formal invitation. • Solution and platform instruction, training and dissemination documents review. • Trial scenario building. • Utilisation local platform facilities. • Detailed trial management planning including output specification, selection of triggering events.	• Technical and organisational setup finalisation.

(*Continued*)

TABLE 5.2 (Continued)

	Dates	*Description*	*Needed For*
Execution	**Dry run 1** – testing and training, maturation of scenario, final concept for solution utilisation in trial.		
	February 18–22, 2019	• Training and verification for participants (solutions, testbed, other). • Solution readiness. • First solution deployment and scenario run. • Finalisation of the scenario development. • Trial management.	• Final technical requirements. • Conclusion on practitioners' participation. • Risk analysis, safety plan, impact assessment, contingency plan for the trial. • Validation of solution readiness (GO, NO-GO).
	Dry run 2 (April 12, 2019)		
	March–April 2019	• Finalisation of integration processes. • Sub-scenarios maturation process. • Maturation and further development.	• Full integration achieved. • Session scenarios frozen.
	Dry run 2 (April 2019) – team readiness test, playing the full trial without external participants.		
	April 8–12, 2019	• Integration review. • Internal dry run with roles and functional solutions utilised as it is planned for trial. • Safety review. • Final acceptance of exercise evaluation processes. • External stakeholders – confirmation.	• Final decision for each solution. • Final validation of integration. • Last modifications of "exercise – scenario," "technical teams plan," timing. • Confirmation of the evaluation tools.
	Trial run – final preparation stage, ended with the trial itself; short, one-to-two-week period for final arrangements.		
	April–May 2019	• Final logistical arrangements. • Freezing non-frozen sections of TAP. • Detailed contagious planning for the trial.	• Achieving full readiness for trial deployment.

	Dates	*Description*	*Needed For*
	Conduction the trial (May 20–24, 2019)		
	May 20–24, 2019	• Solution training. • Observer briefing. • Safety briefing. • Conducting the trial.	• Trial.
Evaluation	**Evaluation**		
	May–October 2019	• Evaluation of the trial results. • Lesson learnt. • Documentation.	• Trial evaluation. • Closing.

trial participants were trained on May 21, 2021, on the solutions. The technical, functional, and organisational set and clean-up of trial 4 required a duration of five days.

In total, 140 people were involved in trial 4. The share of internal and external participants was distributed as follows: 45% trial internal attendees and 55% trial external attendees. The European diversity was reflected in the trial execution by participants from 13 different countries (e.g. Netherlands, Germany, Poland, Austria, France, Sweden, the UK, Israel, and others). The majority of the organisational staff naturally represented the hosting country – the Netherlands. There was a broad representation of Dutch practitioners who represented ten different emergency action centres/crisis teams of the Netherlands. However, there was significant contribution from other states like Germany and Poland reflecting the trans-European network and cooperation within the trial.

Trial 4 was addressing the three gaps and research questions described previously (see "Objectives"). The research questions were further deconstructed to sub-research questions. Logical link between a gap, research question(s), and sub-research questions were agreed upon with the practitioners (work group) participating in trial 4 during the preparation and execution phase. In the next step, for each these issues, result indicators and KPIs (target) were defined and an adequate target value given. In order to conduct the measures for the correctly, qualitative and quantitative indicators demonstrating the level of achievement reached were defined. Key performance indicators (KPIs) to quantify the added value of the testbed environment, and the respective solutions taking into account the EU added value, usefulness, scalability, modularity, reliability, innovation, affordability, cost-effectiveness, usability, and validity.

Trial 4 considers the three dimensions (crisis management, solution, trial), specified by the trial guidance methodology and guidance tool specification in order to assess the innovative potential introduced by the selected solutions. For each dimension, the corresponding KPI was defined, and the adequate technique identified. The evaluation process of trial 4 used the following techniques for the different questions, tasks, and dimensions:

- Observers, including observer support tool.
- Semi-structured interviews.
- Focus group discussion.
- Questionnaires.
- Collection of feedback.
- KPIs measurement using the testbed.

Outcomes

Looking at the overall trial results, it is reasonable to state that the trial has been successfully conducted according to its preliminary planning based on the TGM. However, comparing the recommended timeframes and underlying tasks per trial phase, minor delays and obstacles in its implementation could be identified and handled in a successful way.

Respectively for trial preparation, execution and evaluation phases showed (among others): a great importance of effective information exchange among all groups involved in the trial (trial committee, practitioners, solution providers, observers, trial staff); need of very close cooperation between trial host and trial owner; need of precise planning including the monitoring of all running processes especially during preparation and execution phases; importance of precise and strict data collection plan: tested in practice during dry run 2 as a pilot test of this process and being understand in the same manner by trial committee, observers and the trial staff; and strong need for having reliable reference data.

Trial Dimension

The evaluation of the trial dimension of trial 4 has been conducted mainly through the participants' questionnaires (distributed among all trial "active participants" – i.e. practitioners and observers) and through observation sheets and moderated debriefing discussions. A specific focus was put on the feedback received from the external stakeholders participating in the trial (as observers, external solution providers and practitioners). To evaluate the trial dimension performance a questionnaire was designed for all involved in the trial. This questionnaire was independent of the questionnaires for the trial preparation and execution phases.

The major outcomes related to the trial dimension confirm that the participants' number, background, and commitment supported the trial adequately. The scenario and the simulated environment were deemed realistic enough for the practitioners' immersion. The attendees agreed that they were satisfied with its organisation and the split responsibility of trial owner and trial host; however, the results also show room for improvement with respect to the technical setup, the training, and the scenario building. The complete testbed technical infrastructure was running without any problems throughout the complete trial run.

The key findings of the trial 4 dimension are as follows:

- One hundred forty (140) people were involved (trial internal attendees [45%]; trial external attendees [55%]).
- Evaluation results of the trial are based on the probe of medium and high experienced practitioners.
- All attendees agreed that they were satisfied with its organisation; with respect to the technical setup, the training, and the scenario building, the results also show the opportunity for improvement.
- The majority of the practitioners have already participated in dry run 2. This was seen as advantage, so they were well prepared for the trial as such.
- The evaluation questionnaire for practitioners, observers and solution providers had completed. The feedback received was mostly positive with an average of 89% satisfaction indicated, meaning positive response to measurable questions and yes to the question if the respondent would recommend participating in DRIVER+ trials to others.
- The complete testbed technical infrastructure was running without any problems throughout the complete trial run. All the testbed infrastructure tools were used extensively
- The respondents rather agreed that they were satisfied with the organisation; hereby, most respondents emphasised the trial setup, the aspect of safety and security and the overall satisfaction with the trial as particularly positive. The slightest agreement – but not worse than neutral agreement – was found for the language aspect, which might be due to the fact that Dutch language was allowed in the trial since most of the practitioners and observers were Dutch. It was deemed necessary to allow the local language to ensure that language barriers do not cause major consequences on the trial results and further on the solution evaluation.

Solution Dimension

Solution utilisation and evaluation requires planning and observation as well as measurement of KPIs defined separately for each of the selected solutions. This evaluation process is a part of the three-dimension evaluation of the solutions-in-trial activities (including trial, solutions, and crisis management), which has to be carried out using the trial guidance methodology approach. Therefore, all dimensions were considered with measurements conducted accordingly.

Data for the solution dimension was collected in two ways, both using the OST (observer support tool):

1. For each solution there was a questionnaire dedicated to the use of the solution in that particular block (also accounting for changes in the AC/CT [action centre / crisis team] over the blocks).
2. Checklists were prepared per AC/CT for the observers to specifically track the use of the solution for particular tasks and assignments. Furthermore, the so-called walking observers observed the interaction of solution use between AC/CT (i.e. the CT Water board providing output to another AC/CT).

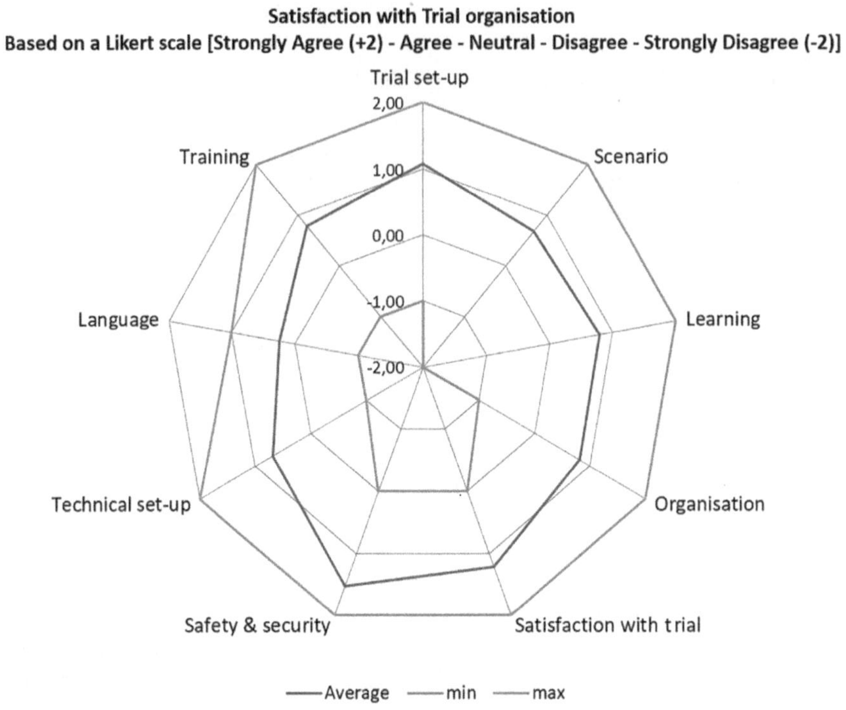

FIGURE 5.7 Trial organisation perception.

In addition to the questionnaires, the digital communication between AC/CT and the solutions was monitored and stored, including results from the individual solutions (i.e. results of flood calculations).

The key findings for the solution dimension (per solution) can be summarised as follows:

- 3Di: The solution was primarily used in the threat phase and significantly facilitated the situational awareness of the practitioners. The provided flood mask gave detailed information on the extent and depth of the flooding. Solution 3Di provided quick simulations of the possible flood extent, based on the actual information available of the area. Especially it calculated the effects of proposed mitigating measures, supporting decision-making.
- ATSA, module KeepOperational: The solution was useful to substantiate traffic circulations plans made by AC police and was able to cope with blockages. It provided useful input to solution HumLog.
- ATSA, module ZKI: the solution is only for the impact phase but provided an objective and accepted (by the practitioners) flood extent for dealing with the flood. However, the practitioners lacked information on the water depth. The interactive PDFs provided in block 4 were hardly used, mainly because

handling of the PDFs was considered difficult by the practitioners in order to complete a given task.

- CrisisSuite: The solution fully covered gap 2 (information exchange). Practitioners considered the solution as a valuable support in multi-agency communication.
- HumLog: The solution has provided objective quantification of the effects of an evacuation strategy, improving decision-making, closing to a large extent gap 3 on large-scale evacuation. The solution was helpful for resource management, regarding evacuation, herewith partly closing gap 1 on resource management.
- SIM-CI: The solution added to the situational awareness of the practitioners, especially since it objectively quantified possible cascade effects of flooding. Effective handling of SIM-CI takes more training but is a valuable addition to assessing the flood threat.

Overall, the feedback to solution providers was useful to further improve their solutions.

Crisis Management Dimension

The results in this section are based on the responses of the practitioners, documented in the crisis management solutions (LCMS/CrisisSuite), the output of the (other) solutions, and the observations made during the trial 4 execution. For this purpose, the observers were handed checklists based on the assignments, including free observations. Also, first impressions from observers and practitioners in the debriefing meeting in the trial were used to analyse the crisis management dimension. Finally, the results were cross-referenced with the questionnaire results of the solutions and the questionnaire on the trial dimension.

Actions taken in a realistic operational environment by the practitioners during the trial were based on currently available means for the chosen country, crisis management plans, standard operating procedures, and good practices of the practitioners.

The main findings are captured here:

- The gap for resources planning (gap 1) was only partly addressed in trial 4: The focus in resource management was redirected to resource planning to support evacuation execution, using solution HumLog.
- Information exchange is of high importance in a crisis situation. In a specific situation where the safety region involves external organisations, information sharing is an important topic. Internally, the safety region uses the legacy system LCMS. For their needs, it is fine, but – for non-technical reasons – LCMS is a closed and restricted information system. Closing the gap 2 on (net-centric) information sharing proved to be feasible with the usage of solution CrisisSuite during trial 4. The experiences in the trial even led to initiatives to formally connect both solutions (LCMS/CrisisSuite).

- The other solutions added significantly to the situational awareness, providing the practitioners with a more accurate and more detailed insight in the (potential) consequences of flooding. Solution 3Di provided objective and actual information on flooding for the threat phase, solution ATSA, module ZKI provided this in the impact phase. Solution ATSA, module KeepOperational facilitated the traffic circulation plan, and solution HumLog provided insight on the feasibility of proposed evacuation strategies.
- Particularly, the solution SIM-CI provided new insights in the cascade effects of flooding enabling the Safety Region to quantify these effects.
- Using solution ATSA, module KeepOperational, to determine effective evacuation routes together with solution HumLog, closed the gap in evacuation (gap 3) and substantiated the proposed evacuation strategies.

Main Takeaways for Preparation, Execution, and Evaluation of a Trial

At the end of the trial journey, many lessons were identified which can be useful to CM organisations interested in organising trials in the future:

- Discuss the current *gaps in crisis management* intensively. Select only a few (one to three) main gaps for selecting adequate solutions; otherwise, it will become very hard to evaluate the single solutions and their possible improvements. Focus on not too complex gaps and keep it simple.
- *Preparation processes* always take longer than expected. Start your planning as early as possible.
- Keep an overview and document all details. Atrial action plan was used for this purpose.
- The *dry runs* were very helpful in order to find the gaps of the trial or to clarify misunderstandings and test the system. It was also valuable for all participants to get to know the procedure of a trial.
- Plan enough buffer time around the real *trial execution* in order to have enough time for unexpected breaks, technical problems, questions, and so on.
- The crisis managers need to be *well trained in using the new solutions* so that they can easily use it during the trial execution. This is important for the evaluation and comparison of the baseline execution with the trial execution.
- All participants should be involved in the *evaluation* in order to get a full view of the tested solutions. Evaluation takes a lot of time, but it is necessary to come to a well-informed decision The trial guidance methodology handbook offers practical guidance through the preparation, execution, and evaluation.

6

THE TRIAL IN AUSTRIA

Testing New Technologies for Increasing Situational Awareness and the Management of Spontaneous Volunteers

Camilo Palacio Ramirez

Introduction

The trial in Austria was the fourth trial organised by the DRIVER+ project for testing new technologies for disaster management. It was carried out in September 12–14, 2019, in Eisenerz, Austria, and the overall objective was the assessment of solutions, which could help to overcome shortcomings and limitations in the management of spontaneous and affiliated volunteers as well as using non-traditional sources of information for decision-making.

The following section introduces the reader to the definition of spontaneous and affiliated volunteers and the components of situation awareness for decision-making, followed by the trial organisation.

Spontaneous Volunteers and Their Role in Disasters

Large-scale disasters usually encourage spontaneous volunteers to take action and respond to the population's needs. Large groups of individuals and self-organised groups converge to the scene to offer assistance (Lauren, 2006). However, this additional workforce is usually acting by their own will and not in line with current coordination activities, bringing additional challenges regarding coordination (Orloff, 2019).

Emergency response organisations have been always dealing, in a way or another, with spontaneous volunteers. In fact, the study made by Lauren (Sauer, 2014) shows that 79% of addressed organisations have encountered spontaneous volunteers in the aftermath of a disaster while about 63% of the organisations have incorporated the volunteers into response activities.

Even though they can support the relief efforts, their lack of disaster-related training puts burden on the emergency response organisations since they need

DOI: 10.4324/9781003256977-11

TABLE 6.1 Summary of Involvement of Spontaneous Volunteers (Sauer, 2014)

Summary Data Table of Survey Responses		
Item	*No. of Yes Responses (n = 19)*	*Percent of Yes Responses*
Use spontaneous volunteers (SVs)	15	79
Have turned SVs away during a past response	11	58
Find SVs generally useful	13	68
Do not consistently perform background checks on SVs	18	95
Consistently credential and verify licensure of SVs	4	21
Provide just-in-time training for SVs	10	53
Sort SVs by skill set, job, or background	12	63
Evaluate SV performance	5	26
Provide benefits	4	21
Accept legal liability for SVs and their actions	6	32
Were sued due to the actions of a SV	3	16
Were sued by an SV	1	5
Reported injuries of SV in the field	8	42
Reported the death of an SV during a past response	2	11

to provide a system to properly manage spontaneous volunteers, such as coordination, registration, personnel, protective equipment, food, and shelter, among others (Sauer, 2014).

These findings highlight the importance for organisations to develop systems for coordinating and managing volunteers effectively in order to provide a safe environment not only to responders but also to affected persons and volunteers. Similarly, the high number of responses (58%) that claimed they have turned spontaneous volunteers away shows that organisations are probably not well prepared to integrate them into the operations and therefore not taking advantage of the added value spontaneous volunteers could bring.

Classification of Spontaneous Volunteers

When referring to spontaneous volunteerism, different definitions and concepts can be found depending on the context and background (Whittake, 2015). For instance, Whittake refers to spontaneous volunteerism as informal voluntarism having these last two types: emergent volunteerism and extending volunteerism (Whittake, 2015).

The association of state and territorial health officials (USA) refers to spontaneous volunteers with the same meaning as self-deployed volunteers (ASTHO,

TABLE 6.2 Definition of Individuals and Groups as Spontaneous Volunteers (Auferbauer, 2019)

Individual Actors	
Unaffiliated volunteers	Persons intending to participate spontaneously in relief efforts without prior affiliation to formal emergency organisations.
Pre-registered volunteers	Persons who have signed up for a programme or initiative and conduct their activity within its organisational boundaries.
Affiliated supporters	Persons who have membership status in a formal CDM organisation, who conduct their activity along its organisational structure.
Affiliated decision-makers	Persons who have membership status in a formal organisation, who are afforded decision-making in its organisational structure.
Composite Actors	
Emergent groups	Formed through the self-organisation efforts of persons as reaction to a crisis or disaster event.
Supporting organisations	Established organisations that participate in relief efforts, but do not engage in such activity as part of their regular operations.
Formal CDM organisations	Established organisations whose regular objectives include crisis and disaster management activities.

2019). Auferbauer (Auferbauer, 2019) makes a more comprehensive classification not only for spontaneous volunteers but for all actors in community interaction and engagement.

A similar approach was followed by Batard (Batard, 2019) where a two dimension approach for voluntarism types is suggested:

1. Unaffiliated volunteers composed by individuals and groups.
2. Affiliated volunteers composed by individuals who belong to an organisation and are organised and formalised.

During our trial, we have used the classification made by Auferbauer for individual actors, specifically the unaffiliated volunteers, and the classification for composite actors: emergent groups.

Situation Awareness Technologies Used in Ddisasters (Level 2)

Situation awareness during disasters is the perception of the environment and events happening in the aftermath of a disaster, which is important to consider for decision-making and for improving the common picture of all agencies responding to an event.

A good understanding of situational awareness includes some challenges due the potential difficulties faced for collecting and disseminating complete, reliable, and high-quality information from first responders to coordination centres (Kedia, 2020).

Ashish et al. (2008) has identified some specific challenges when designing IT solutions to deal with information needed for situational awareness:

1. Diversity of information sources: This relates to the different sources of information relevant to decision-making. These sources can be found at different management levels and provided by different sources, such as verbal communication, satellite imagery, and remote sensing.
2. Diversity of information users: Considering that disasters include the participation of different stakeholders and organisations, the information gathered by each of those actors depends on their role in the emergency and the objectives they have when responding to an emergency. Some organisations may not have specific guidelines on how to share the information among organisations, thus making it more difficult to establish a common operational picture.
3. State of the infrastructure: This relates to how information must be transferred across different structures. This is clear when analysing the tools for information management. Responding organisations have different systems and databases which makes sharing information difficult.

Similarly, the research from Steen (Endsley, 1995) identifies three levels of situation awareness with some roles linked to them:

Level 1 is the perception of elements. At this level, stakeholders receive inputs from technological tools (geographical information systems [GIS], onsite reports, and other information systems). In level 2, practitioners analyse the data from level 1 and make sense of it. This includes causes of the situation and its risks and hazards. Similarly, level 3 is the level that leads to a decision based on future developments, which is done by understanding what is really happening and what the best course of action can be.

Considering the findings from Ashish (2008) and Endsley (1995), a system used for situational awareness should properly address the three challenges described earlier, while providing sufficient information for decision-making.

We now turn to describe the DRIVER+ trial organised in Austria.

Trial Setup, Objectives, and Scenario

The trial was carried out in the framework of the European civil protection exercise (EUCP-EX) IRONORE. The EUCP-EX was organised as a large-scale European civil protection exercise. National and international emergency organisations were present with their volunteers and experts while making use of equipment, vehicles, and tools in simulated disaster scenarios.

The trial's main objective was to evaluate a selection of solutions contributing to international or national CM processes addressing the crisis management dimension, especially in the following fields:

- Volunteer management: In the sense of management of spontaneous as well as affiliated volunteers on the scene in terms of location, tasking, capabilities, and duration of operations.
- Real-time data and information fusion to support incident commander decision-making: Ability to merge and synthesis disparate data sources and models in real-time (e.g. visualisation of resources, spreading models, tactical situation, critical assets map, damaged objects/infrastructure, etc.) to support incident commander decision-making and exchanging crisis-related information among agencies.
- Incorporating information from multiple and non-traditional sources: Reporting of dangerous areas and situation overview from multiple and non-traditional sources (e.g. crowdsourcing and social media) into response operations.
- Psychosocial support: Having the capability to measure stress and/or improving the communication and the awareness of psychological stress of those affected, especially spontaneous and affiliated volunteers.
- Interaction with the population: Including, for example, (1) micro-learning capabilities to communicate to the population safety information and recommendations what they can do during a crisis, (2) registration of affected people, and (3) delivering information from the public to the emergency management authorities.

The trial aimed at finding and testing solutions to improve the decision-making process and the management of spontaneous and affiliated volunteers. It was deemed as important to understand what the components of decision-making are and what influences the management of spontaneous volunteers, in order to collect the data that allows assessing the performance of the solutions and how they help to fulfil the disaster management gaps addressed in trial Austria (trial AT).

A scenario revolving around an earthquake was designed as follows: On September 11, 2019 (day 0), at 22:26, the town of Eisenerz (Austria) was affected by an earthquake with a magnitude of 6.8 (Richter scale) with the epicentre close to town (Figure 2.1, Figure 2.2). The initial damage assessments indicated heavy damages to a range of structures, including residential and industrial buildings as well as critical infrastructure. Considerable damages have been reported in a radius of 60 kilometres including the cities of Vordenberg but detailed reports on the local situation in the urban area and in several remote valleys are not available.

In the region of Eisenerz, 200–300 people are suspected to be trapped alive under the debris and 28 people have been confirmed dead. Almost all of the 4,000 inhabitants are in need of shelter, water, sanitation support, and critical and basic health care. All local emergency services have been deployed.

Because of the massive impact, the Styrian regional authorities filed a request for additional capacities stating an urgent need for USAR[1] teams, medical teams, and WASH capacities. Additional national capacities have been mobilised and the Austrian Federal Ministry of the Interior sent an official request for assistance via EUCPM. All suitable offers were immediately accepted.

Trial Structure

The TGM identifies a group of roles that need to be considered when organising a trial (Fonio & Widera, 2020) and such roles were included in trial AT. In addition, more roles needed to be allocated due the coordination with IRONORE to synchronise the participation of spontaneous volunteers and the use of the technologies into the exercise locations.

In the following figure is the structure that was used during the trial and how the coordination with IRONORE was.

The Application of the TGM

The Trial Preparation Phase

Within the context described earlier, practitioners identified gaps in relation to volunteer management, including spontaneous volunteers, with an additional focus on improving the decision-making process through solutions with ability to provide reliable information to incident commanders. For the selection of solutions, a call for application was launched in October 2018, and 19 solutions from European providers in industry and research. After a review by internal and external practitioners, ten solutions were invited to present their features in a workshop, resulting in the selection of five solutions, including one non-technical approach.

As mentioned, relevant aspects with regards to decision-making were deemed as important:

1. The decision environment is given by the scenario and influenced as well by the expertise of practitioners.
2. The quality of information is the measure of how the information is processed; in other words, low-quality information generally generates data (of information) rejection while good quality information leads to a decision.
3. The flow of information and how it can affect further decisions forms the decision stream.

This component of "quality of information" was of special interest in the frame of the trial. The team organising the trial needed to define clearly what was going to be measured in regard to decision-making – how the information is processed and quality of information.

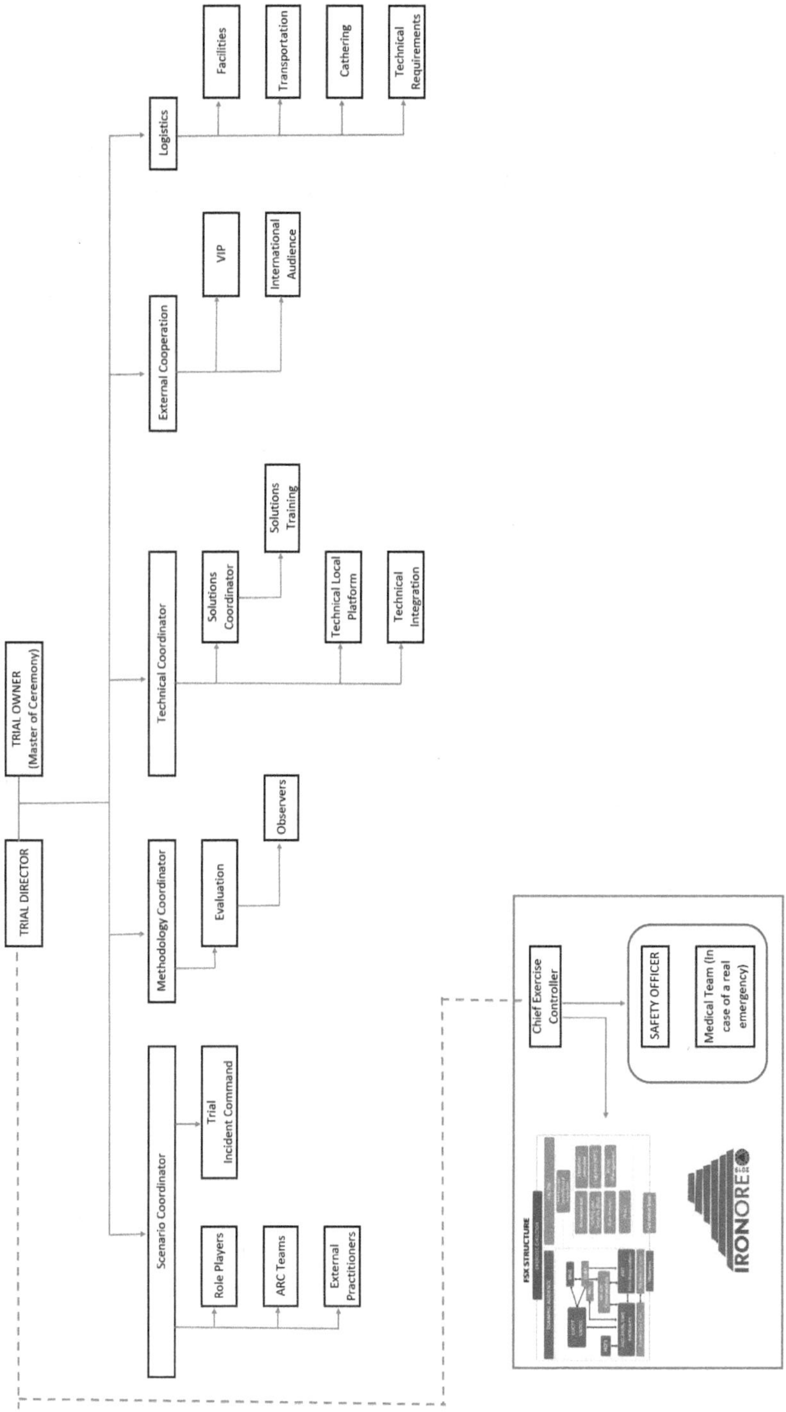

FIGURE 6.1 Trial structure.

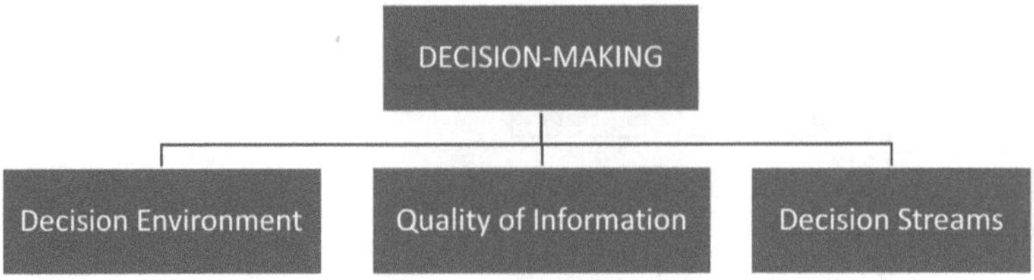

FIGURE 6.2 Components of decision-making (based on Piu, 2013).

The quality of information given to the spontaneous volunteers has also an impact on the tasks they are able to perform. For a proper management of spontaneous volunteers, practitioners should continuously supervise volunteers and assign tasks according to their capabilities, and therefore, the identification of volunteer's skills should be done as early as possible. Therefore, the trial aimed to quantify two characteristics of spontaneous volunteers (SV): (1) how SV performed and (2) how the information (tasks or alerts) is perceived

Data Collection Plan

A data collection plan that allowed to collect feedback from different stakeholders was considered key in this trial.

The first approach was the collection of feedback by means of checklists. These were used by observers in order to collect observations about the expected behaviour of participants. Similarly, a set of predefined questionnaires were created for the practitioners that have used the solution, and the aim of those questionnaires was to collect the first impressions about functionality and use of the solutions. Last, the trial committee has also carried out focus group discussions for better understanding of the added value of solutions in disaster management.

Given that the trial AT had a field component, five different profiles were envisaged to collect data depending on the participants' location as in the control room or in the field. Those profiles were practitioners, field practitioners, field participants, observers, and field observers.

Observers in general had the task to collect observations on how practitioners and participants were using the solutions, as well as the behaviour of participants once they have received a specific task (a dedicated checklist was created for this). Another important task for the observers was the identification of organisational difficulties of trial conduction, external constraints that may influence trial results (solution assessment measurements).

Practitioners and field participants were focused on providing feedback about the hands-on experience with the solutions and the quality of the trial execution. This feedback was collected by a semi-structured questionnaire directly following each session in the trial.

Another way the trial committee has used for collecting data was the development of dedicated focus groups that allowed observers, practitioners, and participants to express freely their impressions about their impressions when participating on each of the testing blocks and how the solutions could be improved according to their own understanding of the crisis management gaps.

The following table is a summary of how the data was collected during the trial.

Sample Plan

While the previous table summarises how the data was collected, Table 6.4 shows a more detailed plan that links specific data collected with the impact of the solution on the crisis management gaps. It also shows the where the data was collected by whom and on which specific testing block.

TABLE 6.3 Data Collection Summary

	Who Will Evaluate?	*What Type of Information Is Collected?*	*How Is the Information Collected?*
Solution Evaluation	Observers	• Behaviour • Actions performed	• Checklist • Evaluation forms (questionnaires)
	Commanders and participants	• Usability of solutions	• Evaluation forms (questionnaires) • Focus groups/ interviews
	Testbed	• Interaction between solutions • Actions performed by operators	• Logs

TABLE 6.4 Specific Data Collection Plan

Testing Block Name	*What?*	*Where?*	*Who?*
Emergent Groups	How the information is processed	Operations room	Observers
	Quality of information		Observers and practitioners
	How the SV performed	Field	Field observers
	How the information (tasks or alerts) is perceived		Field participants

(Continued)

TABLE 6.4 (Continued)

Testing Block Name	*What?*	*Where?*	*Who?*
Situation Assessment	How the information is processed	Operations room	Observers
	Quality of information		Observers and practitioners
Confirmation	How the information is processed	Operations room	Observers
	Quality of information		Observers and practitioners
	How the SV performed	Field	Field observers
	How the information (tasks or alerts) is perceived		Field participants
Chemical Spill	How the information is processed	Operations room	Observers
	Quality of information		Observers and practitioners
	How the information (tasks or alerts) is perceived	Field	Field observers
USAR SitRep	How the information is processed	Operations room	Observers
	Quality of information		Observers and practitioners
	How the information (tasks or alerts) is perceived	Field	Field observers

Solutions to Be Tested

The last step in the preparation phase, once identified the gaps, formulated the scenario, and created a data collection plan, is the selection of relevant solutions based on crisis management gaps (described in trial setup). The following solutions with a focus on the listed functionalities were selected in the solution selection process and assessed during the trial AT.

CrowdTasker

CrowdTasker is a solution for citizen involvement and community interaction. It supports informing citizens, eliciting contributions to the common operational picture by pre-registered parties and integrating efforts of self-organisation. This is achieved by issuing assignments and situational information to a selected crowd of citizens based on their location and skill set, as well as offering a CHATBOT interface for emergent groups to participate using their own organisational infrastructure (such as social media groups).

The gaps addressed by the solution are as follows:

- Coordination of pre-registered volunteers.
- Evaluation of computer-supported interaction with spontaneous volunteers.

Airborne and Terrestrial Situational Awareness

This solution provides real-time aerial images. Real-time aerial imaging significantly enhances situational awareness during major and large-scale disasters. DLR's solution Airborne and Terrestrial Situational Awareness comprises different modules to provide such a real-time aerial imaging and analysis system.

Module 1 is the ground control station U-Fly, used to plan, engage, and monitor aerial missions. The full-size research aircraft Do-228 "D-CODE," which is operated as a drone demonstrator with safety pilots on board, allowing drone-based missions to be executed without regulatory restrictions or safety concerns, executes the missions.

Module 2 is the 3K aerial camera system, specifically developed to acquire and evaluate aerial photographs in near real-time. In addition, it can transfer aerial imagery via data link directly from the aircraft to a mobile ground station to provide the data to decision-makers and rescue forces immediately.

Module 3 is the Center for Satellite-Based Crisis Information, which analyses aerial imagery and generates crisis information maps.

The gaps addressed by the solution are as follows:

- Delivery of near real-time information for situational assessment to improve situational awareness (damage assessment, localisation of routes, volunteers, and equipment).
- Delivery of aerial imagery (GeoTIFF) and reference map products (JPG, GeoPDF).

vieWTerra Evolution

vWTE is a 4D Earth viewer and data integration platform allowing free navigation and querying of the world in different crisis management phases. It presents an ellipsoidal model of the Earth, allowing its users to integrate their own precise data and assets anywhere on the globe, without any area coverage limitations (imagery, cartography layers, 3D objects/buildings, geotagged information, photos, etc.). It can be used to model any type of 3D scene on Earth and create scenarios at their real-world location to simulate events in the crisis preparedness phase and to serve as a global repository for building a custom Earth-wide GIS, either used perfectly offline or ported on an online architecture in order to allow the sharing of multiple information, data, and assets from disparate sources between all stakeholders in the crisis response phase (3D entities, icons, shapefiles, geotagged reports, photos, videos, sound, multiple overlays such as disaster maps, heat maps, tactical situation etc.).

To address the gaps, the solution provides an interactive GIS environment to improve situational awareness/situational assessment and decision-making using the data collected from different sources (other solutions connected through the testbed).

Asign

Asign is a solution that helps reduce critical emergency and disaster response time. It is a complete all-in-one disaster assessment software tool for easy collection, optimal communication, and effective management of operationally relevant critical information. Asign supports collection and communication of photos, videos, geo-texts, tracking, geo-zones, geo-alerts, and assessment forms in a very bandwidth-efficient manner. Specifically, it can communicate photos and video with 99% bandwidth reduction, enabling communication even through low-bandwidth cellular and satellite communication networks while maintaining full precision and accuracy.

The gaps addressed by the solution are as follows:

- Communication support by sharing different information means: photos, videos, geo-texts, tracking, geo-zones, geo-alerts.
- Reduction of critical emergency and disaster response time through optimisation of communication and effective management of operationally relevant critical information.

Psychological First Aid (PFA) Training

The psychological first aid (PFA) training for spontaneous volunteers is a one-day training course, in which you learn and get to practise the main skills needed to give good PFA in a crisis. You will learn the internationally recognised principles of Look, Listen, Link, developed by the World Health Organisation (WHO). The training includes sessions on these three principles and role-plays, discussion sessions, sharing knowledge and experience between participants. Organisations responding to a crisis can implement the training to leverage the resources that spontaneous volunteers bring to a crisis in a positive and safe way.

The leadership seminar for engaging with spontaneous volunteers in crisis response combines the WHO's Look, Listen, Link principles for PFA, knowledge of caring for volunteers, and theory on power relations to build the skills of field-level leaders working for crisis response organisations to engage with spontaneous volunteers during crisis response. Through a series of exercises, analytical sessions, role-plays, and discussion and reflection sessions, the participants activate their own experience and knowledge and learn from each other to be better placed to engage positively and constructively with spontaneous volunteers.

The gaps addressed by the solution are the low awareness and lack of ability to address the risks of adverse mental health effects and decreased psychosocial well-being in spontaneous and trained volunteers following response operations.

The following table captures the correlation between the functions of the solutions and the crisis management functions:

TABLE 6.5 Correlation Between Crisis Management Functions and Solutions

Crisis Management Function	*Solution*
Volunteer management	CrowdTasker
	Psychological first aid (PFA) training
Real-time data and information fusion to support incident commander decision-making	CrowdTasker
	vieWTerra Evolution
	Airborne and Terrestrial Situational Awareness
	Asign
Incorporating information from multiple and non-traditional sources	CrowdTasker
	vieWTerra Evolution
	Airborne and Terrestrial Situational Awareness
	Asign
Psychosocial support	Psychological first aid (PFA) training
Interaction with the population	CrowdTasker
	Psychological first aid (PFA) training

Trial Execution Phase

Trial AT was divided into five testing sessions called blocks (blocks 1 to 5), which can be seen in Figure 6.3. These blocks allowed the trial committee to test the solutions in a logical and chronological way as it would happen in a real disaster situation.

However, some assumptions needed to take place in order to fully test the solution's capabilities, even when the assumptions would not reflect what can and cannot be done in a similar situation without the solutions and with the resources available in Austria. For example, assumptions like no local response capacities in place or tasking spontaneous volunteers to collect information of damaged infrastructure were taken into account. Under normal circumstances, additional response capacities can be deployed from nearby towns or cities and the assessment of damaged infrastructure is relied on trained emergency response staff.

The first testing session was block 1, which had the context that after the earthquake all local response capacities were tasked to different locations and due the insufficient number of responders, inhabitants of the town started to self-organise by social media, and in this block, it was tested the social media functionality of CrowdTasker to analyse how self-organised volunteers can be remotely coordinated in order to perform specific activities aimed to support the immediate need of the community.

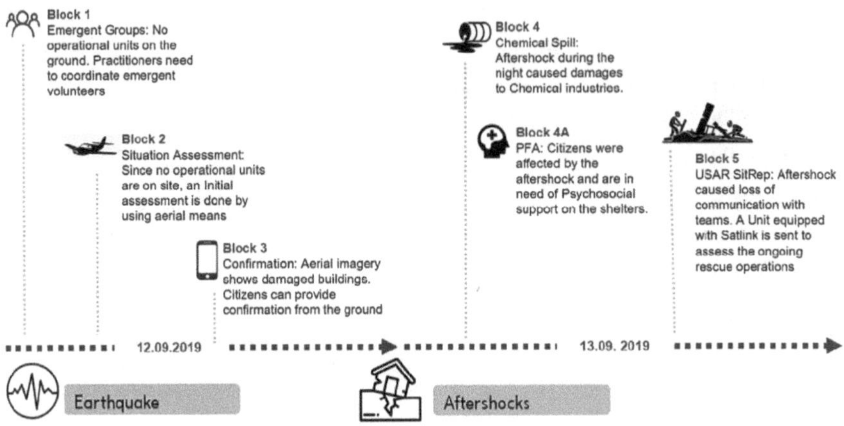

FIGURE 6.3 Scenario sequence.

Considering there were no local capacities in place, it was considered the use of the airborne situational awareness solution that was tested in block 2 and had the goal to get aerial images from specific areas, some of those with clear damages. The identification of damaged areas was the initial point to continue with block 3 where the solution CrowdTasker was used again to task volunteers nearby those areas and request from them to send ground photos from those locations to perform a remote damage assessment and its visualisation was enhanced by the 3D model created on vieWTerra.

The second day of the trial started with block 4 having the assumption that aftershocks during the night caused damage to local industries working with hazardous materials and causing some damages to shelters where the affected population were staying at. Asign was used to create a "geo-fenced" safety zone, alerting the incoming response teams about the potential hazards in the area nearby the industries, whereas CrowdTasker was used to alert the population and request from them information about potential exposure to chemicals. The trained group of volunteers by the PFA training performed some psychosocial support actions to the populations affected at the shelters.

Block 5 was about coping with communication problems when the communication network is compromised or unavailable. The solution Asign and its features about creating links via satellite was implemented during this block where a team was sent to locations to perform some ground assessments and to send imagery to the command post.

Results

The main outcomes of the trial are presented here, following the three-dimension approach of the trial guidance methodology.

Trial Dimension

As indicated in the data collection summary, the feedback for this dimension was collected by questionnaires answered by observers, practitioners, and participants. The focus of this feedback was regarding roles and responsibilities, scenario, and trial organisation.

The results of the evaluation of this dimension reflected the positive perception of trial organisation by all stakeholders who had answered the questionnaires. They found very positive that division of tasks, communication, and safety and security were communicated properly. The duration of the blocks and the storytelling of the scenario were considered as realistic and appropriate and those are probably two of the most important factors when testing new technologies. While long testing sessions may lead to practitioners losing interest, unrealistic scenarios may not allow to fully compare solutions to legacy methods used day by day by practitioners.

The technical infrastructure and support given by solution providers was also scored positively. One of the strategies followed by the trial AT was the implementation of solution's operators which were staff of the solution providers that were fully aware of the solution capabilities and were using the solution under practitioners' commands. Practitioners received specific training about the features of each solution, so they were aware of what the solution was capable of, and having a person who was manipulating the solutions for them gave them the opportunity to focus on fulfilling the tasks.

Only two aspects in the evaluation that were perceived rather low by practitioners. The first one relates to complexity of the tasks to be performed during the trial because they claimed they did not feel that challenged in comparison to their day-by-day activities. These sorts of situations could be analysed from different perspectives; from one side, practitioners may not see the full potential of the solution and its contribution to support complex situations, but on the other side, practitioners may have more time to reflect on the solution features and analyse its potential. The other aspect is that practitioners claimed they did not feel the participation in the trial had enhanced their capabilities to respond to future disasters. This most likely relates to their expectations when taking part in the trial as if they would participate in a training exercise. In the end, the focus was to test the solution and not to train practitioners on novel disaster response methods.

Some aspects during the trial execution have also influenced in some way the data collection plan. For instance, finding practitioners who were available for joining the dry run 2 and the trial may have had an impact on the evaluation of the solutions and its contribution to the crisis management gaps.

Some circumstances caused minor changes on the development of the trial, which cannot be attributed to poor technical performance of solutions. One example is the use of some backup imagery taken the day before by the Airborne solution because the weather conditions were not appropriate for the airplane to take off. Similarly, in another session, the imagery between the Airborne solution and vieWTerra was different due the fact one of the solutions was using a previous dataset, mainly attributed to a human mistake.

Solution Dimension

CrowdTasker

CrowdTasker was used in three blocks, all of them dealing with volunteer management and interaction with the population. According to the evaluation of practitioners, observers, and participants, it can be said that CrowdTasker is a robust technology and helps practitioners to better coordinate spontaneous volunteers.

One of the remarks for this solution is the need of developing a sort of intermediate translator, which is nothing more than a component who can rephrase practitioners needs into a wording easily understandable by spontaneous volunteers. However, most of the tasks were fulfilled by participants as expected.

Airborne and Terrestrial Situational Awareness (ATSA)

ATSA was used in two blocks to collect aerial imagery from the ground to support the crisis management gap of assessing the improvement of situational awareness and to enhance the decision-making process.

The quality and amount of information received in the collected imagery were the most important points scored very positively by practitioners on the questionnaires and in the focus group discussions. The sites prepared beforehand showing damages on them were easily identified by practitioners thanks to the good quality of photos collected by the solution.

The details of the photos were very impressive, and practitioners claimed that its quality was higher than satellite-based images. Similarly, it offers the possibility to fly under some weather conditions which might not be suitable for UAVs.

When these kinds of solutions are used, a good coordination between the national air traffic control and the emergency coordination unit is required. The air space may be closed due arriving help (e.g. helicopters or aerial firefighting operations) and, therefore, may affect the deployment of the solution.

Asign

Asign was also seen as a technology very developed and robust. It aimed to support the crisis management gaps for providing real-time data to support decision-making and incorporating information from multiple and non-traditional sources

Asign was used to create a geo-referenced safety zone, indicating any team crossing that fence that they have entered to a dangerous zone. This is a positive feature to increase team's safety when having several teams deployed on the ground.

The satellite capabilities of the solution were also tested by collecting imagery from remote locations where the communication network was shut down. One of the features is the possibility to reduce photo size to be sent via satellite. Some confirmation options about messages received by teams are missed in the solution

One of the advantages of this solution was the possibility to send photos through the smartphone app, with the possibility to reduce image quality in those areas of the photo where there is no need to have a better definition. The area of the photo that shows most of the information to be shared is sent with a higher quality, which reduces considerably the size of photo files to be sent via satellite.

Considering some of the messages did not reach the teams on the ground, it needs to be further validated if this situation can be explained by a hardware problem or a lack of training or adequate protocols on how to use these technologies. The costs associated with the use of satellite connectivity can be a drawback for the acceptance of this technology by emergency response organisations.

Psychosocial Support First Aid Training (PFA)

The PFA training supported volunteer management and interaction with the population gaps. It was reported that students and trainers had a very good communication and that the training is a good way to raise awareness.

The trained spontaneous volunteers were exercising their newly acquired skills in a realistic scenario (when getting in touch with victims), supported by a field-level volunteer leader from the Austrian Red Cross that attended the leadership seminar.

It was mentioned by the volunteers that they got a better understanding of psychosocial support after the training and were able to know how to react and support others in a difficult situation.

The combination of scenario and virtual reality (VR) created time and space for spontaneous volunteers to practice, reflect, and learn at a high level; however, it also takes time for participants to learn how to engage with VR. It may be that training with VR will take a little longer (one hour per day of training) but the outcomes are deemed good as participants would most likely acquire deeper knowledge and skills and be more skilled at using this sort of training.

vieWTerra Evolution

vieWTerra was used in four out of the five blocks and supported the gaps about providing real-time data to support decision-making and incorporating information from multiple and non-traditional sources.

Practitioners scored positively that vieWTerra was able to provide complementary information. The 3D models of the solution helped to have a more immersive understanding about the solution and the perception of practitioners to implement the tool is that such technology needs a high amount of training and therefore not doable for volunteer-based organisations.

Crisis Management Dimension

Based on the results of the evaluation and data collected during the trial it was apparent that the volunteer management gap was partly achieved by the two solutions that provided some degree of innovation, CrowdTasker, and the PFA training. CrowdTasker has successfully demonstrated that smartphone applications can be used effectively to manage and coordinate volunteers to perform specific actions such as collecting data from the ground to be used for situation awareness. Similarly, the PFA training has the potential to increase awareness about the stress faced during emergencies by increasing the skills of the participants who joined the training showing that these trainings can be a good option to cope with the crisis management gap about psychosocial support for both spontaneous and affiliated volunteers.

CrowdTasker has also demonstrated its potential to be used as a channel for early warning for supporting the gap about communicating with the public during large crises. During the trial, participants reported they have received the alerts in a clear way. The scalability and deployment of the solution shows that it can be used effectively for interacting with the population. The downside of this sort of application is that they need to be handled by a governmental body to guarantee its wide implementation and reliability.

Solutions such as ATSA, vieWTerra, and Asign had different impacts on the gap between real-time data and information fusion to support incident commander decision-making. While the solutions are in a stable technical development, it was not clearly evidenced their real contribution to the decision-making process. This is due the fact the high level of training practitioners may require to fully understand the level of damage shown by the data collected by the ATSA solution or to fully exploit the 3D capabilities of the vieWTerra model. It seems such technologies can be better placed outside of a control room and used by a sort of supporting team which with the help of experts on data/image analysis can process the information provided by the solutions and just deliver information to the command and control centre after a depth analysis.

The contribution to the gap "incorporating information from multiple and non-traditional sources" by solutions such as CrowdTasker, Asign, vieWTerra, and ATSA shows that different technologies can be used in parallel and simultaneously to collect different type of data, and with a correct tool to gather all information (like vieWTerra for example), the information can be presented in a common operational picture tool that allows the visualisation from different levels. While the information collected from citizens by tools such as CrowdTasker seems to be more subjective and not that reliable in comparison to the objective data collected by Asign or ATSA, the analysis allows practitioners to have a better understanding from the public perception so that they can implement actions for addressing population needs.

Lessons Learnt

Since the trial AT was the last trial out of four, organised by the DRIVER+ project, the outcomes and experiences of previous trials were considered throughout all different phases allowing the application of main steps of the TGM in a logical way.

The gaps in preparation phase were quickly identified and defined by practitioners based on previously known needs. The most challenging part was the definition of proper research questions and KPIs since practitioners were familiar with those more research-oriented tasks. This could be overcome by establishing tight collaboration with academic or research organisations.

It is important to note that the selected solutions offered some degree of innovation, which might not be sufficient to close the gaps. For that reason, we created synergies between different solutions and defined testing cases that allowed us to challenge the features that the project was interested in. At the moment of preparing the scenario, one should know very well the features of the technologies to be tested.

The execution phase started with the trial integration meeting, which was the first meeting between all stakeholders with the aim of learning both practitioner needs and solution features. This meeting has proven to be a good practice since solution providers were able to talk with one another and visualise potential synergies.

From this experience, it is important to highlight the following:

1. Prepare a TIM meeting for allowing all stakeholders to get to know each other.
2. Inform solution providers beforehand why their solution was selected and what your expectations about the solution are.
3. Define a detailed working plan with the provider.
4. Be familiar with the features of the solution in order to develop appropriate use-cases.

For the collection of data, the project has used a tool, which was developed by the project itself to collect data not only from the solutions but also from the testbed and observers. This solution, called observer support tool-OST, was used previously in other trials and improvements were done along the project. For trial AT, it was important to have an offline version of the tool since several activities were foreseen to be outdoors. OST proved to be a useful way to collect data for further analysis because it is complemented by the information on the testbed. Nevertheless, some issues developed at the moment of executing the trial and last-minute changes were hard to implement in this tool.

Similarly, the offline version has shown some repeated data when synchronising with the server; in this regard, it was easy to detect the repeated data by comparing the comments within the observations, but it was not the same when having only inputs. For collecting data, one should consider the following:

1. Make use of the data collection, digital tools that facilitate the registry of observations and that provide the results in a format that can be easily processed by statistical tools.
2. Involve in all discussions, including the discussions regarding use-cases, the person(s) in charge of developing the questionnaires and managing the tool.

3. Make sure to differentiate each of the observations according to the profile and the session where it was tested

Observers are very useful as they collect data and provide feedback about the use of the solutions in crisis management – of course, if they have the expertise to do so. One of the difficulties we faced in the trial was the difference between observers and evaluators. Observers are meant to collect behavioural data or identify actions happening at specific moments. While evaluators do essentially the same as observers, there is a conceptual difference since evaluators provide a judgement according to protocols or standardised guidelines. Select the ones appropriated for your trial and be familiar with the expectations of the observers/evaluators.

Having the dry runs 1 and 2 at the same location of the trial revealed possible logistical constraints and one needs to have a plan for that beforehand. While dry run 1 helped to check the technical preparations and communication among solutions, dry run 2 was used to test the real implementation of the TGM and improve the requirements that practitioners need to perform during the trial.

This involved aspects such as training of practitioners, training of observers and evaluators, use-cases tested, logistical needs, data collection, and assignment of roles and responsibilities, among others.

The last step was executing the trial, which ran with no major issues thanks to the structured methodology the TGM provides. The testbed was able to support the exchange of information among solutions, the data were collected according to the evaluation plan and no session has suffered any major loss of data. As lessons learnt, we can highlight better preparation for last minute changes that were needed in order to get proper tests of the solutions, referring specifically to observers' preparation and data collection. In one of the sessions, the solution to be tested required certain weather conditions to be operative. The conditions were defined for a different timeframe, and an OST was not able to include the questionnaires needed to collect the data. However, the data collected to assess the solution was already taken in another session; thus, this situation did not affect the evaluation of the solution. The observers were also confused since they were not briefed properly about this change and their observations about the new session were partially collected.

Focus group discussions as first impression evaluation were very useful not only for the evaluation of the solutions but for solution providers as feedback to improve their solutions.

Conclusions

The TGM has proven to have potential for performing sessions for testing new technologies for disaster management. The methodology guides users to consider what is needed in order to perform these sessions. However, there are some areas where the TGM lacks more information; for instance, it is not mentioned how many observers are needed for a certain observation, how practitioners' skills are matching with solutions features, and how many solutions can be tested in a single trial.

Emergent groups and spontaneous volunteers are very likely to be present after a disaster. As pointed out by practitioners, it is important to have proper methods for coordinating their actions since the emergency response organisations can benefit from the additional work force spontaneous volunteers represent. This also helps to increase citizen's empowerment by engaging them into the emergency response in an organised way while giving them the opportunity to help when they feel the motivation to do so.

In general terms, the trialled solutions contributed to provide nearly real-time data and information, which supported the incident commander in decision-making. The exchange of information and raw data among the solutions was also useful for increasing the situational awareness because it provided different perspectives from the same location.

The interaction with the population, including the management of volunteers, can be improved with some of the tested solutions. In particular, an improved psychological support to volunteers can contribute to a more effective and responsible involvement of these citizens during crisis management situations.

Considering the results of our tests, there is a good future for smartphone applications to serve as tools for the remote coordination of spontaneous volunteers and for improving situational awareness. Further research is needed to define proper guidelines on how to interact remotely with volunteers. Situational awareness seems to be a better use-case to focus on, while performing remote risk assessment with non-trained staff brings additional challenges not only for the decision-making process but for the technical level.

Note

1 In this section, USAR stands for urban search and rescue; WASH for water, sanitation, and hygiene; and EUCPM for European Union Civil Protection Mechanism.

References

Ashish, N., Eguchi, R. T., Hegde, R. M., Huyck, C., Kalashnikov, D. V., Mehrotra, S., Smyth, P., & Venkatasubramanian, N. (2008). Situational awareness technologies for disaster response. In H. Chen, E. Reid, J. Sinai, A. Silke, & B. Ganor (Eds.), *Part of the Integrated Series In Information Systems Book Series* (ISIS, volume 18). Boston, MA: Springer.

ASTHO. (2019). *Emergency Volunteer Toolkit.* www.astho.org/Programs/Preparedness/Public-Health-Emergency-Law/Emergency-Volunteer-Toolkit/Types-of-Volunteers-Fact-Sheet/.

Auferbauer, D., Ruggenthaler, C., Czech, G., & Gojmerac, I. (2019). Taxonomy of community interaction in crises and disasters. *Proceedings of the 16th ISCRAM Conference.* Universitat Politecnica de Valencia, Spain.

Batard, R., Rizza, C., Montarnal, A., Benaben, F., & Prieur, C. (2019). Review-based taxonomy of postimpact volunteerism types to improve citizen integration into crisis response. *ISCRAM 2019 – 16th International Conference on Information Systems for Crisis Response and Management*, May 2019, València, Spain. ffhal-02173967.

Endsley, M. R. (1995). Toward a theory of situation awareness in dynamic systems. *Human Factors Journal*, 37(1), 32–64.

Fonio, C., & Widera, A. (Eds.). (2020). *Trial Guidance Methodology Handbook. DRIVER+ (Driving Innovation in Crisis Management for European Resilience)*, Brussels, 2020. https://tgm.ercis.org/.

ISO. (2017). *ISO 22319:2017(en) Security and Resilience – Community Resilience – Guidelines for Planning the Involvement of Spontaneous Volunteers 3.1.* www.iso.org/obp/ui#iso:std:iso:22319:ed-1:v1:en.

Kedia, T., Ratcliff, J., O'Connor, M., Oluic, S., Rose, M., Freeman, J., & Rainwater-Lovett, K. (2020). Technologies enabling situational awareness during disaster response: A systematic review. *Disaster Medicine and Public Health Preparedness*, 1–19.

Lauren, F. (2006). Spontaneous volunteer response to disaster: The benefits and consequences of good intentions. *Journal of Emergency Management*, 57–58.

Orloff, L. C. (2019). Emergency management: Spontaneous volunteers in response to disaster. In L. Shapiro & M. H. Maras (Eds.), *Encyclopedia of Security and Emergency Management*. Cham: Springer. https://doi.org/10.1007/978-3-319-69891-5_280-1.

Piu, P. (2013). *Decision Making Process: Components and Time Effect*. https://statsneuro.blogspot.com/2013/02/decision-making-process-components-and.html.

Sauer, L. M. (2014). The utility of and risks associated with the use of spontaneous volunteers in disaster response: A survey. *Disaster Medicine and Public Health Preparedness*, 8(1), 65–69.

Whittake, J. (2015). A review of informal volunteerism in emergencies and disasters: Definition, opportunities and challenges. *International Journal of Disaster Risk Reduction*, 13, 358–368.

7

NEW APPROACH TO SELECTION OF INNOVATIVE SOLUTIONS TAILORED TO THE PRACTITIONERS' NEEDS

Marcin Smolarkiewicz, Tomasz Zwęgliński, and Paweł Ogrodnik

Introduction

In the globalised world, response to disasters on a national level can be facilitated by a multinational approach (Zwęgliński, 2020). This type of civil protection and humanitarian aid is provided by many international governmental and non-governmental organisations on the international level, such as United Nations, European Union, International Federation of the Red Cross and Red Crescent, and so on (Zwęgliński & Stefańska, 2021). Such systems significantly broaden the pool of civil protection and humanitarian assets which can be used by a disaster-stricken country in any place of the world. However, widening the coping capacity able to be used by the spectrum of these organisations makes disaster management more complex. Therefore, the growing number of national and international organisations responding to major disasters requires efficient and effective communication and information management protocols. Such protocols are to be supported by technological solutions, such as those being used every day, such as mobile phones, consumer mobile application software, and so on. However, in the communication and information management domain there are still many needs being declared by disaster managers (DRIVER+, 2018a). Such needs were explored in the frame of many research projects (Dimitriadis, 2020). One of them is a need for facilitating the vertical information management and communication in disaster response. It was confirmed by disaster managers working for the European Union Civil Protection Mechanism. Therefore, a set of solutions has been juxtaposed with the participation of disaster managers in order to trial it towards its potential added value in covering the identified need (Fonio & Widera, 2020). The trial has provided data that helped to evaluate whether the set of solutions might improve the situation in the respective domain of disaster managers' performance in a forest fire scenario or not.

DOI: 10.4324/9781003256977-12

Conclusions formulated on the basis of previous trials have not taken into consideration an uncertainty level of the collected data during a particular trial. To tell the truth, it might not always be the case to make such an analysis since it heavily depends on the data collection plan feasible for a given trial and, as such, on the type of data being recorded. To make this possible, there is a strong need for a data collection plan capturing quantitative data (e.g. in interval scale) measuring key performance indicators. Thus, if the data allows uncertainty analysis, it is strongly recommended to do it. Basically, it provides additional information on the reliability of the conclusions formulated on the basis of a trial. As such, uncertainty analysis and error estimation are an important part of scientific research (Palmer & Hardaker, 2011). Error estimation methods used for modelling and engineering approaches in mathematical and physics science (Ayyub & Klir, 2006; Hansson, 1980) have the potential to be adopted and utilised in trials.

This chapter presents results generated by implementation of a newly proposed Performance Indicators Uncertainty Estimation Method (PIUEM), which might be an extension to the proposed evaluation approaches utilised in trials according to the trial guidance methodology (Fonio & Widera, 2020). The method was applied and checked for analysing data captured during the previously mentioned tabletop trial based on a forest fire scenario reflecting the forest fires in Sweden in 2018 (DRIVER+, 2019b, 2020c). The trial was held in November 2019 in two parallel locations, Warsaw (Poland) and Hague (the Netherlands). The key end-users involved, expecting to benefit from the trial, were the Union Civil Protection Mechanism (UCPM) stakeholders, including experts from the Emergency Response Coordination Centre (ERCC, 2020). ERCC is the European Commission (EC) coordinating cell, facilitating disaster management efforts worldwide. The mechanism's overall efficiency is generated by creating synergies between its stakeholders, meaning the European Commission and 35 UCPM participating states (European Court of Auditors Curia Rationum, 2016). The UCPM's main aim is to strengthen cooperation between the system participating states in the field of civil protection, in order to respond globally to disasters as efficiently as possible (Parker et al., 2019). For this reason, the ERCC monitors potential and materialized disasters, facilitating deployment of European-based civil protection assets to a disaster-stricken region. The assistance offered by the UCPM also foresees deployment of an emergency response coordination experts' team called European Union Civil Protection Team (EUCPT) (De Pooter, 2020). The main role of the team is to facilitate coordination of the disaster response efforts by liaising between the EC, UCPM-deployed assets (e.g. civil protection modules), the Local Emergency Management Authority (LEMA) of the stricken country (in the surveyed case, the National Disaster Management Authority, or NDMA), and other key stakeholders during a given disaster. The main aim of the team is to improve common understanding and to provide the most optimal and relevant response to the disaster needs of the affected population (Gestri, 2012). Since 2001, the UCPM has been activated over 400 times (European Commission, UCPM). This number highlights the importance of efficient and timely communication and cooperation between the key components, working on different levels of the system like the ERCC (strategic), EUCPT (operational), and civil protection modules (tactical). Therefore, optimisation of these processes gives the chance for finding and strengthening the synergies between these components in order to reinforce the system as a whole.

One of the trial's key objectives was to demonstrate the trial approach as a pertinent and applicable method for searching for an added value coming out of new solutions tailored for disaster management purposes. The general purpose was to find out how the solutions might support disaster management by (1) limitation of existing shortcomings in information management and communication flow between key components of the UCPM and (2) improvement of common operational picture by merging and synthesising disparate data sources and models in (near) real-time perspective. The trial addressed these aspects by searching for an answer on the research question formulated as follows: Does the set of new solutions (innovation line) provide an added value through improving vertical communication and information management between practitioners working on strategic, operational, and tactical level of the UCPM, in comparison to the currently working protocol (baseline)? Already published results (DRIVER+, 2020a, 2020c; Foks-Ryznar et al., 2019) have shown that information management and communication between different levels of disaster response structures could be reinforced by the set of new solutions in accordance with the currently legacy systems. However, these conclusions were based mainly on measured indicators, not including uncertainty estimation.

Therefore, the assumption for this trial was as follows: The results figured out in the course of the trial require an uncertainty estimation method in order to increase reliability of its conclusions.

Materials and Method

Trialling new solutions, such as protocols and technologies, should be realised in the closest possible environment to the one in which the trialled solutions will ultimately work in practice (Weigel et al., 2010). Therefore, real-life disaster environments might be perceived as the best for this purpose. However, such conditions are highly unpredictable and not easy to be implemented into controllable and methodological research processes. Therefore, simulating a disaster environment is the second-best option to conduct such a type of research.

The trial guidance methodology provides comprehensive guidance in order to support disaster management practitioners in searching for an added value of innovative solutions towards facilitating crisis management (Fonio & Widera, 2020; Foks-Ryznar et al., 2019). Implementation of innovative solutions into regular use represents a significant challenge as it must find an evidence-based answer to this question: How do we improve operational efficiency with proven reliability? Trials are prepared, conducted, and evaluated with broad and in-depth involvement of disaster management practitioners on each of its stages in order to ensure validity and reliability of the trial findings (Fonio & Widera, 2020; DRIVER+ 2018b, 2019a). A trial is always focused on a precisely identified gap or need of a given disaster management organisation. This is usually either stemming from specific requirements of a disastrous situation, or it is related to shortcomings in operational, procedural, and/or administrative protocols. The relevance of a trial's results is highly dependent on the reliability of the trial setup. The more adequately and realistically the scenario, context, and facilities of the trial reflect the real disaster environment, the more reliable results can be expected. Therefore, the scenario of the surveyed trial was tightly based on the forest fires in Sweden (2018), adapted and extended to the trial purposes in close cooperation with UCPM practitioners involved in the response to this particular disaster. The highlighted frames indicate the elements (Figure 7.1) are of significant importance in the scenario from the process-wise perspective. These are the steps of the scenario flow that were mainly explored in the trial.

Reflecting the entire spectrum and dynamics of real disaster conditions is simply impossible. Neither correlation study nor true experiments are usable for this natural phenomenon. This is due to the multispectral and non-repeatable character of disasters. This especially takes into account research on the broad socio-technical domain of decision-making supported by technological tools. Notably, disaster management is a highly complex socio-technical process, closely binding human behaviours corresponding to social sciences with technical interaction corresponding to technical sciences. A trial provides the means to address the challenge in an appropriate manner. Using a trial approach is justified since it enables running of the innovation line and baseline without a random distribution of participants. Furthermore, a trial is used to assess the impact of an intervention, meaning a new solution, introduced to currently working protocols. This intervention is to be observed and measured during a trial, evaluating the performance of trialled solutions towards decision-making facilitation.

In case of this trial, there was a complete set of following solutions trialled (DRIVER+, 2019b):

- A disaster management network aimed at improving the process of sharing situation awareness among stakeholders involved in disaster response, as well as facilitating disaster managers in taking optimal decisions by supporting them with real-time exchange of information between tactical, operational, and strategic levels.
- "GIS and simulation" suite of products which allows practitioners to build a virtual 4D representation (3D synthetic environment plus time dimension) of any potential disaster area facilitating a common operational picture (COP).
- Drone dedicated software providing maps from the disaster-stricken area.
- A reporting tool capable of sharing information in-between different levels of the response operation.

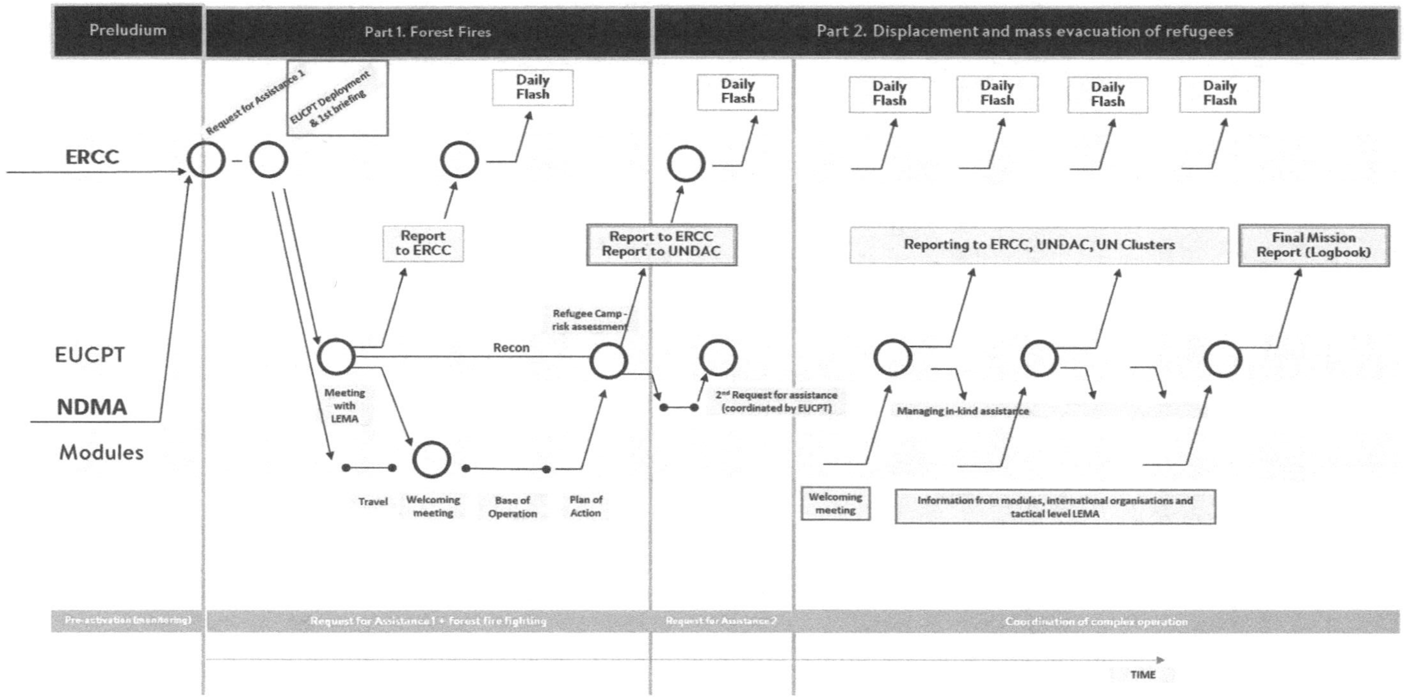

FIGURE 7.1 Trial scenario workflow. Frames indicate the elements which are of significant importance for the research done during the trial (information process and products evaluated).

Source: DRIVER+ report (DRIVER+, 2020c).

The practitioners participating in the trial had been trained on the solutions used, including hands-on training, prior to the trial (DRIVER+, 2020b). In line with the trialling rules, the practitioners operated the solutions by themselves during the trial in order to limit potential biases. However, during the trial the practitioners operating the solutions were being silently observed by the solutions providers who intervened only on request of the practitioners in order to be facilitated in case of urgent need concerning the solutions' performance.

There were information management and communication protocols trialled among strategic and operational, as well as operational and tactical levels of the UCPM. The teams of disaster managers constituted an independent variable on a nominal scale. For comparison purposes they generated reference data on the basis of their experiences of working with a set of current, legacy solutions. This was defined as the baseline. Data to feed the key performance indicators was generated and collected while working with the set of trialled solutions. This was defined as the innovation line. The trialled interactions between strategic (ERCC) and operational (EUCPT) levels were designed in dependable groups research settings. The same group of participants for ERCC and EUCPT generated both reference data and, later, the data for the innovation line run during the experience of working with the trialled solutions. While the interactions between the operational (EUCPT) and tactical (civil protection modules' team leaders) levels were trialled in research settings involving undependable groups. In this case, there were two separate groups of participants with similar disaster management profiles. Both approaches allowed the analysis of how the new solutions were experienced by the trial participants and, as such, how the solutions influenced the dependent variables, including collecting and analysing data relevant for determining the values for the key performance indicators. This data was collected by structured questionnaires and reviews. This was further translated to the efficiency level of the overall decision-making process.

Figure 7.2 presents the trial concept. In total, there were four different protocols of information management and communication flow trialled. There were two for each inter-level communication in the period of a two-day-long trial. Processes 1 and 2 were related to the communication between the strategic (ERCC) and the operational (EUCPT) level (dependable groups setting); processes 3 and 4, between the operational (EUCPT) and the tactical (UCPM modules) level (undependable groups setting). Each process was evaluated in half-day sessions. The sessions were logically linked by the scenario flow, guaranteeing a real continuum of disaster management. The emergency response tasks assigned to the participants during each session were tightly bound with the needs previously revealed by the practitioners on the base of their experiences in working with currently used legacy systems. The experiences related to past UCPM missions, including the forest fires in Sweden (2018), constituted background knowledge which allowed the participants to appoint elements requiring improvements in the existing legacy system. The designed and simulated process of the trial mirrors realistic protocols which always take place while all the three levels are involved in an international civil protection mission of the UCPM.

For evaluation purposes, a modified version of cost-benefit analysis (Lebow, 2020; Reniers & Noël Van Erp, 2016) was implemented. One is modified since typically, in such a method, both the cost and the benefit of an action are owned by the same stakeholder who simply invests some effort to achieve some benefit. Since the surveyed cases concerned communication protocols, the method had been adapted only to a certain extent. It means that the incurred cost of creating a message was assigned to one stakeholder, while the gained benefit was assigned to another one; cost incurred by a sender (a team generating an information product, such as EUCPT producing daily situational report on operational level), and a receiver (a team receiving the information product, such as ERCC getting the situational report for its own analysis on a strategic level). Thus, the utilised modification is more accurate for researching communication efficiency. For example, in the case of a communication flow process, somebody who produces and sends information bears the cost of the information production and transmission. In consequence, the information receiver, not the sender, is benefiting from having this information and using it for his/her own purposes. In practical words, the modification boils down to the fact that the sender/producer of an information product (e.g. the EUCPT preparing a daily situational report relates to process 2 on Figure 7.2) devotes some effort (cost e.g. time, energy, wisdom, etc.) to generate this particular product, while another actor, a receiver (e.g. in case of process 2, the

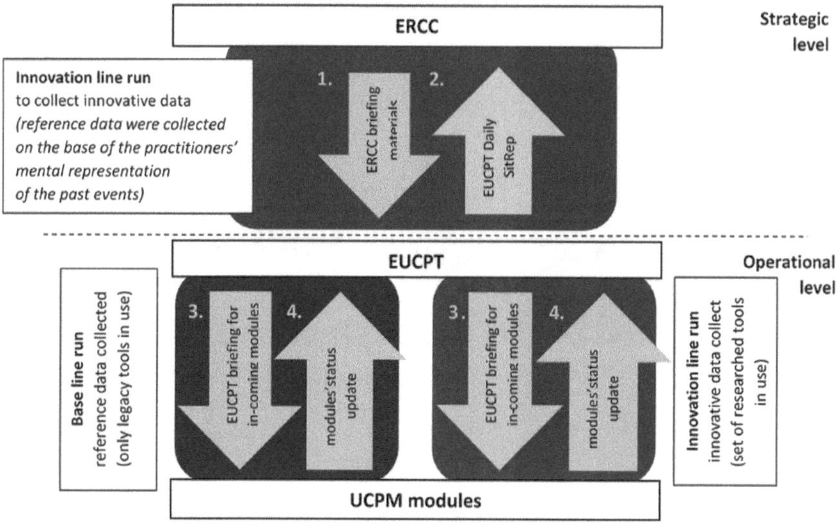

FIGURE 7.2 Concept of the trial – correlation between baseline and innovation line assessment.

Source: DRIVER+ report (DRIVER+, 2020c).

beneficiary/receiver of this situational report is ERCC), benefits from the information as a result (e.g. a good quality report is easier to be analysed by the receiver). As such, the modified method applied to all four surveyed cases of the information management and communication flow processes (Figure 7.2).

Detailed descriptions of the trialled information management and communication flow processes are presented in Table 7.1, where I-n (n = 1, 2, 3, 4) information product corresponding to the communication process n (respectively for these ones presented on Figure 7.2) is indicated.

The Performance Indicators Uncertainty Estimation Method (PIUEM), as a new method, was implemented for the first time in this particular trial in order to analyse the results of the trial. PIUEM assumes that data obtained by the questionnaire method has uncertainty, which can be measured as dedicated questions addressed in effort and result indicators in an adequate manner using interval scale. Even for a trialling approach, where low response statistics of the research is typical, uncertainty of mean values of effort and result indicators can

TABLE 7.1 Information Products Evaluated During the Trial Sessions

Product	*Description*	*Trial Reference*
I-1	Process: ERCC briefing materials Research group setting: Dependable Communication: Top-down Levels: Strategic-operational Between: ERCC-EUCPT Info package of briefing materials (including support, such as visualisations in innovative solutions) generated by the ERCC (sender) and provided to the EUCPT (receiver) being deployed for the Union Civil Protection Mechanism mission	The task in the session referred to preparing and sharing (ERCC) as well as receiving and digesting (EUCPT) the initial documents on the mission and operational tasks assigned to the EUCPT during ERCC briefing done prior to the UCPM mission with the trialled solutions. Before the session started, each ERCC (N = 3) and EUCPT (N = 4) practitioner taking part in the trial was surveyed on his/her opinion based on own experiences (including forest fires in Sweden, 2018) relating to preparing and sharing (ERCC) as well as receiving and digesting (EUCPT) information during the ERCC briefing (the captured data represented the baseline data). The experiences of the practitioners were collected in a form of questionnaires on the trialled process. For comparison purposes, the questionnaire was done by the same practitioners (ERCC and EUCPT) after the trial session with the experience of working with the trialled solutions in simulated forest fires scenario based on Swedish forest fires (innovation line). Since the same group of practitioners (working for ERCC and EUCPT) produced data for the baseline (legacy system) and innovation line, the survey was based on dependent groups setting.

Product	*Description*	*Trial Reference*
I-2	Process: EUCPT Daily SitRep Research group setting: Dependable Communication: Bottom-up Levels: Operational-strategic Between: EUCPT-ERCC Daily situational report (SitRep) generated by the EUCPT (sender) and provided to the ERCC (receiver)	The task in the session referred to preparation and sharing (by EUCPT) as well as receiving and digesting (by ERCC) two daily situational reports (SitReps) from the disaster (each for one day of the trial). These SitReps, created by the EUCPT (N = 4) were used by the ERCC (N = 3) to figure out the latest update of the situation in the forest-fire-stricken country and consider potential options for further UCPM measures responding to the newly identified disaster needs. For the assessment, the same setting of dependent groups has been applied as in the above row (I-1).
I-3	Process: EUCPT briefing for incoming civil protection modules Research group setting: Undependable Communication: Top-down Levels: Operational-tactical Between: EUCPT-UCPM civil protection modules Briefing materials (including support, such as visualisations used in legacy solutions in the baseline run and in the innovation line run) generated by the EUCPT (sender) and	The task in the session referred to the preparation and sharing of briefing documents by the EUCPT (N = 4) to the incoming UCPM modules (N = 8). There were four independent teams (each included two practitioners) working on the tactical level as UCPM modules. Two of them in the baseline setting run (N = 4). The teams worked with legacy systems covering many different tools (like emails, MS Office, or Virtual OSOCC) and formats (like .doc, .pdf, .jpg, or .bmp files). The two other teams worked in the innovation line setting run (N = 4), meaning they used the trialled solutions. After the trial experience of working with the solutions (two teams) and with legacy systems (two other teams), all of the practitioners completed a questionnaire in order to generate data for the baseline and the innovation line comparison purposes. The setup implemented for this part of the trial was based on independent groups (there were different practitioners generating the baseline and different groups for generating the innovation line data). However, it is important to underline that the profiles of the practitioners in all groups were very similar since on their daily bases they represented the same type of modules – Forest Fire Fighting with Vehicles Module (FFFVM).

(*Continued*)

TABLE 7.1 (Continued)

Product	*Description*	*Trial Reference*
	provided to the team leaders of UCPM civil protection modules (receivers) incoming to the disaster-stricken country	As such, despite the fact that these practitioners represented four different EU countries (Austria, Bulgaria, Romania, Poland), they still were trained in the same European training programme of the UCPM, which guarantees the same qualifications of all of them, making the teams well comparable (as independent variables).
I-4	Process: Civil protection modules' status update Research group setting: Undependable Communication: Bottom-up Levels: Tactical-operational Between: UCPM civil protection modules – EUCPT Status update materials on the emergency response (including support, such as visualisations used in legacy solutions in the baseline run and the outcomes of the trialled solutions in the innovation line run) generated by the team leaders (senders) of the UCPM modules and provided to the EUCPT (receiver)	The task in the session covered the status updates on the ongoing emergency response from the field, reported by the UCPM modules (N = 8) to the EUCPT (N = 4). The setup of this part of the trial mirrored the one presented in the row above (I-3).

Source: Own elaboration based on DRIVER+ report (DRIVER+, 2019b). N represents a number of practitioners who took part in the trial.

In the table, information products (I-1: ERCC briefing materials, I-2: EUCPT Daily SitRep, I-3: EUCPT briefing for incoming civil protection modules, I-4: civil protection modules' status update) evaluated during the trial are described together with their reference to the trial.

be estimated by using t-test (Pfanzagl & Sheynin, 1996) for responses distribution based on the questionnaires' outcome. The questionnaires, used for capturing the data on the trialled protocols' efficiency, covered eight criteria: usability, editability, formatting, searchability, structure, visualisation, and relevance of the existing legacy system protocol (the baseline) and the newly proposed protocol emerging trialled solutions (the innovation line). For every criterion, dedicated questions have been formulated, addressing both effort (for sender) and result (for receiver) indicators measured on an interval scale of 1–10. To find a relation between the baseline and the innovation line, pairs of the sets of captured data (baseline, innovation line) for each information product had to be analysed, taking into account that the worst situation is represented by the highest value of effort indicator (equal 10) and the lowest value of result indicator (equal 1). This basically means a maximum effort invested to generate an information product, followed by a minimum result on the information product addressee. The best situation is represented by the lowest value of effort indicator (equal 1) and the highest value of result indicator (equal 10) – in short, minimum effort used and maximum result received.

For each pair of the sets of captured data (baseline, innovation line) presented in the effort-result coordination system, a two-dimension vector, starting from the baseline point and ending at the innovation line point, is a graphical representation of the added value generated by the innovative solutions package. There are three main interpretations possible in the model:

- The vector with any effort component (increasing or decreasing) and a decreasing result component (marked by dash line and dark grey colour) represents a situation where introducing the solutions set does not bring added value to the trialled protocols, such as getting a lower result in comparison to the existing legacy system is always negative.
- The vector has a decreasing effort component, such as less work dedicated to generating an information product and an increasing result component (marked by solid line and medium grey colour). This represents a situation where introducing the new solutions set makes the trialled protocol more efficient (lower effort, higher result) in comparison to the currently used protocol. This type of positive impact was not observed in the case of the surveyed solutions.
- The vector with an increasing effort component (more work) and an increasing result component (marked dotted line and light grey colour) represents a situation where introducing the solutions set makes disaster management more laborious from the perspective of practitioners (higher effort). However, at the same time, it brings better results. Therefore, in certain cases, it may be worth investing a higher effort. Secondly, it shows a potential value of the trialled solutions which should have been further explored in order to bring new knowledge on how to improve it. Such knowledge is of special importance for the solutions producers and developers.

To estimate which setup (the innovation line or the baseline) is perceived as more useful by the practitioners (within the uncertainty interval), additionally a single result-effort indicator (REI) can be defined as

$$\sqrt{(Result)^2 + (m - Effort)^2} \quad (1)$$

where
REI is a result-effort indicator,
Result is the value of the result indicator,
Effort is the value of the effort indicator, and
m is the maximum effort indicator + 1 (for the scale of 1–10, m = 11).

As well as result and effort, the value of the REI indicator is measured on an interval scale from $(2)^{½}$ to $(200)^{½}$. The higher value of the REI indicator means that the setup (innovation line or baseline) is more suitable for the practitioners. This approach allows direct comparison of the innovation line with the baseline.

The error of the REI can be calculated with use of the general derivatives formula (Braga-Neto, 2020; Ku, 1966; Rouaud, 2013) for error propagation:

$$\delta Q = \left|\frac{dq}{dx}\right| \delta x \quad (2)$$

where Q = *Q(x)* is any function of variable *x*.

Results

Results generated by application of the PIUEM during the trial are presented in Figure 7.3 and Figure 7.4. Mean values of effort and result indicators based on the questionnaires' outcome were calculated (together with their error ranges) for the baseline and the innovation line, for each information product (I-1 ÷ I-4) evaluated during the trial.

Figure 7.3 presents result and effort indicators measured for the baseline (open dots), and the innovation line (full dots) information products (I-1, I-2, I-3, and I-4). Error ranges were calculated for 95% confidence interval ($\alpha = 0.05$) by using t-test for response distribution.

It can be noticed from Figure 7.3 that for I-1 and I-2, change of effort and result indicators from the baseline to the innovation line is marked by a red vector (dash line, dark grey). These results mean that in the innovation line in I-1 (ERCC briefing materials for EUCPT) and I-2 (EUCPT daily situational reports to ERCC), there was a higher effort devoted by the sender to produce a relevant information and the receiver perceived the product as less relevant (more difficult to use and understand) in comparison to the solutions used for the same purpose in the currently working baseline.

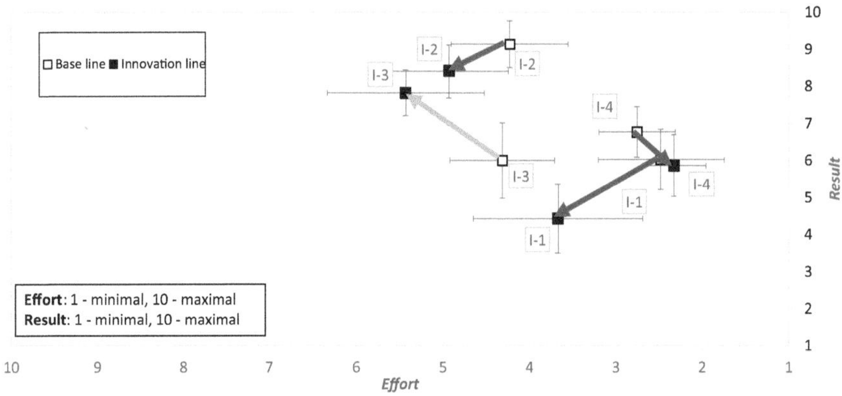

FIGURE 7.3 Result and effort indicators measured for baseline and innovation line information products (I-1, I-2, I-3, and I-4). Error ranges represent a 95% confidence interval ($\alpha = 0.05$) calculated by using t-test for response distribution. Arrows represent the transition from the baseline to the innovation line in the result-efforts frame of reference for each information product. The meaning of the arrow colours (different line types in grey scale) is explained at the end of section Materials and Method of this chapter. Practitioners N: I-1, I-2 (N = 7); I-3, I-4 (N = 12).

Source: Own collaboration based on the results of the trial (DRIVER+, 2019b).

For I-3, a change of effort and result indicators from the baseline to the innovation line is marked by a yellow vector (dotted line, light grey). This result means that for the innovation line in I-3 (EUCPT briefing materials for incoming civil protection modules), there was a higher effort devoted by the sender to produce a relevant information; however, the receiver perceived the product as more relevant (contains more valuable information) in comparison to the solutions used in the baseline.

For I-4, a change of effort and result indicators from the baseline to the innovation line is marked by a red vector (dash line, dark grey). This result means that in the innovation line in I-4, the practitioners invested less effort to produce the information product (civil protection modules status update to EUCPT). This is good; however, in consequence, the receiver perceived the product as less relevant (contains less valuable information) in comparison to the baseline.

It has to be noticed that taking into account errors of the findings presented on Figure 7.3, interpretations formulated seem to be more reliable for I-1 information products than for I-2, I-3, and I-4 (where confidence intervals are highly overlapping). This finding was observable only due to the fact that PIUEM was applied to the trial's results analysis. Thus, the comparison of effort and result indicators, without the uncertainty estimation in situations where difference between absolute indicators' value is minimal, may lead to wrong conclusions.

On Figure 7.4 REI, values calculated for the baseline (open squares) and the innovation line (filled squares) information products (I-1, I-2, I-3, and I-4) are presented. Error ranges were calculated by applying formula (2) to equation (1).

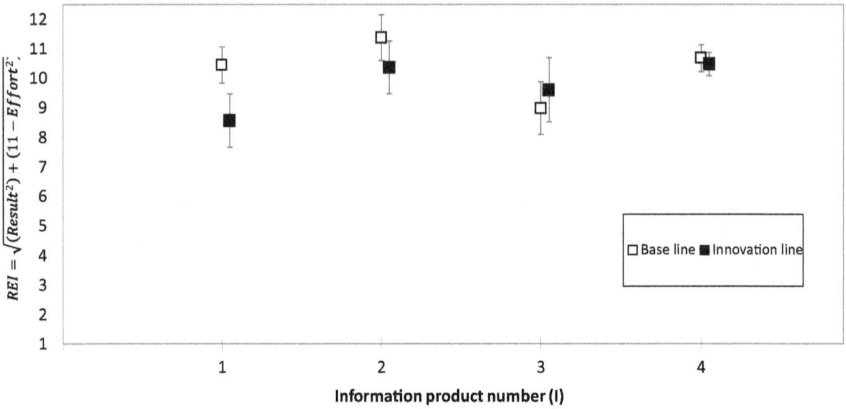

FIGURE 7.4 Result-effort indicator (REI) calculated for baseline and innovation line information products (I-1, I-2, I-3, and I-4). Error ranges represent a 95% confidence interval (α = 0.05); innovation line results were shifted on the I-axis by 0.05 to prevent errors from overlapping. Practitioners N: I-1, I-2 (N = 7); I-3, I-4 (N = 12).

Source: Own collaboration based on the results of the trial (DRIVER+, 2019b).

It can be noticed from Figure 7.4 that for I-1, I-2, and I-4, REI calculated for the baseline has higher values than for the innovation line. These results mean that the set of solutions currently used by the practitioners as the baseline were perceived by them as more useful and more appreciated than the proposed set, used as the innovation line during the trial. REI calculated for the I-3 has a higher value for the set of solutions used as the innovation line than for the set of solutions used as the baseline. These results are fully corresponding to findings that have been noticed from Figure 7.2. They allow the formulation of an interpretation that the innovation line generating the I-3 information product (EUCPT briefing materials for incoming civil protection modules) was perceived by practitioners as more useful in comparison to the legacy set (the baseline) in the trial.

It has to be noticed that taking into account errors of results presented on Figure 7.4, interpretation of the revealed results seems to be more reliable for I-1 information products than for I-2, I-3, and I-4.

Discussion

The solutions, as shown at Figure 7.3 with the red vectors (dash line, dark grey), were assessed by the practitioners as those that do not improve the process of decision-making a lot. This is figured out from the data. While using the trialled solutions the effort dedicated to produce and communicate information to the practitioners was higher (in two out of three cases) than they usually devote to using the currently working protocols. In addition, what is more important, the increased effort has not brought positive results on the information receiver's end when compared with existing protocols. Taking into account results of uncertainty estimation for the trial, it can be found that for the communication between the ERCC and the EUCPT, there is no added value observed on the EUCPT site. The trialled solutions did not demonstrate extra potential in case of reusing the briefing materials received from the ERCC before the EUCPT deployment in the innovation line (Figure 7.3 and Figure 7.4, results for I-1 product). It is possible that technical barriers typical for new solutions, as well as a desire to use known tools is a reason for this result. The effort of the ERCC might be reduced while the ERCC members would have more time to better familiarise themselves with the new solutions (DRIVER+, 2020c) and in that case the result-effort indicator (REI) probably would be higher.

In the communication flow from the EUCPT to the ERCC being realised in a form of situational report, the new solutions also did not show extra potential (Figure 7.3 and Figure 7.4, results for the I-2 product). Using the solutions, the ERCC found it less usable than the currently used legacy tools. At the same time, the EUCPT had to dedicate more effort to use the solutions in order to produce the SitRep. As it comes to the effort dedicated by the EUCPT, the new solutions did not reduce the EUCPT effort on structuring their SitRep. At the same time, the SitRep is perceived to be worse structured by the ERCC than it used to be with the legacy tool usage (DRIVER+, 2020c). However, it is worth to underline that (1) the SitRep in a current status of its legacy tool is perceived by the ERCC as the one which meets the ERCC expectations and (2) values of result-effort indicators calculated for the baseline set of solutions in comparison to those calculated for the innovation line are close to equal, taking into account error ranges.

Preparation of the status update reports by the UCPM modules for the EUCPT demonstrated the lower potential of the solutions (the innovation line) to facilitate the communication between the EUCPT and the UCPM modules (Figure 7.3 and Figure 7.4, results for I-4 product). As proven with the I-4, the trialled set of solutions demonstrated lower potential in bottom-up communication between tactical and operational levels than in the analysed top-down channel. However, values of REIs calculated for the baseline set of solutions, in comparison to those calculated for the innovation line, are close to equal, taking into account error ranges.

The yellow vector (dotted line, light grey) (Figure 7.3) informs that this particular process has been facilitated by the trialled solutions. This is especially the case from the information receiver point of view with the more important stakeholder of the communication process, benefiting from the analytical efforts done on the tactical level. When looking at the information exchange between the EUCPT and the UCPM modules, it can be stated that the information flow to the UCPM modules (Figure 7.3 and Figure 7.4, results for I-3 product) is perceived as more beneficial for the innovation line than for the baseline. However, applying PIUEM allows the conclusion that the difference is small and could be neglected due to error ranges.

Unfortunately, there are no green vectors (solid line, medium grey) (Figure 7.3), representing the information management and communication processes which would have been recognised by the disaster managers as those being improved by the trialled solutions. The trialled set of solutions did not allow any decrease of dedicated effort in producing and communicating, and also receiving and making use of information reported both ways between strategic and operational levels.

The trial showed that neither the ERCC nor the EUCPT found the set of trialled solutions bringing them added value. It means that the trial results indicate that the solutions in their present version do not reinforce the communication between strategic and operational levels as demonstrated in the trial conditions. This observation was strengthened by implementation of the PIUEM, which proved with its results the reliability of these findings, especially for the I-1 information product (the ERCC briefing for the EUCPT).

Nevertheless, the trialled solutions revealed some added value in communication between operational and tactical levels. For example, in the I-3 (EUCPT briefing for incoming civil protection modules), the civil protection modules leaders recognised the information product generated by the trialled solutions brought them some benefit (the result increased) in comparison to the currently used protocol. However, the EUCPT had to devote more effort to produce and communicate it which constituted the cost of receiving a better information product. In this case, the trialled solutions demonstrated their potential. However, further work needs to be invested by the solution providers in order to decrease the level of effort dedicated by the practitioners to work with the solutions, especially while producing an input to the information management process. This conclusion cannot be taken as fully granted (as it was in I-1) since the PIUEM results for I-3 demonstrate the results are in the scope of error. However, this gives an indication at least providing a clear direction for further research and development towards making the solutions more accurate to the practitioners' operational needs.

Conclusions

The trial demonstrated that the set of the trialled solutions, in three out of four cases, does not bring an added value to the UCPM vertical information management and communication processes. It was proven especially for both directions of communication between the strategic and the operational levels. Nevertheless, the communication between the operational and the tactical level proved some potential. However, this requires more investment from the solutions providers in order to make the solutions better suited for the practitioners' needs.

These results were verified by the PIUEM, as the uncertainty estimation method designed to increase the trials' results reliability. The method proved its value in the evaluation of new solutions' potential to facilitate disaster management, providing new knowledge on the reliability check for the achieved results during a trial. The PIUEM method could be successfully used for research data acquired from a trial with an acceptable error level. In the case of the surveyed trial, the relative error for the REI was roughly 4–11% (Figure 7.4). These results allow the conclusion that implementation of PIUEM in trialling of disaster-management-oriented solutions makes the revealed results more reliable by indicating strengths and weaknesses of the trialled solutions with a proven error assessment.

Finally, the key challenge for the new solutions' providers is their willingness to successfully commercialise their products, especially in a specific sector, such as disaster management. This could be significantly burdened by strong psychosocial habits and preferences of the disaster management practitioners to believe in and use legacy systems already known by them by heart. Perhaps, paradoxically, not the technical aspects but psychological ones could be the main obstacle to be overcome by the providers of new solutions in order to find a bridge for introducing innovations into the disaster management domain.

Acknowledgements

Appreciation to David Lund for checking the English grammatical correctness of the manuscript.

This chapter has received funding from financial resources for science in 2018–2023 granted by the Ministry of Science and Higher Education of the Republic of Poland, grant no. 3851/7.PR/17/2018/2.

References

Ayyub, B. M., & Klir, G. J. (2006). *Uncertainty Modelling and Analysis in Engineering and the Sciences*. New York: Chapman and Hall/CRC. https://doi.org/10.1201/9781420011456.

Braga-Neto, U. (2020). Error estimation for classification. In *Fundamentals of Pattern Recognition and Machine Learning* (pp. 151–183). Cham: Springer. https://doi.org/10.1007/978-3-030-27656-0_7.

De Pooter, H. (2020). The civil protection mechanism of the European union: A solidarity tool at test by the COVID-19 pandemic. *InSights*, 24(7). www.asil.org/insights/volume/24/issue/7/civil-protection-mechanism-european-union-solidarity-tool-test-covid-19.

Dimitriadis, D. (2020). *Civil Protection Needs More and Targeted EU Involvement*. www.euractiv.com/section/climate-environment/opinion/mon-civil-protection-needs-more-and-targeted-eu-involvement.

DRIVER+. (2018a). *D922.11 – List of Crisis Management Gaps*. www.driver-project.eu/wp-content/uploads/2018/08/DRIVERPLUS_D922.11_List-of-CM-gaps.pdf.

DRIVER+. (2018b). *D922.21 – Trial Guidance Methodology and Guidance Tool Specifications (Version 1)*. www.driver-project.eu/wp-content/uploads/2019/01/DRIVER_D922.21_Trial-guidance-methodology-and-guidance-tool-specifications-version-1-compressed.pdf.

DRIVER+. (2019a). *D922.41 – Trial Guidance Methodology and Guidance Tool Specifications (Version 2)*. www.driver-project.eu/wp-content/uploads/2019/03/DRIVER_D922.41_Trial-guidance-methodology-and-guidance-tool-specifications-version-2-1_120319-FINAL.pdf.

DRIVER+. (2019b). *D942.25 – Report on the Application of the Solutions in Final Demo*. www.driver-project.eu/wp-content/uploads/2020/01/DRIVER_D942.25_Report-on-the-application-of-solutions-in-Final-Demo-Final.pdf.

DRIVER+. (2020a). *D941.31 – SP94 Overall Evaluation of the Trials and Final Demo*. www.driver-project.eu/wp-content/uploads/DRIVER_D941.31_SP94-Overall-evaluation-of-the-Trials-and-Final-Demo.pdf.

DRIVER+. (2020b). *D942.31 – Report on Trainings for the Selected Solutions*. www.driver-project.eu/wp-content/uploads/2020/03/DRIVER_D942.31-Report-on-trainings-for-the-selected-solutions-Final.pdf.

DRIVER+. (2020c). *D947.12 – Report on Trial Evaluation – Final Demo* (Internal Project's Report Restricted for Public Dissemination). https://www.driver-project.eu/wp-content/uploads/Summary-Final-Demo_final-1.pdf.

Emergency Response Coordination Centre (ERCC). (2020). *European Civil Protection and Humanitarian Aid Operations*. https://ec.europa.eu/echo/what/civil-protection/emergency-response-coordination-centre-ercc_en.

European Court of Auditors. (2016). *Special Report no 33/2016: Union Civil Protection Mechanism: The Coordination of Responses to Disasters Outside the EU Has Been Broadly Effective*. www.eca.europa.eu/en/Pages/DocItem.aspx?did=40303.

Foks-Ryznar, A., Ryzenko, J., Trzebińska, K., Tymińska, J., Wrzosek, E., & Zwęgliński, T. (2019). Trial as a pragmatic and systematic approach for assessing new solutions in crisis management and rescue operations. *Proceedings of SPIE 11176, Photonics Applications in Astronomy, Communications, Industry, and High-Energy Physics Experiments 2019*. https://spie.org/Publications/Proceedings/Paper/10.1117/12.2537925. https://doi.org/10.1117/12.2537925.

Fonio, C., & Widera, A. (Eds.). (2020). *Trial Guidance Methodology Handbook. DRIVER+ (Driving Innovation in Crisis Management for European Resilience)*, Brussels, 2020. https://tgm.ercis.org/.

Gestri, M. (2012). EU disaster response law: Principles and instruments. In A. de Guttry, M. Gestri, & G. Venturini (Eds.), *International Disaster Response Law* (pp. 105–128). The Hague: T.M.C. Asser Press. https://doi.org/10.1007/978-90-6704-882-8.

Hansson, C. B. (1980). Error estimation. In C.B. Hansson (Eds.), *Prepare for Science, Introductory Mathematics for Physical and Engineering Sciences* (pp. 58–74). Oxford: Pergamon. https://doi.org/10.1016/C2013-0-10155-9.

Ku, H. H. (1966). Notes on the use of propagation of error formulas. *Journal of Research of the National Bureau of Standards – C. Engineering and Instrumentation*, 70C(4), 263–273. https://doi.org/10.6028/jres.070c.025.

Lebow, R. N. (2020). Crisis management and beyond. In R.N. Lebow (Eds.), *Between Peace and War* (pp. 345–393). London: Palgrave Macmillan. https://doi.org/10.1007/978-3-030-43443-4_8.

Palmer, T. N., & Hardaker, P. J. (2011). Handling uncertainty in science. *Philosophical Transactions of the Royal Society A*, 369, 4681–4684.

Parker, C. F., Persson, T., & Widmalm, S. (2019). The effectiveness of national and EU-level civil protection systems: Evidence from 17 member states. *Journal of European Public Policy*, 26(9), 1312–1334. https://doi.org/10.1080/13501763.2018.1523219.

Pfanzagl, J., & Sheynin, O. (1996). Studies in the history of probability and statistics XLIV A forerunner of the t-distribution. *Biometrika*, 83(4), 891–898. https://doi.org/10.1093/biomet/83.4.891.

Reniers, G. L. L., & Noël Van Erp, H. R. (2016). Cost-effectiveness analysis. In *Operational Safety Economics: A Practical Approach Focused on the Chemical and Process Industries* (pp. 219–234). West Sussex: John Wiley & Sons, Ltd. https://doi.org/10.1002/9781118871546.

Rouaud, M. (2013). *Probability, Statistics and Estimation: Propagation of Uncertainties in Experimental Measurement (short edition)*. www.incertitudes.fr/book.pdf.

Weigel, C., Schäffner, G., Kattwinkel, P., Viehweg, P., Hehle, M., & Bergmann, D. (2010). Technologies for exhaust aftertreatment testing under real conditions. *MTZ Worldwide*, 71, 48–53. https://doi.org/10.1007/BF03228001.

Zwęgliński, T. (2020). An overview of the European union instruments contributing to the internal security of the organization and its member states. *Scientific Papers of the Main School of Fire Service*, 74(2), 181–196. https://doi.org/10.5604/01.3001.0014.3178.

Zwęgliński, T., & Stefańska, M. (2021). From mobilization to lessons learnt – Civil protection and humanitarian aid operations in Mozambique 2019. *Security Dimensions*, 36(36), 94–121. https://doi.org/10.5604/01.3001.0015.0488.

SUB-PART 3.2

Simulation as Decision Support

8

DYNAMIC FLOOD MODELLING IN DISASTER RESPONSE

Tomasz Zwęgliński, Cor-Jan Vermeulen, Marcin Smolarkiewicz, Anna Foks-Ryznar, Karolina Bralewska, and Bernard Wiśniewski

Introduction

Floods, drought, wildfires, and low-flow events, as well as water scarcity situations and extreme storms, are likely to become more intense, longer, and more frequent. Such types of disasters are perceived as a worldwide increasing phenomena which seriously impact human life. Van Aalst proves that there are several reasons for this, however, the key one is the climate change that results in severe atmospheric and hydrologic events (van Aalst, 2006). An intensification of floods, hurricanes, droughts, and other disasters obviously leads to high losses in human life. Furthermore, it has devastating consequences for private and public property, agriculture, forestry, industry, and critical infrastructure having a specific feature of multiplying negative impact through cascading effects (Lawrence et al., 2020; Setola & Theocharidou, 2017; Zwęgliński et al., 2020). The increase in air and water temperature, combined with changes in precipitation, water availability, and water quality and the increase in extreme events, may lead to changes in ecosystems, life cycles, and biodiversity in the disaster resistance business (DRB) in the long-term. This is frequently mentioned to be one of the most relevant climate change consequences (ICPDR, 2020). Therefore, high-impact, low-probability events, such as major disasters, are recently more and more common. Among them, flood is broadly recognised to be the highest risk event in Europe and worldwide.

Since the dynamically changing situation naturally triggers efforts of disaster managers, scientists, politicians, and citizens, the European Parliament and the Council have introduced new regulations which are tailored towards facilitating the EU member states with enhanced flood risk management protocols and flood response (EU, 2007, 2019). Furthermore, the EU has established a common alert system for rapid intervention prior to the mobilisation of the European civil protection potential, called rescEU (Zwęgliński, 2015). This and other aspects are

DOI: 10.4324/9781003256977-14

also studied by Brinke (ten Brinke et al., 2008), de Jong and Helsloot (de Jong & Helsloot, 2009), Kolen (Kolen et al., 2009), and Zwęgliński and Gawroński (Zwęgliński & Gawroński, 2014a, 2014b), who made a survey on a broad arsenal of policies, regulations, procedures, and good practices which are being implemented in order to reduce flooding risk to as low as reasonably practicable (Puppim de Oliveira, 2009).

In parallel, special attention is put on the development of new technologies in service to address this particular risk reduction and counteraction. As Kulmala et al. prove, the dynamically developing technology sector almost every day provides new solutions which are claimed to be supportive in the decision-making process in disaster management (Kulmala et al., 2016). Many of them are already implemented in flood-prone countries, the others are still achieving excellent progress, covering, exploring, and improving functionalities optimally tailored for decision-making protocols. These solutions include, for example, a toolbox for effective evaluation of flooding (Vermeij-Van den Braak et al., 2009), methodologies for risk-based evacuation planning (Kolen & van Gelder, 2018), use of space applications in crisis management activities carried out on the national level (Ryzenko & Smolarkiewicz, 2010), or real-life training on disaster management (Vermeulen et al., 2012; Vullings et al., 2011).

In particular, flood simulation software solutions are rapidly developing technologies, which have already significantly influenced disaster management (Liu et al., 2012; Nkwunonwo et al., 2020). This type of software is used for different purposes in disaster prevention, preparedness, and response. This opens up new possibilities for decision-making by providing short-term prognosis on flood development in its initial phase. Burston, Horita, and Izumi postulate that efficiency and effectiveness of decisions, taken prior to a disaster or in its very initial phase, play a key role in mitigating its impact and are more and more dependent on supportive technologies (Burston et al., 2015; Horita et al., 2018; Izumi et al., 2019). All this should lead to generating relevant, accurate, and timely information that limits uncertainty to which disaster managers are naturally exposed while taking decisions (Shannon, 1948).

Since disaster management systems are financed by public funds, the budget allocations in these fields need to be rationally planned and spent in the most efficient and transparent way. However, regarding new solutions, due to the lack of objective results of the real performance or tests, the public stakeholders have to decide *a priori*. Thus, often they made decisions that are mainly based on commercial argumentation of solution providers. As Fonio and Widera state, testing new technologies in a realistic setup prior to their purchase is a critical measure for effective investments and their successful implementation in disaster management (Fonio & Widera, 2020). In this respect, Foks-Ryznar et al. highlight that creating a testing environment that mirrors real conditions is a challenging task, however still possible (Foks-Ryznar et al., 2019).

Kolen and van Gelder prove that one of the requirements for pertinent decisions during disaster management is possessing an access to data which is a reliable and timely contribution to the decision-making process (Kolen & van Gelder, 2018). Furthermore, data needs to be processed into a form of usable and appropriate information for the disaster managers' analytical work. This is the final effort which should result in an optimal decision. Therefore, the overall process is non-linear, highly complicated, and exhausting for disaster managers, who are in addition mentally burdened with the human life related consequences of their decisions. Thus, the disaster environment represents a highly uncertain, complex, and dynamic reality that makes decision-making extremely challenging (Wiśniewski et al., 2017; Zhou et al., 2017). Hence, there is no doubt that disaster managers benefit from getting access to the relevant examination results of solution performance being thoroughly trialled prior to being used in a real disaster response.

Hadimlioglu and Schröter confirm the results of the literature review. There is a significant need for trialling flood simulation software towards finding new protocols to be implemented in standard operating procedures, with a special attention to measuring parameters of their potential usability in real conditions, such as decision-making time consumption, quality of decisions, and so on (Hadimlioglu et al., 2020; Schröter et al., 2020). Most of the available research involves qualitative study confirming the need and claiming efficiency and efficacy of flood simulation, although not providing valid measurements on these performance indicators (Ghazali & Kamsin, 2008). Furthermore, there is a deficiency of meaningful research analysing the use of such products for decision-making purposes in an early stage of disaster. Hence, this chapter is reflecting on this gap. It is aimed at providing disaster managers with the findings on trialling of a dynamic flood simulation software in two sequential flood related trials in Poland and the Netherlands. The trials addressed a crisis management function that is the decision-making process in an early stage of a major disaster. The research questions were based on the initial assumption that the dynamic flood modelling software (innovation line) supports disaster managers by improving the decision-making process in comparison to currently used legacy systems, such as static flood modelling solutions (baseline).

Materials and Method

Weigel postulates that trailing the performance of solutions, including protocols and technologies, should be realised in the closest possible conditions to those in which the particular solution will ultimately be used (Weigel et al., 2010). Real-life disaster environments seem to be the best testing ground; however, they are highly unpredictable and, as a consequence, not easy to be implemented into a controllable process of a trial design that would ensure collecting reliable and valid data on the solution performance. Due to the dynamic, uncertain, non-repeatable, and costly characteristics of a disaster, there is no sense, even more, it is simply impossible to mirror the full spectrum and character of such phenomena for trialling purposes. Therefore, simulating a real environment, being a part of the broader concept of a trial, is the optimal option for achieving pertinent results on solution performance and on its potential to cover an identified crisis management gap.

Thus, the important thing for the trial purposes is to imitate in an utmost realistic way the key crisis management functions within comparable settings, which are the baseline (the process as it works now in the legacy system – i.e. without the trialled solution) and the innovation line (i.e. the process as it is proposed to work when the trialled solution is part of the system). Obviously, the trial scenario is to be designed in a way that enables the identified crisis management gap to be covered by the new solution functionality. As the gap appears in the flow of the scenario during the trial, it triggers the particular crisis management function to be realised, such as in the given case decision-making process on flood development. And finally, such a designed trial environment should be configured and assessed by the disaster management practitioners, who are taking part in the trial, as the ones for whom disaster reality is a daily business and to whom the trialled solution is expected to serve. This approach guarantees an acceptable level of reliability and validity of the trial findings.

Trials are conducted to evaluate a kind of an intervention, such as a new solution implemented in the currently working crisis management function. The added value should be demonstrated by covering the identified crisis management gap of the system the practitioners work in. It might be done through specifying and evaluating key performance indicators (KPIs), such as efficiency and/or efficacy of the trialled solution in the trial setup. It might be assumed that the modernisation of a protocol by using the trialled solution in the given crisis management function is considered as a manipulation of an independent variable in order to influence the dependent variables that are measured with key performance indicators fed with the observed, collected, and analysed data generated during the trial run. Since trial participants are not randomly assigned, and disasters are not repeatable in their character, the trial method does not eliminate the problem of confounding variables from the research point of view. However, it provides results generating new knowledge on highly probable answers to a given research question since these are crisis management practitioners who are following a rigorous methodological approach proposed in the trial guidance methodology (Fonio & Widera, 2020) – so-called TGM. For this purpose, disaster management practitioners were directly involved throughout both trials at different stages of their realisation, starting from the preparation phase, through execution, up to the evaluation phase. Their contribution started with an identification of the gaps and operational needs to be addressed by new solution, and it was continued with formulation of data collection and evaluation plan tailored to check the gaps coverage by the trialled solution performance during the trials. Furthermore, the practitioners consulted and tested scenarios and the trials setup in order to verify if they mirror the practitioners' working environment. Afterwards they participated in the selection of solution(s) to be trialled – that is, by assessing which solution in the market might have potential to cover the identified gaps and support the operational needs. Finally, disaster management practitioners trialled the solution and participated in the interpretation of the collected data for evaluation purposes. Thus, the trials were triggered by the formulated gap and the needs related to it.

Thus, the trials were led mainly by and for the disaster management practitioners, following a pragmatic and systematic approach proposed in the TGM, including close supervision of the overall process by researchers experienced in implementation of this method (Fonio & Widera, 2020).

Trials

Trial Poland (Trial PL)

As it is suggested by Fonio and Widera (Fonio & Widera, 2020), setting up the trial required disaster managers to identify a potential crisis management gap they want to address. In the analysed case it was the lack of real-time information on flood development in the initial phase of the disaster. Furthermore, the disaster managers identified three core innovative features of the trialled solution that have been perceived as these ones that might most likely improve the crisis management function that in the trialled case was the decision-making process. The identified features of the solution are as follows:

- Short computation time in combination with a high spatial resolution and an accurate prediction of floods.
- Ease of model adaptability to test deliberate measures.
- A realistic visualisation of model outputs.

Furthermore, according to the information shared by the solution provider, the tested solution claimed to be able to provide rapid flood development forecasting (prediction of flooding locations, water depths, and water arrival times, among others), constructing a common operational picture of floods and exploring quickly various flooding scenarios depending on the measures undertaken as well as decisions made.

Hence, the objective of the trial was to evaluate an interactive, dynamic flood simulation software in terms of its efficiency and efficacy (e.g. by such criteria as decision-making time). For this reason, an indoor tabletop trial was conducted in Warsaw (Poland) in May 2018. The trial scenario was based on a massive chemical spill release resulting in a major flood. This scenario was drawn from the historical data of a real incident – the caustic waste release in Ajka Alumina Plant in Hungary (2010) (Milacic et al., 2012; Toth, 2013). As shown by Hendriks (Hendriks et al., 2019) and Zwęgliński (Zwęgliński, 2019), in a real disaster, the decision is worked out in complex internal group discussions, within a disaster management team, being worked out in a human-technology interactive environment. Therefore, to avoid biases, the decision-making environment was reproduced in an optimal manner – for example, by a realistic scenario deeply consulted in advance with Hungarian experts involved in the emergency response to the real case, generation of time pressure on the disaster managers during their performance of the assigned operational tasks in the trial, appropriate practitioners' involvement, and so on.

The trial was attended by disaster managers from 13 countries including Bulgaria, Estonia, Latvia, Lithuania, Croatia, Hungary, Israel, Italy, Poland, Slovakia, the Netherlands, Sweden, and the United Kingdom. By profession, there were eight fire officers, four emergency medical technicians (EMTs), two police officers, and two hydrological experts. All involved practitioners possessed more than ten years of experience in disaster management on a regional and/or national level, being responsible for operational and strategic decisions. Each group consisted of eight people, in non-equivalent but almost similar profile designs as suggested by Price (Price, 2015). Each group consisted of the same number of people mirroring the same professions. This led to the construction of two independent, however utmost similar groups in their profiles (e.g. years of experience, profession, etc.).

One group was working with legacy tools (*the baseline*), while the other one operated in a newly proposed protocol implementing the trilled solution (*the innovation line*). These two groups constituted *independent variables* defined in *nominal scale*. The team working in the *baseline* used currently available tools dedicated to flood risk management, such as elevation maps, Google Maps, and the Polish IT system dedicated to facilitate disasters risk analyses with static models – ISOK (each group included one Polish hydrological expert working with ISOK on a daily basis) (ISOK, 2020). The group working in the *innovation line* operated the dynamic flood modelling software being trialled. Both groups worked in parallel runs in order to generate data relevant for comparison purposes. As such, the groups constituted disaster management teams that were tasked the same – to decide on a sequence of evacuation of four public buildings in the context of a flood scenario run. This decision was to be preceded by collecting, processing, and analysing available data from the different technological solutions available for each of the groups.

Such an approach enabled the trial to provide an adequate environment for generating data required to assess key performance indicators, such as decision time. Furthermore, it was possible to interpret the data in order to figure out how the trialled solution influenced the efficiency and efficacy (*dependent variables*) of the overall decision-making process. Hence, efficiency was determined by the following quantitative indicators measured in ratio scale (minutes):

- Time taken for decision-making.
- Time dedicated to each sequential phase of the decision-making process.

On the other hand, efficacy was assessed by qualitative indicators measured in a nominal scale. They were as follows:

- Proposed sequence of the buildings evacuation.
- Factors taken into account in decision-making.

Thus, while the *baseline* and *innovation line* were running, appropriate data was collected in order to provide relevant and robust input for evaluation purposes.

There was also an additional *qualitative* process implemented in the trial to ensure reliability of the collected data. This data was collected in frames of the trial and the solution dimensions. This process was conducted by observers who were tasked to register any potential biases which might affect the reliability of the trial findings. Furthermore, both the observers and the disaster managers involved in the trial provided their opinions on trial organisation and the solution performance in the form of questionnaires. The collected data were coded in the Likert scale, from *strongly disagree* (coded −2) to *strongly agree* (coded +2) (Jamieson, 2004). These results reflected the observers' and the practitioners' assessment, such as on the extent to which the simulated environment mirrored a real flood environment or how realistic were the disaster managers' working conditions. As a result, the overall evaluation process allowed the evaluators to control the reliability of the trial findings.

Results

The quantitative and qualitative results observed and recorded during the trial are presented with distinction to efficiency and efficacy of the decision-making process.

The results relevant to *efficiency* of the decision-making process are presented in Table 8.1.

Table 8.1 shows that in total there was a small difference of three minutes in favour of the team working in the *innovation line* (with the solution). However, this group made the initial decision much earlier than the group working at *baseline*, literally after 22 minutes from launching the decision-making process, including

TABLE 8.1 Efficiency of the Decision-Making Process (Trial PL)

Key Performance Indicators				
Independent Variables	*Dependent Variables*			
Type of group	Criteria for decision-making defined	Initial decision	Final decision	Total decision-making time
Values				
Nominal scale	**Ratio scale**			
	Direct measurement (minutes)			
	Setting decision criteria	Analytical process		Sum
Baseline	7	N/A	55	62
Innovation line	2	20	37	59
	Indirect measurement – percentage share of the total time (%)			
Baseline	11.29	N/A	88.71	100.00
Innovation line	3.39	33.90	62.71	100.00

the time for defining criteria which should be involved in the analytical part of the process. At the stage of the initial decision the disaster managers working in the *innovation line* were able to decide which building was to be urgently evacuated (as the first one) due to flooding development, which was not a case in the *baseline* at all. Thus, using the trialled solution the crucial initial decision could be taken 40 minutes earlier than the decision taken in the *baseline* (working with the legacy system).

The analytical process realised in the *innovation line* leading to the decision on the first building to be evacuated (initial decision) encompassed 33.9% of the total decision time. In comparison, the group working in the *baseline* needed 88.71% of their time for analytical purposes while making the decision.

Moreover, time devoted for setting criteria, which should be taken into account during the decision-making process, was also significantly shorter for the *innovation line*. The time slot constituted only 3.39% of the overall process, while the *baseline* consumed 11.29% of the total decision time. The reasoning for this phenomenon is explained here in the part devoted to efficacy.

Decision-making *efficacy* supported by the solution performance is analysed with qualitative indicators. The results relevant to the process are presented in Table 8.2.

TABLE 8.2 Efficacy of the Decision-Making Process (Trial PL)

Key Performance Indicators		
Indicators	*Values (Nominal Scale)*	
Independent Variables		
	Baseline	*Innovation Line*
Dependent Variables		
Quality of the decision – proposed sequence of evacuation	1. Hostel 2. Food factory 3. *Power plant or hospital** 4. *Hospital or power plant**	1. Hostel 2. Food factory 3. *Power plant* 4. *Hospital*
Factors taken into account during decision-making process	**Exposure**	
	1. Distance from the breach. 2. Number of people inside at the given moment. 3. The elevation of the land. 4. *Working hours of the building.*** 5. *Potential economic loss if not evacuated.*	1. Distance from the breach. 2. Number of people inside at the given moment. 3. The elevation of the land. 4. *Potential time to flood a building.*

Key Performance Indicators		
Indicators	*Values (Nominal Scale)*	
Independent Variables		
	Baseline	*Innovation Line*
Dependent Variables		
	Vulnerability	
	6. *Physical and mental status of the people being inside – such as handicapped people, people being asleep, and so on.*	5. *Type of hospital in the context of evacuation challenges.*
	7. *Individual needs of potentially affected people.*	
	Likelihood	
	8. *Probability level that some of the buildings will be semi-evacuated – such as school (maybe students can leave school by themselves).*	N/A
	9. *Time needed for evacuation.*	
	10. *Destination venue of the evacuees.*	
	Internal Team Aspects	
	11. *Competences available in the group – such as knowledge about topography, terrain elevation, and so on.*	N/A

* Depending on self-evacuation capacity
** Values qualitatively different in both groups are presented in *italic style*

The proposed evacuation sequence of the buildings at risk of flooding was almost the same in both groups. First two objects were the same in the *baseline* and the *innovation line*. What is more, the order of the third and the fourth buildings was also, although conditionally, the same.

Table 8.2 presents qualitative data on the factors the teams took into account while making the decision. These data were grouped into three risk related criteria (exposure, vulnerability, and likelihood) and one criterion concerning teamwork. The results demonstrate that the team working in the *innovation line*, in comparison to the *baseline*, took a lot less factors into consideration while deciding. This explains the time difference in defining criteria for decision-making (see Table 8.1). The visualisation offered by the dynamic modelling software allowed the group working in the *innovation line* to not assess the likelihood aspects of the flood scenario development (flooding of particular areas) by themselves. Possibly, the reduction of factors resulted in shorter decision time with the same quality of the decision (equal evacuation sequence of the buildings). In addition, it was observed that the team working with the trialled solution (*innovation line*) did not consider the competences they had in the team. This reduced the number of discussions resulting in shorter time as well. Probably, the algorithm being a part of the dynamic flood modelling software constituted an alternative for the process of identification and proper usage of the competences of individual team members.

In addition, there was also a survey conducted on the opinion of the practitioners' experience with the solution gained during the trial. It was done to assess potential impact of the solution on disaster management, including facilitation of the following aspects: creation of an integrated common operational picture, cross-border resource management, modelling cascading effects, and modelling chemical threat dynamics. All opinions on the surveyed aspects were positive (average scores 1.13, 0.64, 1.29, and 1.20, respectively), as presented in Figure 8.1.

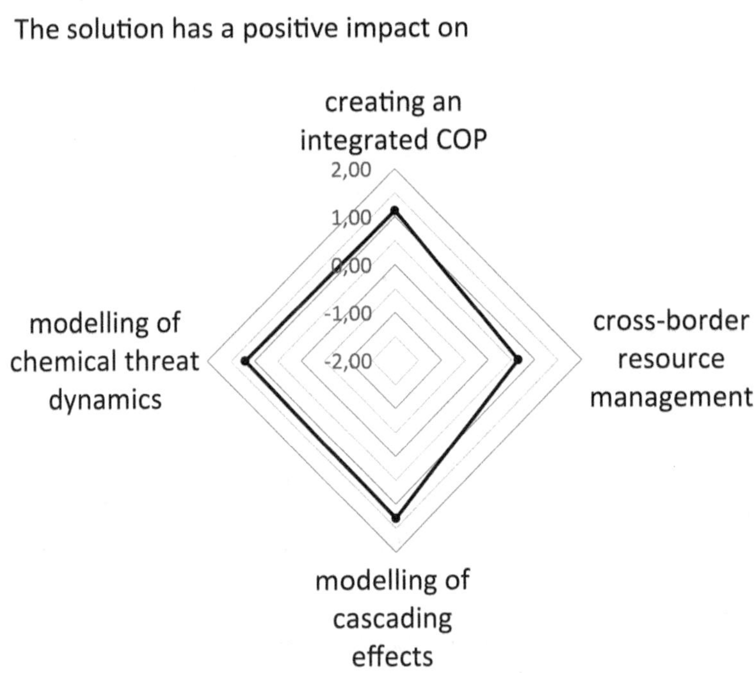

FIGURE 8.1 The impact of the solution on broader disaster management context. (trial PL). Practitioners: $n = 16$.

Trial and Solution Dimensions Evaluation

The trial and solution dimensions evaluation were aimed at identifying factors affecting the trialling conditions and the solution performance, which could potentially bias the results. In order to monitor these areas, a special questionnaire survey was conducted. Both the practitioners and the observers were requested to express their opinions on the trial organisational aspects. The results are presented in Figure 8.2.

In general, the practitioners' and the observers' perception of the trial dimension was positive. According to the practitioners, the flooding scenario, the trial setup, safety precautions, and security measures were adequate for testing purposes. The disaster managers satisfactorily understood and realised the assigned tasks, which was important to ensure the validity of results. Furthermore, they received sufficient training on the solution prior to the trial and did not identify any major technical hiccups during the conduction of the trial. Thus, these results suggest that the practitioners' experiences have not been biased. Moreover, positive opinions on the overall trial organisation, the data generated in the course of the practitioners' experience, and the quality of work with the trialled solution advocate for validity and reliability of the trial findings.

As it comes to the solution dimension, the functionalities were recognised as having potential to meet the practitioners' operational needs (Figure 8.3). The disaster managers' opinions were positive on most of the questioned aspects of the solution's performance. Nevertheless, the solution was perceived as not very easy to learn how to use. It might mean that the organisations, which are planning to purchase the solution, should foresee special training sessions prior to implementing it into their system.

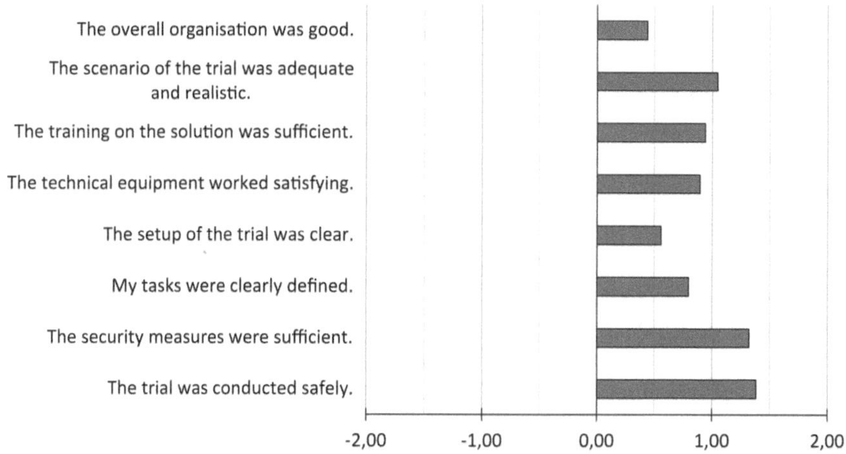

FIGURE 8.2 Trial dimension results on the trial aspects (trial PL).

Practitioners: $n = 16$. Observers: $n = 8$.

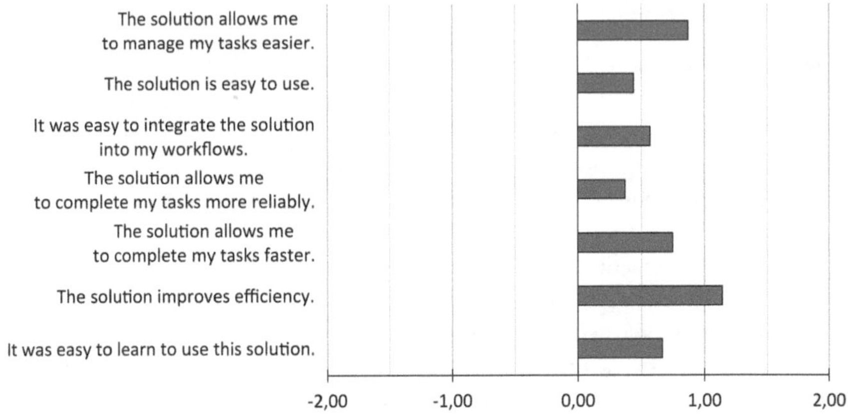

FIGURE 8.3 Solution dimension results on the solution aspects (trial PL).

Practitioners: $n = 16$. Observers: $n = 8$.

Key Findings of the Trial in Poland

The team working in the *baseline* did not have any tool that could help them in foreseeing a time horizon when the area will be flooded. This situation influenced a longer time to come to their final decision. In comparison, the team working in the *innovation line* was able to assess the time horizon for flooding of the respective area which was provided by the trialled solution. This gave them more confidence for structuring the decision process and to come up with a more efficient decision, by dividing the process into two sub-decisions. The initial decision concerned the building requiring an urgent evacuation, while the final decision included the complete evacuation sequence for three other buildings. It proves that implementation of the solution changes the way of decision-making, providing more initial information on the flood development much faster than in the *baseline*. The conclusion is also evidenced by a reduced number of criteria (only six factors) taken into consideration in the process of decision-making in the *innovation line*, while in the *baseline*, 11 factors were considered. Moreover, the time-oriented visualisation generated by the solution supports internal team communication within the team and external information exchange resulting in shortening of the decision-making time and making the decision easier. Finally, the *innovation line* team did not consider any aspect related to the likelihood of the flood development since it was simply automatically covered with the simulation results generated by the solution. These aspects positively influenced the decision efficiency by decreasing decision processing time. As a consequence, the limitation of uncertainty makes the overall decision process more organised and thus less stressful.

All positive aspects in the *innovation line* have been observed regardless of the efficacy of the decision-making process which occurred to be basically the same for *baseline* and *innovation line* (the sequence of the buildings evacuation). However, information generated by the dynamic flood modelling software may lead to more objective conclusions, which can improve the quality of the decision, especially in more complex cases or while the decision-makers are less experienced.

The evaluation of trial and solution dimensions shows that the presented results of the trial are reliable and valid. The statement is rooted from and based on the positive perception of all the surveyed organisational and solution-oriented aspects of the trial. The practitioners and observers involved in the trial did not identify any circumstances which might have significantly biased the collected data.

Trial Netherlands (Trial NL)

Similar to the trial in Poland, in the Netherlands a tabletop trial was carried out in The Hague in May 2019. The same dynamic flood modelling solution was introduced to the decision-making protocols (*innovation line*) and compared to the protocols used on a routine basis (*baseline*) in the region of The Hague.

The disaster managers involved in this trial, although not the same as those involved in the trial in Poland, were willing to explore the same features of the solution. They recognised its functionalities as very promising for covering the same gap as in the Polish trial, which was a lack of the real-time information on flood development in the initial phase of the disaster.

The trial scenario referred, as much as possible, to the local geographical conditions and disaster situations appealing to the practitioners' experience. Therefore, the trial was based on a realistic scenario of a major flood caused by an extreme high tide at the coastal area, coinciding with a storm. Such a flood threat is one of the key scenarios considered in the Dutch risk assessment (ten Brinke et al., 2008; van Duuren & Zwęgliński, 2019). The disaster management triggered by the scenario involved a deployment of evacuation assets and mitigation of impact of cascading effects.

In total, 13 practitioners were involved in testing the solution within the trial, representing the primary Dutch disaster management entities on a tactical level. The vast majority (eight) had between five and ten years of professional experience, while three practitioners had between one and five years of experience and two of them more than 15 years.

In the Dutch situation, a regional operational team (ROT) coordinates the flood response on a tactical level. Their tasks primarily consist of obtaining the operational picture of the flood (situational awareness) and advising the strategic level represented by the mayor of The Hague. Since the interagency cooperation is the key aspect in disaster management (Zwęgliński & Morgado, 2018), the ROT is supported by so-called action teams (or crisis teams, AC/CT) composed of first responders (municipalities, fire services, medical services, police service, and military services) and other stakeholders, such as the Water Board, power company, and public transport. The tested dynamic flood modelling software was used by the Water Board (a crisis team of seven disaster managers), which got hands-on experience operating the solution in a realistic setting of the trial. In accordance with the Dutch operating procedures, the Water Board distributed the results of simulations (flood forecasts) to other involved stakeholders, such as the ROT, as it would have been done in a real case of the trialled scenario.

The trial involved two phases: the flood threat phase (a high probability of flooding in the near future (days) and the flooding phase (after the dike breach). Each phase consisted of two parts: creating a common operational picture (situational awareness) followed by implementation of mitigation measures.

The crucial aspects validated during the trial were as follows:

- Mitigation of cascade effects: The flooding exposes a large part of the population (one million) to the risk of drowning/flooding. Furthermore, electricity/telecom failure affects the communication between parties (public and private) so that a clear understanding, and information exchange becomes almost impossible. In addition, the flooding has also had an impact on both resource planning and transport logistics. Hence, coordination, information exchange, enhancement of the common operational picture, situation assessment, resource management, and communication are the crucial challenges to deal with during this large-scale disaster.
- Information analysis: The national crisis management system (LCMS) is used to build the common operational picture of the flood (threat), to share the information and support decision-making. This system is only accessible to the first responders. Other stakeholders, like the Water Board, power companies, and communication providers use similar but unconnected systems. The communication is therefore indirect, using mail or telephone communication (*baseline*). The dynamic flood modelling tool was tested to assess its potential to improve the overall process and its ability to provide additional information to the system (*innovation line*).

During the trial, all communication between ROT and AC/CT was recorded for evaluation purposes. Additionally, each session of the trial was followed by observers (experienced disaster managers other than those testing the solution), who have been given instructions on which aspects should be observed, in terms of the role and tasks of each team. For example, observers in each group observed team dynamics and recorded the critical time in the processes of decision-making and information sharing. Finally, questionnaires were used to collect practitioners' opinions after having an experience in working with the solution in each session of the trial. In order to accomplish the survey with *qualitative data*, the questionnaires assessing the impact of the flood modelling tool on the disaster management aspects (e.g. facilitating the process of building a common operational picture, assessing cascading effects) were completed. The efficiency and efficacy of the response using the tested solution was compared to the baseline meant as the response executed using the legacy system represented by contingency plans and operational procedures in case of flooding.

The objective of the trial was threefold:

- Evaluate the usefulness of the flood modelling tool in resource planning activities for large-scale and long-term disaster operations.
- Exchange net-centric data to improve information flow between relevant parties and thus improve the shared understanding of the current situation.
- Support the planning and management of a large-scale evacuation, taking into account the real-time traffic information.

Similar to the trial in Poland, a *qualitative* process of trial and solution dimensions evaluation was implemented to ensure the quality of the trial. It was conducted by the observers who were asked to register any potential biases which might affect the outcome of the test. Furthermore, both groups, the observers and the practitioners, completed additional questionnaires on their assessment of the trial organisation and the solution-related aspects. The results of these surveys represent the perception of the observers and the practitioners how the simulated environment reflected a real flood situation and the disaster managers' working conditions. Such a process enabled the organisers to assess the reliability of the trial findings.

The solution performance was researched towards its potential to bridge specific disaster management gaps (Horita et al., 2017), which had been specified by the practitioners prior to the trial. All results presented in the figures have been again calculated as the average value of data coded in the Likert scale (Jamieson, 2004).

Results

Figure 8.4 presents the average rates of the practitioners' opinions (Dutch Water Board representatives, n = 7) on the solution towards their four operational needs: supporting situational awareness, supporting quantification of mitigation measures, facilitation of decision-making, and supporting the command, control, and communication (3C). The highest rate was given to the indicator coded as *supporting situation awareness in flooding* tested in the very initial phase of the flood scenario development (1.2 points). Three other features were also assessed positively, with bigger value added in creating a common operational picture (average scores of 0.5, 0.6, and 0.8, accordingly) than in implementation of mitigation measures (scored in average 0.0, 0.5, and 0.5).

The solution functionalities were further surveyed in the context of their relevance and maturity levels (Figure 8.5). The relevance of all features has been assessed as "good" (average score of 1.00 in creating a common operational picture, 0.71 in the implementation of mitigation measures). However, the survey results with regard to the maturity suggest that the solution still requires some further investment and development (average score 0.35 and 0.18 accordingly).

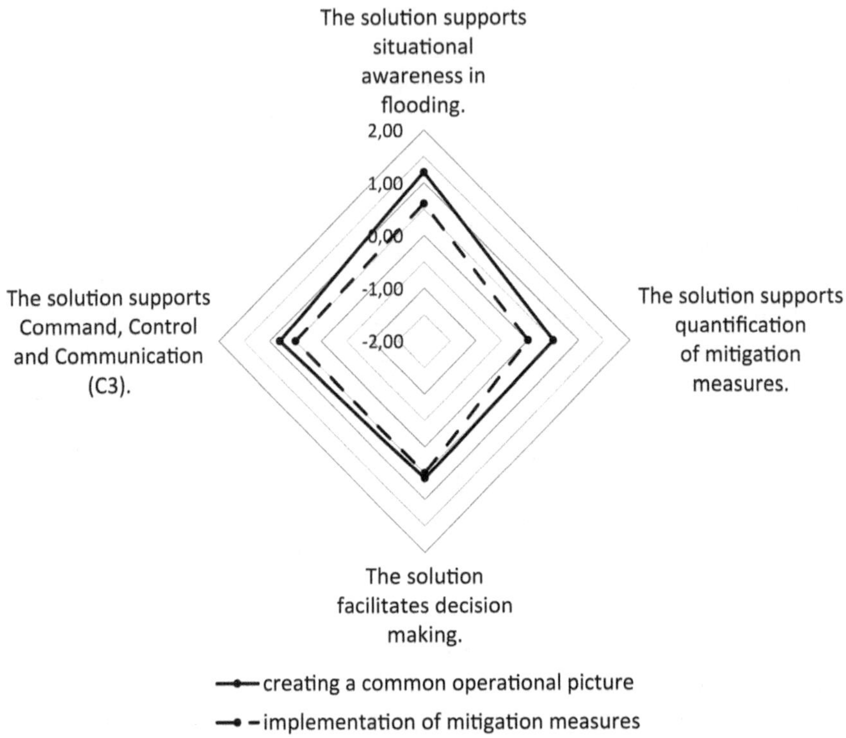

FIGURE 8.4 Average results on the solution potential in creating a common operational picture and implementation of mitigation measures (trial NL).

Practitioners $n = 7$.

Finally, the results suggest that the solution is more appropriate to be used for situational awareness (in the very initial stage of a flood development) than mitigation measures (in the later stage of a flood), what was concluded based on the higher ratings given by the practitioners during the first stage of the disaster scenario development. This may be explained by the fact that at the initial stage the disaster managers have to deal with uncertainty on the status and evolution of the flood threat, while afterwards they can rely on the actual data and their own experience.

Trial and Solution Dimensions Evaluation

The trial and solution dimensions evaluation process was aimed at identifying biases on the trailing conditions and the solution performance which could potentially impact the results. In order to monitor these areas, another questionnaire survey was conducted. Both groups, the practitioners and the observers, were requested to express their opinions on the trial organisational aspects. Data was collected in the Likert scale (Jamieson, 2004). The results are presented in Figure 8.6.

In general, the practitioners' and the observers' perception of the trialling conditions was positive. According to the practitioners, the flooding scenario, the trial

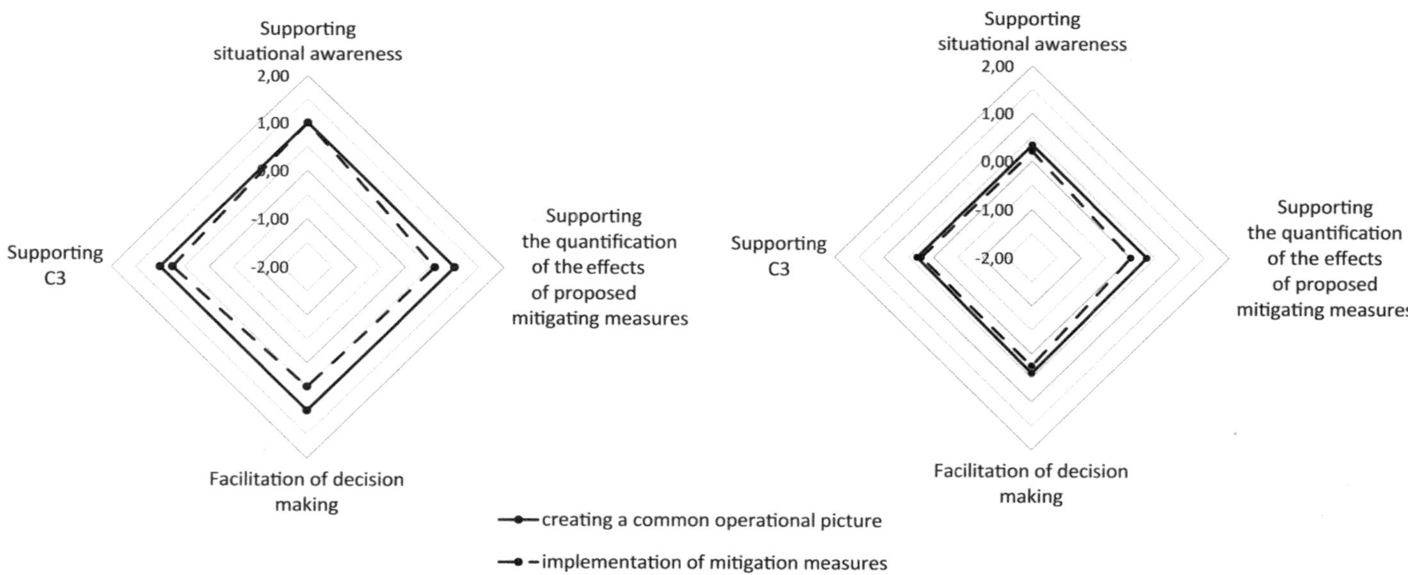

FIGURE 8.5 Average results on the solution performance: the solution (a) relevance and (b) maturity (trial NL).

Practitioners: $n = 7$.

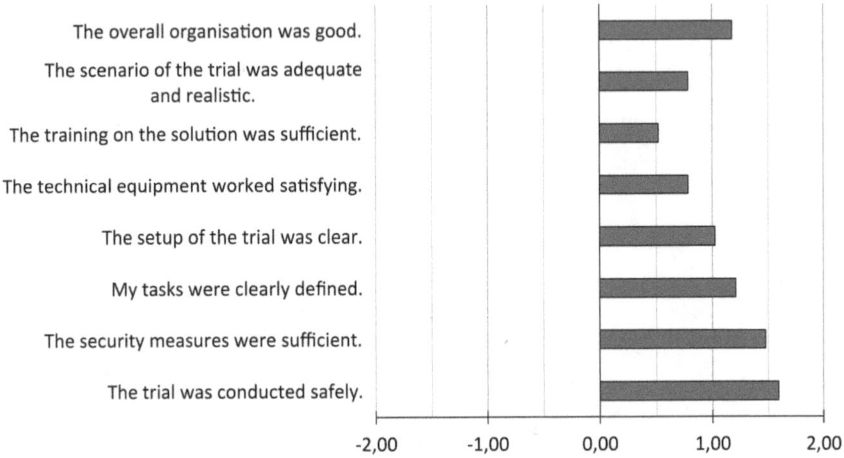

FIGURE 8.6 Trial dimension results on the trials aspects (trial NL).

Practitioners: n = 13. Observers: n = 10.

setup, the safety precautions, and the security measures were adequate for trialling purposes. The disaster managers sufficiently understood and realised the assigned operational tasks. Moreover, the stakeholders attending the trial stated that they received sufficient training on the solution prior to the trial and did not identify any major technical hiccups during the conduction of the trial. Thus, the trial and solution dimensions findings suggest that the practitioners' experiences have not been biased. Furthermore, positive opinions on the trial organisation, the data generated in the course of the practitioners' experience, and working with the tested solution substantiate the reliability and validity of the trial results.

The tested tool functionalities were assessed as having potential to meet the practitioners' operational needs (Figure 8.7). The disaster managers' opinions were positive on most of the questioned aspects of the solutions' performance. However, like in the previous trial, the solution was perceived as not very easy to use.

Key Findings of the Trial in the Netherlands

The discussion on the results generated by the trial is structured around the three main objectives.

Tools to Improve Resource Planning Activities in Large-Scale and Long-Term Disaster Operations

Activities in a large-scale and long-term disaster operation require a significant amount of assets and resources coming from various areas, such as public, private, and non-governmental sectors. During a long-term and multi-agency disaster response, usually responders face two types of problems. There is either a very large number of resources which are expected to be distributed at the right time to the

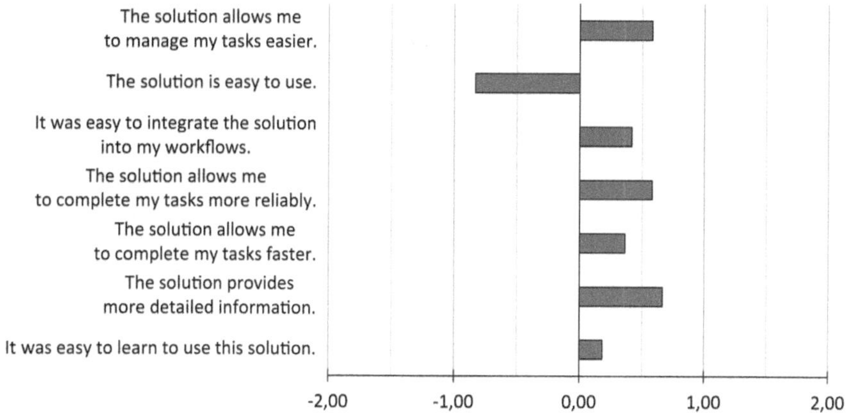

FIGURE 8.7 Solution dimension results on the solution aspects (trial NL). Practitioners: n = 13. Observers: n = 10.

right place or shortages of specific resources which should be managed with appropriate care in order to address the most urgent needs in an effective manner. This generates a need for simulation tools that would allow improvement of planning activities. From the disaster managers' perspective, the most desirable functionalities of such a solution are, for example, facilitation of decision-making process to optimise deployment of civil protection assets, ability to monitor operational timing of different rescuers/resource activities and the number and location of those in action. Moreover, in large-scale and long-term disaster operations the improvement of resource planning activities could be more valuable in case of solutions, which can forecast dangerous zones, zones being at risk of flooding impact, or zones potentially exposed to the cascading effects (e.g. loss of power).

The trial demonstrated that the flood forecasts generated by the trialled solution, especially in terms of flood threat modelling, improved decision-making on deployment of specialised equipment and human resources. This facilitated the organisation of logistics; for example, the disaster managers benefitted from the detailed information on the flood forecast for planning of mitigating measures (like emergency dikes or pumps).

In the flood phase, an actual flood mask (aerial image of the flooded area) was provided. The flood mask supported decision-making on the deployment of human resources and equipment. In future cases, aerial or satellite imagery of flooding can improve simulation performance.

Net-Centric Data Exchange Improves Information Between Relevant Parties and Thus Improves the Shared Understanding of the Current Situation

Net-centric data exchange is an important factor determining the effectiveness of the disaster management process. Situational awareness is obtained by sharing information in real-time between all actors involved. The lack of a net-centric information exchange necessitates bilateral contacts, which, as demonstrated during baseline in the trial, are time-consuming and error-prone processes. The use of the new solution resulted in more detailed information, based on the best (actual) data presented in an objective manner. The net-centric information exchange provides a shared situational assessment, thanks to the use of such results of modelling as flood maps and cascade-effects prediction. The advantages of net-centric information exchange in the trial are as follows:

- The information was shared instantaneously and continuously; all organisations used the most up-to-date information.
- The information between safety regions and the stakeholders involved (i.e. Water Board, power company, public transport) were exchanged faster. Information was shared digitally, especially the flood maps (in contrast to phone or mail communications, followed by importing this information into the systems).
- All organisations used the same data (source), so no mix-up of different versions of information influenced the crisis management action.
- The information was unambiguous since the organisations shared their information directly. There was no person in between that might distort the information.

The Modelling Tool Supported the Planning and Management of a Large-Scale Evacuation Taking into Account Real-Time Traffic Information

The flood simulation was highly useful for indicating collection points for the evacuees, locating evacuation assembly points in areas not prone to inundation. As demonstrated during the trial, the practitioners based their decisions on the available flood modelling. The main advantages are the detailed information on the forecast flood and substantiation of the effects of proposed protective measures (like emergency dikes or pumps).

Synthesis

The trials in Poland and the Netherlands both tested the same solution for flood simulation to close the gap of the lack of real-time information on flood development. Real-time information is deemed crucial to decision-making in case of flood. The trial in Poland used a chemical spill scenario, while the trial in the Netherlands used a scenario of large-scale flooding in a densely populated urban area. The important features in the solution were (1) fast flood modelling with a high spatial resolution, (2) simulation of mitigating measures, and (3) visualisation of model outputs. In the Polish trial, the solution was used for determining a strategy on evacuation of buildings at risk of flooding and measuring the process with key performance indicators valid for the efficiency and efficacy of the newly proposed innovation line. These parameters have been analysed in comparison with the baseline run results. The Netherlands trial dealt with flooding on a much larger scale and used the trialled solution in combination with other innovative tools like net-centric information management and evacuation modelling. Both trials were tabletop exercises on a tactical level with a realistic scenario.

For the Polish trial, the flood modelling solution proved its efficiency since decisions on the evacuation strategy were made in a shorter time, without any discrepancy on its efficacy in comparison to the baseline run. In the Dutch trial, the solution was operated by accepted experts (namely, the Water Board). The evacuation strategy decision was made on more detailed flood information, including water arrival times and water depth to determine safe havens during flooding and collection points for evacuation. Furthermore, the effects of mitigation measures (like emergency dikes) were assessed to protect vital infrastructure in the areas at risk of being flooded. In both trials, the visualisation of the flood area was considered a huge benefit of using the solution. The flood ensured a shared understanding of the flood and prevented miscommunication on the flooded area (common operational picture).

In both trials, the practitioners' and the observers' perception of the trailing conditions (trial dimension) was positive. The Netherlands trial was scored a little bit better, probably because it benefitted from the experiences of the Polish trial that was conducted earlier. The positive opinions of the practitioners indicate that in both trials there have not been identified any serious biases that might significantly influence the trials' findings. It advocates for the reliability of the trial findings.

With respect to the solution dimension, the results were positive and quite similar in both trials. The one exception was on the ease of use of the flood modelling software. Opposed to the Polish trial, the crisis managers in the Dutch trial scored negative, although the involved practitioners were well trained in using the solution. The explanation for this is the extensive use of modelling mitigation measures in the trial which necessitates a thorough understanding of hydrodynamics to comprehend and explain the flood modelling solution outcomes.

Participants in both trials emphasised the added value of dynamic flood modelling in disaster management, including fast provision of the solution products such as dynamic flood models and their usefulness for decision-making purposes in case of a massive flood.

Conclusions

The trials proved the potential of the dynamic flood modelling software solution to facilitate disaster managers in their key role, which is efficient and effective decision-making in case of a major disaster such as flood. The solution implemented in the decision-making protocols as the innovation line demonstrated its superiority over the legacy tools (the baseline), such as static models. In quantitative research, the solution proved to have the potential to shorten the decision-making time, positively influencing teamwork by making the decision process more structured and less stressful through the provision of data that significantly reduces uncertainty. Moreover, the dynamic flood modelling software positively influenced resource planning of the civil protection assets used in the flood response, facilitated information management among involved stakeholders, and improved efficiency of evacuation measures.

Acknowledgements

This paper has received funding from financial resources for science in 2018–2023 granted by the Ministry of Science and Higher Education of the Republic of Poland, grant no. 3851/7.PR/17/2018/2.

References

Burston, J., Ware, D., & Tomlinson, R. (2015). The real-time needs of emergency managers for tropical cyclone storm tide forecasting: Results of a participatory stakeholder engagement process, *Natural Hazards*, 78, 1653–1668. https://doi.org/10.1007/s11069-015-1794-7.

de Jong, M., & Helsloot, I. (2009). The effects of information and evacuation plans on civilian response during the National Dutch crisis exercise "Waterproef." *First International Conference on Evacuation Modeling and Management ICEM2009*, The Hague.

EU (European Union). (2007). *Directive 2007/60/EC of the European Parliament and of the Council of 23 October 2007 on the Assessment and Management of Flood Risks*. https://eur-lex.europa.eu/legal-content/EN/TXT/PDF/?uri=CELEX:32007L0060&from=EN. Accessed 30 August 2020.

EU (European Union). (2019). *Decision (EU) 2019/420 of the European Parliament and of the Council of 13 March 2019 (Amending Decision No 1313/2013/EU on a Union Civil Protection Mechanism)*. https://eur-lex.europa.eu/legal-content/EN/TXT/PDF/?uri=CELEX:32019D0420&from=EN. Accessed 30 August 2020.

Foks-Ryznar, A., Ryzenko, J., Trzebińska, K., Tymińska, J., Wrzosek, E., & Zwęgliński, T. (2019). Trial as a pragmatic and systematic approach for assessing new solutions in crisis management and rescue operations. *Proceeding of SPIE 11176, Photonics Applications in Astronomy, Communications, Industry, and High-Energy Physics Experiments 2019, 111763T*. https://spie.org/Publications/Proceedings/Paper/10.1117/12.2537925. https://doi.org/10.1117/12.2537925.

Fonio, C., & Widera, A. (Eds.). (2020). *Trial Guidance Methodology Handbook. DRIVER+ (Driving Innovation in Crisis Management for European Resilience)*, Brussels, 2020. https://tgm.ercis.org/.

Ghazali, J. N., & Kamsin, A. (2008). A real-time simulation and modeling of flood hazard. *Proceedings of the 12th WSEAS International Conference on Systems*, 22–24 July 2008, Heraklion, Greece. https://dl.acm.org/doi/proceedings/10.5555/1513605.

Hadimlioglu, I. A., King, S. A., & Starek, M. J. (2020). FloodSim: Flood simulation and visualization framework using position-based fluids. *ISPRS International Journal of Geo-Information*, 9(3), 163. https://doi.org/10.3390/ijgi9030163.

Hendriks, M., Vullings, E., van Campen, S., & Hameete, P. (2019). An interoperability framework for trials and exercises. *Proceedings of the 30th International Forum for the Military and Civil Simulation, Training and Education Community (ITEC)*, 13–16 May 2019, Stockholm, Sweden.

Horita, F. E. A., Albuquerque, J. P., & Marchezini, V. (2018). Understanding the decision-making process in disaster risk monitoring and early-warning: A case study within a control room in Brazil. *International Journal of Disaster Risk Reduction*, 28, 22–31. https://doi.org/10.1016/j.ijdrr.2018.01.034.

Horita, F. E. A., Albuquerque, J. P., Marchezini, V., & Mendiondo, E. M. (2017). Bridging the gap between decision-making and emerging big data sources: An application of a model-based framework to disaster management in Brazil. *Decision Support Systems*, 97, 12–22. https://doi.org/10.1016/j.dss.2017.03.001.

ICPDR (International Commission for the Protection of the Danube River). (2020). *Climate Change Adaptation Strategy*. www.icpdr.org/flowpaper/viewer/default/files/nodes/documents/icpdr_climatechangeadaptationstrategy_2.pdf.

ISOK. (2020). https://wody.isok.gov.pl/imap_kzgw/.

Izumi, T., Shaw, R., Djalante, R., Ishiwatari, M., & Komino, T. (2019). Disaster risk reduction and innovations. *Progress in Disaster Science*, 2. https://doi.org/10.1016/j.pdisas.2019.100033.

Jamieson, S. (2004). Likert scales: How to (ab)use them. *Medical Education*, 38(12), 1217–1218. https://doi.org/10.1111/j.1365-2929.2004.02012.x.

Kolen, B., Kutschera, G., & Helsloot, I. (2009). A comparison between The Netherlands and Germany of evacuation in case of extreme flooding. *Urban Flood Conference*, Paris. https://citeseerx.ist.psu.edu/viewdoc/download?doi=10.1.1.567.3827&rep=rep1&type=pdf.

Kolen, B., & van Gelder, P. H. A. J. M. (2018). Risk-based decision-making for evacuation in case of imminent threat of flooding. *Water*, 10(10). https://doi.org/10.3390/w10101429.

Kulmala, I., Salmela, H., Kalliohaka, T., Zwęgliński, T., Smolarkiewicz, M., Taipale, A., & Kataja, J. (2016). A tool for determining sheltering efficiency of mechanically ventilated buildings against outdoor hazardous agents. *Building and Environment*, 106, 245–253. https://doi.org/10.1016/j.buildenv.2016.06.034.

Lawrence, J., Blackett, P., & Cradock-Henry, N. A. (2020). Cascading climate change impacts and implications. *Climate Risk Management*, 29. https://doi.org/10.1016/j.crm.2020.100234.

Liu, Y., Zhang, W., & Cui, X. (2012). Flood emergency management using hydrodynamic modelling. *Procedia Engineering*, 28, 750–753. https://doi.org/10.1016/j.proeng.2012.01.802.

Milacic, R., Zuliani, T., & Scancar, J. (2012). Environmental impact of toxic elements in red mud studied by fractionation and speciation procedures. *Science of the Total Environment*, 426, 359–365. https://doi.org/10.1016/j.scitotenv.2012.03.080.

Nkwunonwo, U. C., Whitworth, M., & Baily, B. (2020). A review of the current status of flood modelling for urban flood risk management in the developing countries. *Scientific African*, 7, e00269. https://doi.org/10.1016/j.sciaf.2020.e00269.

Price, P. C. (2015). Quasi-experimental research. In P. C. Price, R. S. Jhangiani, & I. A. Chiang (Eds.), *Research Methods of Psychology* (pp. 125–142). Washington, DC: The Saylor Foundation.

Puppim de Oliveira, J. A. (2009). The implementation of climate change related policies at the subnational level: An analysis of three countries. *Habitat International*, 33(3), 253–259. https://doi.org/10.1016/j.habitatint.2008.10.006.

Ryzenko, J., & Smolarkiewicz, M. M. (2010). Space-enabled information environment for crisis management. Scenario-based analysis and evaluation in an operational environment. *Acta Astronautica*, 66(2010), 33–39. https://doi.org/10.1016/j.actaastro.2009.03.088.

Schröter, E., Kiefl, R., Neidhardt, E., Gurczik, G., Dalaff, C., & Lechner, K. (2020). Trialing innovative technologies in crisis management – "Airborne and terrestrial situational awareness" as support tool in flood response. *Applied Sciences*, 10, 3743. https://doi.org/10.3390/app10113743.

Setola, R., & Theocharidou, M. (2017). Modelling dependencies between critical infrastructures. In R. Setola, V. Rosato, E. Kuriakides, & E. Rome (Eds.), *Managing the Complexity of Critical Infrastructures. A Modelling and Simulation Approach* (pp. 19–42). Cham: Springer. https://doi.org/10.1007/978-3-319-51043-9.

Shannon, C. E. (1948). A mathematical theory of communication. *The Bell System Technical Journal*, 27, 379–423.

ten Brinke, W. B. M., Bannink, B. A., & Ligtvoet, W. (2008). The evaluation of flood risk policy in The Netherlands. *Proceedings of the Institution of Civil Engineers – Water Management*, 161(4), 181–188. https://doi.org/10.1680/wama.2008.161.4.181.

Toth, J. I. (2013). Key actors of the red sludge disaster in Hungary. In L. Wetrs, P. Taylor, & A. Michelot (Eds.), *Confronting Ecological and Economic Collapse. Ecological Integrity for Law, Policy and Human Rigths* (pp. 143–155). London: Taylor and Francis Group.

van Aalst, M. K. (2006). The impacts of climate change on the risk of natural disasters. *Disasters*, 30(1), 5–18. https://doi.org/10.1111/j.1467-9523.2006.00303.x.

van Duuren, C., & Zwęgliński, T. (2019). Analysis of the dutch crisis response system in the context of the common European union resilience. *Internal Security*, 11(2), 55–64. ISSN 2080-5268, https://doi.org/10.5604/01.3001.0013.8205.

Vermeij-Van den Braak, E., Vermeulen, C. J. M., & Birnie, J. A. (2009). Toolbox with ideas to improve the effectiveness of our roads during an evacuation. *1st International Conference on Evacuation Modeling and Management ICEM 2009*, 23–25 September 2009, Scheveningen.

Vermeulen, C. J. M., Bruggers, M., & Nagel, R. (2012). Disaster management training environment. *Flood Risk Conference 2012*, 20–22 November 2012, Rotterdam.

Vullings, E., de Kleermaeker, S., & Vermeulen, C. J. M. (2011). Serious gaming for training of multiple parties in disaster management. *ITEC*, 10–12 May 2011, Köln.

Weigel, C., Schäffner, G., Kattwinkel, P., Viehweg, P., Hehle, M., & Bergmann, D. (2010). Technologies for exhaust after treatment testing under real conditions. *MTZ Worldwide*, 71, 48–53. https://doi.org/10.1007/BF03228001.

Wiśniewski, B., Kozioł, J., & Falecki, J. (2017). *Podejmowanie Decyzji w Sytuacjach Kryzysowych*. Szczytno: Wyższa Szkoła Policji.

Zhou, L., Wu, X., Xu, Z., & Fujita, H. (2017). Emergency decision making for natural disasters: An overview. *International Journal of Disaster Risk Reduction*, 27, 567–576. https://doi.org/10.1016/j.ijdrr.2017.09.037.

Zwęgliński, T. (2015). European civil protection response potential in floods. In T. Zwęgliński, P. Gromek, M. Gikiewicz, & A. Prędecka (Eds.), *The BaltPrevResilience Monograph Report on Awareness Raising, Bridging and Building Community Resilience in the Baltic Sea Region* (pp. 146–168). Warszawa: Wydawnictwo Arka.

Zwęgliński, T. (2019). Metodologiczne podstawy teorii ewaluacji ćwiczeń w zakresie zarządzania kryzysowego i ochrony ludności. In B. Wiśniewski & T. Zwęgliński (Eds.), *Racjonalizacja zarządzania jednolitymi formacjami umundurowanymi odpowiedzialnymi za bezpieczeństwo wewnętrzne. Tom IV* (pp. 252–277). Warszawa: Szkoła Główna Służby Pożarniczej. https://bibliotekasgsp.locloudhosting.net/files/original/d3158537accb2679dfc764929efc9ea6.pdf. Accessed 03 Nov 2020.

Zwęgliński, T., & Gawroński, W. (2014a). Case study incident – river flooding. In E. Obcarskaite & A. J. Olsson (Eds.), *Blue Book Two – Macro-Regional Approach through Sharing Experiences in Flood Prevention in the Baltic Sea Region* (pp. 28–31). Stockholm: Council of the Baltic Sea States. www.bsr-secure.eu/wp-content/uploads/2015/03/Blue-Book-Two.pdf. Accessed 20 Nov 2020.

Zwęgliński, T., & Gawroński, W. (2014b). Identified gaps and future priorities for flood prevention practice in the Baltic Sea Region. In E. Obcarskaite & A. J. Olsson (Eds.), *Blue Book One – Mapping Flooding and Flood Prevention in the Baltic Sea Region* (pp. 40–41). Stockholm: Council of the Baltic Sea States. www.bsr-secure.eu/wp-content/uploads/2015/03/Blue-Book-One.pdf. Accessed 15 Nov 2020.

Zwęgliński, T., & Morgado, C. (2018). Analytical approach to cooperation of the State Fire Service with the Police in Poland in the context of brief comparison with Portuguese reality. *Internal Security*, 10(1), 161–182. https://doi.org/10.5604/01.3001.0012.7498.

Zwęgliński, T., Smolarkiewicz, M., & Gromek, P. (2020). *Efekt kaskadowy współczesnym wyzwaniem zarządzania kryzysowego*. Warszawa: Szkoła Główna Służby Pożarniczej.

9

ON THE PRACTITIONER-DRIVEN USE AND MISUSE OF SIMULATION AS DECISION SUPPORT

Lessons from a Mass Evacuation Trial

Adam Widera, Michael Middelhoff, Niclas Rotering, Claas Cassens, and Bernd Hellingrath

Introduction

Simulation has been widely identified as a tool to tackle the problem of emergency evacuation in response to disasters (Kunz & Reiner, 2012; Leiras et al., 2014). Simulation is used in this context to investigate different aspects of an evacuation, like evacuee behaviour, traffic management, shelter location placement, or the use of transport modes. Following the existing simulation guidelines and best practices (see, for example, Borshchev, 2013; VDI, 2000), the main objective of these simulations is to evaluate different strategies, based on a given scenario that is transferred into a simulation model. While the amount of research in this field is increasing, its relevance in practice is still limited (Kunz et al., 2017). An area that is often overlooked or not documented is the involvement of practitioners in the creation and use of the simulations.

In this paper, we present a critical case study about a practitioner-driven simulation approach for the management of mass evacuations in urban areas. The underlying research question asks, how can speculative simulation foster situation awareness and sensemaking in multi-agent evacuation scenarios?

The structure of the paper is organised as follows: We first present related works on how simulation is intended to be used for supporting decision and sensemaking, followed by an overview of existing simulation approaches for the specific goal supporting evacuation planning and management. On this basis, we describe the novel evaluation approach of the TGM and how the simulation environment has been applied in the respective trial as a case study. In a next step, both the actual simulation results and the evaluation from the involved practitioners will be analysed. The observed phenomena on how the TGM process changed the traditional objectives and intended use of simulation through a systematic practitioner involvement will

DOI: 10.4324/9781003256977-15

be critically reflected in the discussion. The paper closes with a short summary, a discussion of several limitations, and the formulation of a research agenda covering a potential paradigm shift in the area of simulation design and application.

Related Work

The amount of research papers in the area of simulation of humanitarian logistics and supply chains has risen significantly over the previous years. To be more precise, from 2011 to 2017, the amount of papers published increased from 22 in 2011 to 95 in 2017, with more papers being published every following year from 2012 on (Behl & Dutta, 2019). Additionally, the review of Behl and Dutta (2019) states that dynamic modelling and simulation is a prominent area of research in the humanitarian supply chain sector. This is reinforced further by reviews such as Banomyong et al. (2019), who found most of their reviewed papers from 2005 to 2016 to use simulation and modelling. However, the authors conclude that cross-organisational case studies and surveys should be used more often to increase knowledge in the field (Banomyong et al., 2019) and that the needs of stakeholders as of preparedness, response time and usage of resources needs to be further analysed (Behl & Dutta, 2019).

These findings were complemented through a separate literature review covering simulation-based evacuation approaches within the SCOPUS database. The initial search string consisted of a combination of "humanitarian logistics" or "disaster relief" with the terms "simulation" and "evacuation." To limit the results to decision support systems that strive to enhance – in the best case – practitioner's capabilities to make decisions, the terms "decision support" were added as well.[1] The timeframe for the papers was set to 2010 or newer to only include recent publications in the search results. The initial results led to 18 papers, ranging from 2011 to 2020. After screening the results based on title and abstract, ten papers remained as relevant or related to our subject.

Among the literature results are studies that focus on a single aspect of evacuation such as described by Ozdamar (2011) where the organisation and usage for helicopters in disaster relief operations after an earthquake is planned and optimised. In this case, helicopters are used for the evacuation of wounded and the delivery of medicines, vaccines, and medical equipment, like blood bags. A mathematical model is proposed, as well as a route management procedure to simulate the usage of helicopters and aid contingency planning. Similarly, Andreeva-Mori et al. (2015a, 2015b) proposed a particle swarm optimisation in combination with a greedy-search algorithm to optimise utilisation of helicopters in disaster relief operations. The centre of their research is the assignment of helicopters to evacuation of injured or isolated people but extension to goods transportation is planned. The solution is capable of supporting decision-makers in assigning helicopters with different characteristics to missions so that the time of each helicopter can be best utilised according to mission requirements and constraints like flight distance and helicopter capacity.

A second set of studies can be summarised under the category of agent-based simulation studies. Four of the papers utilise this simulation technique in different forms to inspect and analyse different aspects of evacuation. Yang et al. (2020) built a multi-agent system to analyse the behaviour of pedestrians during urban waterlogging in China. The developed simulation tool analyses the impact of shelters, the route choices of pedestrians and further rescue operations once shelters are reached. Their findings showed that the pedestrian herd mentality and the distribution of shelters have a significant impact on the success of post-disaster evacuation. Sopha et al. (2019) built an agent-based model as well. The authors examined different approaches and strategies for last-mile relief delivery, as well as the impact of early and self-evacuation. They have been able to reproduce the general pattern of internal displaced persons behaviour in context of the 2010 Mount Merapi eruption, which served as a data basis for this study. With their findings, they have been able to draw practical implications for the design of evacuation plans and profound decisions on efficient and effective last-mile relief distribution. The third paper utilising agent-based modelling, in this case in a hybrid model in combination with discrete-event simulation, is the study of Na and Banerjee (2019). The authors composed a GIS-based agent-based discrete-event simulation modelling framework to gather insights on the traffic assignment and the configuration and location of resources such as vehicles and relief facilities in the case of no-notice natural disasters. The framework has been tested in various experiments based on the city of San Francisco and great emphasis has been placed on traffic assignment and traffic control. However, the degree to which the system has been developed and tested in collaboration with emergency management personnel is limited. The last paper of this category is the paper of Lee et al. (2017) about the RealOpt real-time decision support suite. RealOpt offers tool-based decision support for mass dispensing, emergency preparedness, and biosurveillance. Primarily, the solutions offer decision support for response facilities and locations, resource needs and supply tools and real-time population flow. The application areas expand over biodefence where it has been applied to H1N1 regional planning and earthquake disaster relief, like in Haiti in 2010, and the evacuation tools have been used extensively in 2011 after the Fukushima disaster. A standalone version of the evacuation module, RealOpt-Evacuate, is in development to support planning and training of decision-makers.

The third category of the identified papers are not directly related to evacuation simulation but can be seen as supportive methods to the analysis of pedestrian and vehicle movement during natural disasters and thus offer great potential to increase the quality of assumptions made for further simulations. The first studies of this subject to be mentioned here are by Wang & Taylor (2016a, 2016b). In these two studies, the authors examined human mobility during and its perturbations under the influence of natural disasters such as hurricanes. In their first study (Wang & Taylor, 2016b), they developed a simulation of the human mobility behaviour, based on animal movements, under three different scenarios, the imposition of evacuation zones, the occurrence of flooding, and the abrupt failure of certain infrastructures. Their second study (Wang & Taylor, 2016a) again focused on human mobility during a hurricane but the data basis was expanded by multiple resources, including the US Census data, Twitter data, and Google Maps to better simulate commuting routes and points of interest in the city of New York. The last paper of this category goes in a similar direction: Osogami et al. (2013) developed a simulator with the X10 language to simulate behavioural models of traffic for entire cities to evaluate the transpiration system regarding quick evacuation, rescue distribution of relief supplies, and other activities reducing the impact of natural disasters. The agent-based approach simulates individual drivers with real probe-car data. The final simulation delivers data and insights into how resilient the transportation system of a city is and offers data for evaluation of its capabilities. These include rescue, distribution of relief supplies, and other mitigation and recovery efforts after natural disasters.

As it can be concluded, the simulation-based specific decision support to mass evacuations in case of floodings are not covered in the current literature. While important insights can be concluded, such as when looking at behavioural dynamics of the evacuees, the developed approaches in particular neglect the decision-makers needs when planning and executing relatively unexpected and the interdependency between agencies, decisions, and consequences. Accordingly, a practitioner-driven and simulation-based decision support for mass evacuations in urban flooding situations is described in the following.

Research Design and Methodology

While the applied simulation environment has been developed in a broader action research project (Widera et al., 2015; Widera et al., 2017; Widera et al., 2019), its specific application followed the recently developed trial guidance methodology (TGM) within the EU-funded project *D*riving Inno*v*ation in Crisis Management for *E*uropean *R*esilience (DRIVER+). The case study was executed in a so-called trial at the safety region of The Hague in the Netherlands (see Chapter 5). The trial was executed in May 2019, and it covered a flash flood scenario simulating a lock breach caused by severe weather conditions. The actively involved crisis management (CM) practitioners covered all relevant authorities and response organisations in the respective safety region, ranging from representatives of the town administration, over the water board, police, or military, up to fire services or the health sector. Thus, all inputs to the simulation model reflect practitioner realities (e.g. all actors, organisational structures and interrelations, and data on the available resources and capacities and the corresponding costs) as they were determined by the respective responsibilities. The overarching goal of the trial was to assess the potential added value of innovative CM solutions, such as the simulation model. It is worth to be noted that the identification of those innovations has been also decided by the involved CM practitioners. The novel assessment approach, the TGM, required a realistic embedding of the simulation model into the current operating legacy systems, both organisation and technology-wise (Fonio &

Widera, 2020). Hence, the simulation model and its application were connected to the relevant systems in place, which ensured a high level of realism not only of the "world" which has been simulated but also of how and by whom the simulation would be used in reality.

Case Study

The simulation applied in the evacuation scenario introduced earlier is modelled in a simulation environment specialised towards the assessment of humanitarian logistics operations (Widera et al., 2017). Given the various autonomous actors and to reproduce the practitioner-driven perspectives, an agent-based discrete-event simulation maps the behaviours following demand and supply relationships. The overall objective is to fulfil demands as efficiently and effectively as possible given a network configuration. Effectiveness describes the fact that logistics is the main driver for delivering the right relief items and services at the right time, amount, and quality meeting the beneficiary's needs, whereas efficiency reflects that logistics can be considered as the biggest cost factor in disaster relief (Schulz, 2008; Tomasini & Van Wassenhove, 2009). The simulation environment focuses on the four logistical challenges of agent collaboration, route planning, prepositioning of facilities and supplies, and bottleneck analysis. Therefore, the model base data maps the city infrastructure in terms of locations and availability of resources of the involved actors. That includes locations and capacities of hospitals, public transport, and road infrastructure; locations of police and fire stations and buildings available as potential evacuation sites; and other locations of interest utilised in urban crisis response. Users of the simulation configure a scenario with the flood-affected city area, available first responder resources, and designated evacuation camps.

Within the trial, it was possible to import flood maps, which forecast the water propagation and use real-time routing services considering road utilisations provided by other trial participants. The model automatically adapts the available infrastructure and evacuation needs based on the population in the affected area, required hospital evacuations with special medical needs, and unavailable sections in the road and transport network. Prior to the execution, the scenario is validated according to the evacuation need and the provided capabilities – for example, if sufficient evacuation sites are available for the affected population. A scenario can be started with insufficient capabilities but will inform about resulting overutilisation in advance. Within a simulation run, the actors aim to fulfil their demand-supply relationships, like the demand of inhabitants to be picked up at various locations in the affected area or the hospitals to transport patients to other medical facilities. Key performance indicators of the simulation are the progress of the evacuation over time up to the required total time to evacuate the affected area, the utilisation of resources and infrastructure, and estimations on transport and food and water supply costs. In detail, a complete record of all activities of agents is logged for further analysis. Both the simulation environment and the model of the presented case were developed in close collaboration with practitioners working in disaster relief and representatives from the actors in the flooding scenario. During the development and application of the simulation, practitioners showed an increasing interest in the potential to model and test speculative scenarios, although these are less likely in reality and only assumptions on required data inputs are available.

The Trial Context

The trial context features a fictional flooding event in the safety region of The Hague, Netherlands. It was executed in 2019 in the city of The Hague as a table-top scenario. The starting point are weather forecasts predicting strong winds and a storm joint with rising sea levels. Crisis management protocols are initiated, and the response forces are informed. The trial was split into two major parts. Starting with these initial forecasts, the preparedness phase dealt with all activities before the actual flood was taking place. The event of the actual flood turned out more severe, leading to a breach of the lock of Scheveningen. The following response phase dealt with all activities during the actual flood. During both phases, several groups were tasked with individual assignments depending on their role in crisis management. The evacuation of the urban area is led by military forces in collaboration with fire brigades, police, medical personnel, and the local government. Military was tasked to prepare possible evacuation actions and to inform other parties in crisis meetings during the preparedness phase. In the response phase, they were tasked to plan evacuation activities to be executed.

The Simulation Environment

The simulation applied in the evacuation scenario introduced earlier is modelled in the simulation environment HumLogSIM, which is specialised towards the assessment of humanitarian logistics operations and built in the simulation software AnyLogic as part of a broader action research approach (Widera et al., 2015; Widera & Hellingrath, 2016, Widera et al., 2017; Widera, 2022).

HumLogSIM is a performance assessment platform that serves logistic processes in crisis management. It can operate on both current operational logistics networks and fictional (planned) network configurations. The functionality comprises strategic planning support and tactical and operational decision support by assessing and comparing the network performance under given situations and realistic crisis management actions. The main target groups are higher-level crisis managers, coordinating field operations, and responder networks in their strategic network design and configuration. The name HumLogSIM refers to the humanitarian logistics context, thereby representing planning, implementation, and controlling of efficient and cost-effective flow and storage of goods, materials, and equipment, as well as related information, from point of origin to point of consumption, aiming to meet the beneficiaries' requirements under given resource capacities. HumLogSIM is an adaptable simulation environment for discrete event-based and agent-based simulations. It represents crisis management activities within and between humanitarian organisations on the way to a defined objective whilst assessing the overall performance.

HumLogSIM is applicable to different settings and operational levels of crisis management. On a strategic level, it supports creating or updating a relief network of one or multiple responder organisations. The simulation environment can analyse different possible network configurations in terms of facility locations, relief goods stock values, warehouse capacities, human resources, and others. By executing so-called "what if" scenarios, the performance of a planned network can be simulated under a fictive crisis event, which can make use of historical crisis event data to represent a realistic scenario.

On a tactical level, these capabilities can also be utilised to compare possible execution plans and to estimate their expected outcomes. Given a relief network of one or multiple responder organisations in an actual crisis event, crisis management commanders can test different strategies in addressing the crisis without any influence on real-world resources. The simulation capabilities can thereby be freely adapted to the needs of an individual organisation, like the construction of sandbag barriers, evacuation of the population, or allocation and transportation of human resources. The simulation will take all available resources and capacity restrictions into account to provide a complete and sophisticated execution protocol, which can then be used by command staff to compare execution strategies.

The simulation environment consists of three parts. A collection of models and data that was built in collaboration with domain experts and partners from the Safety Region Haaglanden, a simulation engine that executes the simulation model with variable input data, and a data analysis and exploration tool that allows easy overviews and deep-dives into the resulting data of an executed simulation run.

FIGURE 9.1 HumLogSIM user interface.

Simulation Model and Data

Given the requirement of the scenario of various autonomous actors and to reproduce the various practitioner-driven perspectives, an agent-based discrete-event simulation model maps the behaviours of actors following demand and supply relationships. The simulation model is focused on the four logistical challenges of agent collaboration, route planning, prepositioning of resources, and bottleneck analysis. Therefore, the model base data maps the city infrastructure in terms of locations and availability of resources of the involved actors.

- Agents
  - Type
    - Stationary
    - Moving
    - Organisational
  - Chain
    - Goods
    - Transport/evacuation
- Organisation/communication:
  - BPMN model

There are three basic types of actors involved in the simulation: stationary, moving, and organisational agents. Stationary agents have a fixed location on the map and provide certain resources or capacities to store goods or supply internally displaced persons (IDPs). These are hospitals, IDP camps, police and fire stations, and other locations of interest that are utilised in an urban crisis response. Moving agents are vehicles or IDPs that move between stationary agents. All of them start at a stationary location and either have certain needs that are provided or provide services that are needed by stationary agents. Besides the IDPs, moving agents are, for example, buses, transport vehicles, and police cars. Organisational agents represent larger organisations that coordinate the evacuation processes. Together, they determine the outline of the evacuation process and can request or deliver additional resources to assist in it. These are the regional crisis response organisations from the safety region that plan the evacuation and the surrounding safety regions and supraregional organisations that can deliver additional resources.

The overall evacuation process consists of three semi-dependent sub-processes: a transport chain, a supply chain, and a communication process. The transport chain models the transportation of IDPs from their origin locations at the start of the simulation to the destinations, which are either camps in or near the region at which IDPs are supplied afterwards or locations outside of the region at which IDPs can be considered safely evacuated. In general, there are two different types of routes on which the IDPs can be transported. Either they evacuate themselves by car or they use the public transportation options provided by the emergency response organisations. These public options are transport by bus, train, or ambulance. For the evacuation by train, the existing train and tram networks with 73 stations are utilised and modelled in the simulation model. IDPs can either leave the region via this network or transfer into a bus to be transported to IDP camps. For the transport via bus, the IDPs gather at collection points, which are placed across the whole region at bus stops and train stations. Buses are provided by the regional transportation organisations and are available at six depots across the region. From here, the buses are allocated by demand to the collection points and transport the IDPs from them to IDP camps and secure locations outside of the region. Another mode of transportation is the transport by ambulance. With ambulances, IDPs are transported to hospitals in and outside of the region. The supply chain models the transportation of resources to IDP camps. At IDP camps, IDPs must be supplied with certain emergency goods. Therefore, a typical humanitarian supply chain with the supplier, the logistic centre, cross-docking points, and storage centre is modelled, to deliver the goods. As moving agents, trucks are used to perform the transportation of goods. The communication process is guiding the whole evacuation process. In it, the crisis response organisation from the safety region, which are the ROT, the RPT, and the municipalities, plan and initiate the evacuation. They can also ask for support from neighbouring safety regions and the military. These organisations can deliver additional resources in the form of transport vehicles. Besides locations and resources, the agents also contain detailed models about their resource consumption. That is especially important for vehicles that consume resources depended on their movement and utilisation based on gas and driver costs. Similar costs calculations are performed for needed goods in the supply chain that must be procured.

Simulation Engine

Users of the simulation configure a scenario with the flood-affected city area, available first responder resources, and designated evacuation camps. With these "what if" scenarios, the performance of a planned network can be simulated under a fictive crisis event, which can make use of historical crisis event data to represent a realistic scenario. Within the trial, it was possible to import flood maps, which forecast the water propagation, and use real-time routing services considering road utilisations provided by other trial participants. The model automatically adapts the available infrastructure and evacuation needs based on the population in the affected area, required hospital evacuations with special medical needs, and unavailable sections in the road and transport network. Prior to the execution, the scenario is validated according to the evacuation need and the provided capabilities – for example, if sufficient evacuation sites are available for the affected population. A scenario can be started with insufficient capabilities but will inform about resulting overutilisation in advance. Within a simulation run, the actors aim to fulfil their demand-supply relationships like the demand of inhabitants to be picked up at various locations in the affected area or the hospitals to transport patients to other medical facilities. Key performance indicators of the simulation are the progress of the evacuation over time up to the required total time to evacuate the affected area, the utilisation of resources and infrastructure, and estimations on transport and food and water supply costs. In detail, a complete record of all activities of agents is logged for further analysis. The simulation is executed with the simulation engine AnyLogic, which provides efficient computations, standard model libraries, and a visual interface to monitor and explore the simulation during run time.

Data Analysis and Exploration Tool

To analyse the data generated during the simulation, the open-source data analytics tool Metabase is used. The data from the simulation run is stored in a database and when the run completes a dashboard is automatically generated in Metabase, that gives an overview over the configuration and performance indicators for the simulation run. Performance indicators that are used here show the development of different measures over the time span of the simulation. Measures here are, for example, the number of successful evacuations by transport mode, utilisation rates of the different vehicle types, and costs that occur with the different parts of the evacuation.

Execution

The execution of the trial was split into two major phases. Before the actual flooding event occurred, meteorological forecasts and a rise in the sea levels suggest an increased risk of flooding. The preparedness phase dealt with the collection of information and taking preventive actions while the infrastructure was not yet affected. In this phase, the most critical tasks were to establish a common situational awareness and to forecast potential preventive actions through, for example, simulations. After the event occurred, the forecasts could be improved using fictional flooding simulation from the trial scenario, which illustrated the imagined flooding. This response phase was driven by updating the common operational picture established in the previous phase and deciding on actions to be taken.

The tasks of the simulation environment covered evacuation planning (during the threat and flooding phases) and organisational logistics (such as planning of personnel or calculation of IDP camp-related demands and costs). During the threat phase, the solution was used by two separate teams, the action centre (AC) police and the AC evacuation table. For practitioners, resource planning took into account following functions: identification of needed personnel, operational and logistics support, identification of all required resources, possibility to compare potential evacuation sites, and options for facilitating selection of resources. The criteria were also assessed under the angle of their maturity and relevance to the practitioners. During the response phase, HumLogSIM was used by the AC police (with respect to resource planning) and was assessed by prism of 12 indicators, such as possibility for easier staff planning, possibility to compare evacuation sites and relevance of results obtained, operation and logistic support, and providing the best selection of resources. The highest value was identified for easier staff planning criterion. The practitioners rated that the solution as useful in terms of operations and logistic support, identification of required resources, comparison of evacuation sites, and needed personnel considered. Most of the criteria were also assessed as mature and relevant to the practitioners what additionally reinforce the positive findings for these criteria. The lowest rating was identified on maturity of the best selection of recourses functionality.

HumLogSIM was also trialled towards its potential to develop and assess evacuation strategies through demonstrating ability to assess road conditions, highlight and comment areas, support to efficient route planning, add obstacles information to the operational map, assist dispatching rescue vehicle, and route decision support. This task was done by the AC evacuation table. Trial 4 revealed that the practitioners were satisfied with overall performance of the solution. As it comes to individual functionalities surveyed, the possibility to highlight and comment areas on the operational map is perceived well (score above 0.5); however, the maturity of this functionality is still an issue to be improved. A detailed presentation of the evaluation results of the solution are presented in Chapter 5.

Results and Discussion

The usage of the simulation by the practitioners changed drastically throughout the execution of the trial, which can be seen from the scenarios being executed and the decisions taken afterwards. The preparedness phase before the actual flood followed a quite standard analysis procedure featuring predicted, best-case, and worst-case scenarios based on the provided forecasts on the situation. For that, the practitioners were mainly interested in the overall evacuation time and number of inhabitants evacuated. The model calculated that it will take at least 18 hours to evacuate around 42,000 people from the endangered districts, while having sufficient resources with additional further unused capacities. Using these additional resources, it would be possible to evacuate up to 190,000 people with only a little more time of 22 hours. This is reasoned by the included setup time for the evacuation response forces and the city network infrastructure. Based on this information, the practitioners were quite confident in dealing with the upcoming event. They prepared the evacuation sites, according to the utilisations derived in simulation and shared the outcomes to the other participants. With the surprising element that the flood breaches the lock and thereby a central flooding prevention, the response phase turned out to be significantly more severe than forecasted. In the same way, the interest of the practitioners in the simulation results changed as well. Due to largely affected infrastructure, the model calculated based on the new information an evacuation time of 31 hours for roughly 100,000 people. The practitioners then continued to test more severe scenarios featuring, for example, an unavailable train network or a cut in available buses, extending the time up to 40 hours. They also included overly extensive scenarios of around 250,000 affected people although these were not to be expected. Due to the increasing degree of speculative inputs, the practitioners were less interested in the actual numbers produced but in the behaviour of the network as such in order to support their decision-making. Their main interest was to identify potential points of failure, in order to guide and prioritise optional actions they were considering. The simulation was useful to provide visualised and quantified impressions that helped to weigh possibilities for response actions. On the one hand, this was a clear misuse of the simulation for purposes it was not designed for; on the contrary, the practitioners were aware of that and still found it valuable to just observe the model, which fostered their situational awareness.

Summary and Outlook

Under the emerging phenomenon of practitioner-driven crisis management, we apply a newly developed methodology, the TGM, to simulation studies in disaster relief and conclude a potential paradigm shift to speculative simulations. Our main findings suggest that leaving the rather strict guidelines on executing proper simulation studies (e.g. VDI, 2000) might lead to serious shortcomings in the analysis of the results – that is, evaluation. Surprisingly, the "misuse" of several simulation recommendations, such as incorporating potentially inaccurate data or choosing for an operational use of intentionally strategic decisions, significantly widens the scope of the simulation objectives. The presented case has shown how speculative simulations had a positive impact on practitioners' sensemaking regarding the situational awareness and the following decision-making being supported through higher data-driven evidence.

Several limitations need to be considered when discussing the case study and our findings. First and foremost, it needs to be stressed that the applied theory, the TGM, follows a mixed-research approach. Although this combination enabled the discovery of the added value of practitioner-driven speculative simulation, the results need to be seen as not generalisable. From a qualitative research perspective, we might conclude a certain transferability, which, however, would highly benefit from other case studies covering different socio-cultural contexts (such as other countries or different setups of stakeholders) and especially other CM and civil protection areas (such as fire services or health emergencies). Another limitation refers to the CM context in general. The focus on the CM sector comes along with a too narrow scope when it comes to the emergence of practitioner-centred design and application of humanitarian information systems (see, for example, Turoff et al., 2004; Comes & Van de Walle, 2016). Here, further applications of speculative simulation approaches in other domains are needed in order to validate the findings as a theoretical contribution on the design and use of simulation. Using estimations in simulations turned out to allow for an improved or "soft" data-driven decision-making in CM, but other operational realities certainly imply different requirements and expectations.

However, within the CM domain, these speculative simulations seem to equip the stakeholder with access to more intuitive "what if" questions and answers, which can be realised with less effort compared to strict simulation studies. This also means that the results are less robust and do not provide accurate forecasts for implementations based on these scenarios. Yet practitioners seem to accept this downside given the advantage of faster long- and short-term decision-making in time critical disaster relief operations. This might be caused by the fact that crisis situations are often characterised by dynamic, uncertain, and complex circumstances, which in turn force the decision-makers to refer to their experiences and sometimes even gut feeling (see Comes, 2016). Exploring alternative solutions in a secure and non-operational context becomes more accessible to practitioners and enables them to identify the most critical influencing factors by passing on exact KPIs due to the speculative nature of the simulation.

Note

1 The complete search string applied in Scopus was defined as following: TITLE-ABS-KEY (((humanitarian AND logistics) OR (disaster AND relief)) AND (simulation) AND (evacuation)) AND (decision AND support) PUBYEAR > 2010.

References

Andreeva-Mori, A., Kobayashi, K., & Shindo, M. (2015a). Helicopter mission assignment in disaster relief based on particle swarm optimization. *Proceedings of the AIAA Infotech at Aerospace*. https://doi.org/10.2514/6.2015-0367.

Andreeva-Mori, A., Kobayashi, K., & Shindo, M. (2015b). Particle swarm optimization/greedy-search algorithm for helicopter mission assignment in disaster relief. *Journal of Aerospace Information Systems*, 12(10), 646–660. https://doi.org/10.2514/1.I010362.

Banomyong, R., Varadejsatitwong, P., & Oloruntoba, R. (2019). A systematic review of humanitarian operations, humanitarian logistics and humanitarian supply chain performance literature 2005 to 2016. *Annals of Operations Research*, 283(1–2), 71–86. https://doi.org/10.1007/s10479-017-2549-5.

Behl, A., & Dutta, P. (2019). Humanitarian supply chain management: A thematic literature review and future directions of research. *Annals of Operations Research*, 283(1), 1001–1044. https://doi.org/10.1007/s10479-018-2806-2.

Borshchev, A. (2013). *The Big Book of Simulation Modeling, Multimethod Modeling with Anylogic 6*. Chicago: AnyLogic North America.

Comes, T. (2016). Cognitive biases in humanitarian sensemaking and decision-making lessons from field research. *Proceedings of the IEEE International Multi-Disciplinary Conference on Cognitive Methods in Situation Awareness and Decision Support* (CogSIMA).

Comes, T., & Van de Walle, B. (2016) Information systems for humanitarian logistics: Concepts and design principles. In I. Haavisto, G. Kovacs, & K. M. Spens (Eds.), *Supply Chain Management for Humanitarians. Tools for Practice* (1st ed., pp. 339–352). Philadelphia: Kogan Page.

Fonio, C., & Widera, A. (Eds.). (2020). *Trial Guidance Methodology Handbook. DRIVER+ (Driving Innovation in Crisis Management for European Resilience)*, Brussels, 2020. https://tgm.ercis.org/.

Kunz, N., & Reiner, G. (2012). A meta-analysis of humanitarian logistics research. *Journal of Humanitarian Logistics and Supply Chain Management*, 2(2), 116–147.

Kunz, N., Van Wassenhove, L. N., Besiou, M., Hambye, C., & Kovács, G. (2017). Relevance of humanitarian logistics research: Best practices and way forward. *International Journal of Operations & Production Management*, 37(11), 1585–1599. https://doi.org/10.1108/IJOPM-04-2016-0202.

Lee, E. K., Pietz, F. H., Chen, C.-H., & Liu, Y. (2017). An interactive web-based decision support system for mass dispensing, emergency preparedness, and biosurveillance. *ACM International Conference Proceeding Series Part F128634*. https://doi.org/10.1145/3079452.3079473.

Leiras, A., de Brito Jr, I., Queiroz Peres, E., Rejane Bertazzo, T., & Tsugunobu Yoshida Yoshizaki, H. (2014). Literature review of humanitarian logistics research: Trends and challenges. *Journal of Humanitarian Logistics and Supply Chain Management*, 4(1), 95–130.

Na, H. S., & Banerjee, A. (2019). Agent-based discrete-event simulation model for no-notice natural disaster evacuation planning. *Computers and Industrial Engineering*, 129, 44–55. https://doi.org/10.1016/j.cie.2019.01.022.

Osogami, T., Imamichi, T., Mizuta, H., Suzumura, T., & Idé, T. (2013). Toward simulating entire cities with behavioral models of traffic. *IBM Journal of Research and Development*, 57(5). https://doi.org/10.1147/JRD.2013.2264906.

Ozdamar, L. (2011). Planning helicopter logistics in disaster relief. *OR Spectrum*, 33(3), 655–672. https://doi.org/10.1007/s00291-011-0259-y.

Schulz, S. F. (2008). *Disaster Relief Logistics: Benefits of and Impediments to Horizontal Cooperation between Humanitarian Organizations*, dissertation at the Technische Universität Berlin.

Sopha, B. M., Achsan, R. E. D., & Asih, A. M. S. (2019). Mount Merapi eruption: Simulating dynamic evacuation and volunteer coordination using agent-based modeling approach. *Journal of Humanitarian Logistics and Supply Chain Management*, 9(2), 292–322. https://doi.org/10.1108/JHLSCM-05-2018-0035.

Tomasini, R., & Van Wassenhove, L. (2009). *Humanitarian Logistics*. Palgrave MacMillan, London.

Turoff, M., Chumer, M., Van de Walle, B., & Yao, X. (2004). The design of a dynamic emergency response management information system (DERMIS). *Journal of Information Technology Theory and Application (JITTA)*, 5(4), 3.

VDI. (2000). *VDI 3633 – Simulation von Logistik-, Materialfluss- und Produktionssystemen Grundlagen*, VDI-Guidelines, Verein Deutscher Ingenieure, Düsseldorf.

Wang, Q., & Taylor, J. E. (2016a). Data-driven simulation of urban human mobility constrained by natural disasters. *Proceedings of the Winter Simulation Conference 0*. https://doi.org/10.1109/WSC.2016.7822366.

Wang, Q., & Taylor, J. E. (2016b). Diffusion and simulation of human mobility using online network data to examine mobility constraints. *Construction Research Congress 2016: Old and New Construction Technologies Converge in Historic San Juan – Proceedings of the 2016 Construction Research Congress*, CRC 2016. https://doi.org/10.1061/9780784479827.150.

Widera, A. (2022, in press). *Performance Management in Humanitarian Logistics: Development of a Process-driven and IT-supported Performance Measurement System*. Logos.

Widera, A., Bubbich, C., & Hellingrath, B. (2019). Visualizing humanitarian logistics performance: Design, application and evaluation of an IT-supported balanced scorecard approach. *International Journal of Information Systems for Crisis Response and Management*, 11(1).

Widera, A., & Hellingrath, B. (2016). Making performance measurement work in humanitarian logistics. The case of an IT-supported balanced scorecard. In Haavisto, I., Kovacs, G., &

Spens, K. M. (Eds.), *Supply Chain Management for Humanitarians. Tools for Practice* (1st ed., pp. 339–352). Philadelphia: Kogan Page.

Widera, A., Hellingrath, B., & Bubbich, C. (2015). Humanitarian logistics dashboards – A design-related requirements analysis. *Proceedings of the IEEE Global Humanitarian Technology Conference* (pp. 92–99), Seattle.

Widera, A., Konradt, C., Böhle, C., & Hellingrath, B. (2017). A multi-method simulation environment for humanitarian supply chains. *Proceedings of the International Conference on Information and Communication Technologies for Disaster Management.*

Yang, Q., Sun, Y., Liu, X., & Wang, J. (2020). MAS-based evacuation simulation of an urban community during an urban rainstorm disaster in China. *Sustainability (Switzerland)*, 12(2). https://doi.org/10.3390/su12020546.

SUB-PART 3.3

Situation Awareness Approaches

10

THREE-DIMENSIONAL MODEL AND ORTHOPHOTOMAP'S QUALITY EVALUATION TOWARDS FACILITATING AERIAL RECONNAISSANCE OF FLOOD RESPONSE NEEDS

Tomasz Zwęgliński and Marcin Smolarkiewicz

Introduction

One of the key determinants of optimal management is a relevant decision-making process which leads to rational decisions. The value of these decisions increases with the impact generated by the decision in its consequences. Simply, the higher value is to be gained or protected as a result of the decision, the importance of running an efficient and effective decision-making process increases adequately. Since a decision is always evaluated by a prism of its consequences, the significance of the decision-making process highly grows if we consider it is done for disaster management purposes (Zwęgliński, 2019). Disasters, by definition, are perceived as events which lead to high losses in human life and other important values, such as environment or property. Therefore, building a common operational picture (Römer et al., 2016), which is shared by a broad spectrum of disaster responding agencies on the scene, facilitates the cooperation between them and finally should have led to optimal response (Zwęgliński & Morgado, 2018).

Therefore, surveying potential innovations which might facilitate the decision-making process in disaster realities is an important aspect in order to make critical decisions more relevant – decisions which will lead to achieving acceptable goals, for example, in respect to disaster response efficiency and efficacy while using new technologies (BBK, 2017). Efficiency might mean, for example, on-time decision-making, while efficacy might be assessed by checking if the decisions lead to meeting real and actual needs of the affected population or fulfilling the operational needs (Oddo & Bolten, 2019). One of the key determinants of making such decisions is having access to data which construct a reliable and on time contribution to the decision-making process (Kurz et al., 2011). This contribution is preceded by data collection, their analysis and processing to the form, which makes them usable for the decision-maker's purposes.

DOI: 10.4324/9781003256977-17

Decision-making in disaster management appears as a highly complicated process which is additionally burdened with high uncertainty of working in a complex (Stufflebeam & Zhang, 2017), dynamically changeable environment. Therefore, new technologies or new concepts dearly require to be introduced to decision-making processes (Römer et al., 2016; Oddo & Bolten, 2019; Kurz et al., 2011; Marturano et al., 2020). However, it is crucial to find out beforehand to what extent new solutions might facilitate the overall process of decision-making. For example, it should be figured out how efficient and effective a solution is towards meeting disaster managers' and first responders' needs (Römer et al., 2016; Zwęgliński et al., 2019). In addition, utmost preferably efficiency and efficacy should be quantitatively measured with respect to validity and reliability of the collected data, resulting in an overall evaluation process being in place (Feher Waltz et al., 2010).

Disaster management is an interdisciplinary process involving many stakeholders and, through that, being determined by many dimensions, such as social, organisational, technological, economical, and others. Furthermore, major disasters, understood as high-impact, low-probability events, are not very common phenomena, which do not follow one recurrent manner and frame. Simply, disasters are comparable one to another only to a very limited extent. Disasters always have different contexts, by happening in different societies and geographical locations and challenging different types of systems representing different emergency readiness and society resilience levels. Therefore, creating one testing environment, which mirrors all disasters in a universal manner, is simply impossible. However, setting a disaster-related simulation environment that is sufficient for trailing purposes is still feasible at least to the extent which enables assessing an added value of a solution in a specific context (Hendriks et al., 2019). Due to the multispectral and not repeatable character of disasters, the appropriate method for doing this seems to be a trial approach (Craig & Hannum, 2006; Fonio & Widera, 2020). The approach provides means to address the challenge in an optimal manner.

One of the rapidly developing sectors, which already significantly influence disaster management, is the usage of unmanned aerial vehicles (UAVs) (Gilman, 2014). Drones, including UAVs, are implemented to support situational awareness – for example, by providing information from aerial observation, sampling and/or measuring aerial contamination, transporting necessary items, and so on (Feltynowski & Zawistowski, 2018a). This technology opens a broad spectrum of new possibilities which contribute to disaster management, besides its initiatory use for military purposes (Meyer, 2015; Hodgson et al., 2018). To be able to fulfil given functionality, UAVs are always equipped with appropriate additional tools like cameras for observation, detectors for measuring contamination, etc. Therefore, an UAV is a universal tool which is able to be adapted and combined with other tools in order to fulfil a given task depending on the operational need (Römer et al., 2016) or an identified crisis management gap. One of the basic functionalities of UAVs is aerial observation which is used for reconnaissance missions in disasters, both in rapid mode or in longer term perspectives (Meier, 2020). Drone flights are already being used for these purposes, providing 2D movies or aerial images which feed up decision-makers with utmost updated data increasing their situational awareness (Eyerman, 2018; Soesilo et al., 2017).

However, in recent years, 3D modelling is more and more used as a technology that brings a lot of value in different fields, such as in medicine or spatial planning. Thus, solution providers of a software that possesses the functionality to transform 2D into 3D models are searching for potential new areas where their product might be used. On the other hand, crisis management practitioners do search for best ways to conduct reconnaissance of needs in flooded areas in the most efficient and effective ways. From air, as a first general approximation, space applications may be used (Ryzenko & Smolarkiewicz, 2010); however, a more precise operational picture is highly required. Therefore, there is a strong need for trailing such a type of solution in order to make an attempt to evaluate potential added value of such technology in disaster management. This was the crucial aim for which the trial has been designed, planned, conducted, and evaluated.

The assumption for the trial was that quantitatively measured quality, expressed by efficiency and efficacy of flooded population needs reconnaissance from air, based on analysing 3D model and orthophoto (the innovation line), is better than traditional way of conducting such reconnaissance with use of 2D movie and flight images (the baseline) (Verykokou et al., 2016). The reasoning for the assumption is substantiated by the fact that a 3D model enables, for example, to investigate all sides of buildings' elevations from different angles, which is not possible with a 2D perspective. By that it might positively influence the aerial assessment of needs during massive floods and other major disasters (Meier, 2015).

The results of the trial lead to completing the aim of the chapter, which is defining an added value of 3D modelling and orthophotomapping in supporting the decision-making process on affected population's needs reconnaissance in flooded urban areas.

Materials and Method

The research was designed as a full-scale trial (Hendriks et al., 2019; Foks-Ryznar et al., 2019; Fonio & Widera, 2020; Schröter et al., 2020). In terms of internal validity, the trial method is generally somewhere between correlational studies and true experiments. Nevertheless, the method is perceived as the optimal one to seek for an added value of trialled solutions in the context of disaster management, including their performance in an unpredictable and unrepeatable environment reflecting major disaster realities. In the surveyed case, 2D movie and flight images from a drone flight were considered as a baseline that was compared with the 3D model and orthophotomapping usage (innovation line) for the purpose of flooded areas needing reconnaissance (Zwęgliński, 2020).

The trial had an outdoor character and was carried out in the area of the firefighting training field in Nowy Dwor Mazowiecki (Poland) (DRIVER+, 2021a). Relatively big area, located in the catchment of Vistula and Narew Rivers. It was an optimal location to simulate flooding conditions and through that reflect the reality the most possible way. The session aimed at recognising the needs of the affected population trapped in flooded buildings. Two independent groups (working in the baseline and innovation line) of disaster managers, taking part in the test, were tasked to analyse data gained from the same drone flight. The data were collected out of the same needs location (so-called ground truth); however, provided in different forms to each of the groups that were the following:

- 3D model and orthophotomap.
- 2D movie and flight images.

There were in total 16 disaster managers taking part in the trial from 11 European countries. Each group was furtherly split into pairs (DRIVER+, 2021b), giving in total eight pairs of practitioners. The baseline group of the disaster managers (consisting of three pairs out of eight) got a 2D movie and pictures (Feltynowski & Zawistowski, 2018b), while the teams in innovation line (five pairs out of eight) worked with a 3D model and orthophotomap generated by the trialled solution. The drone flight lasted 13 minutes. According to a comparison study done as the trial approached both the baseline group (three pairs) and innovation line group (five pairs) were not randomly assigned; however, they were reasonably selected to minimise differences between them, such as on age, experience as crisis manager, and so on (Craig & Hannum, 2006). This approach limited potential biases on the trial findings, increasing reliability of the test (Feher Waltz et al., 2010).

The trial triggered shaping flow charts of the decision-making process for both surveyed cases, the traditional one used nowadays, as the baseline, and the one surveyed, implementing the new solution, as the innovation line. What differentiates the two lines is an additional activity to be taken in the innovation line. This extra step relates to generating a 3D model and orthophotomap. It means that potential implementation of the new solution in the future would require changing a standard protocol concerning running aerial assessments in order to enable producing additional analytical products like 3D model and/or orthophotomap (Zwęgliński, 2020).

The trial reflected a standard protocol used in Poland for signalling visually the needs of the affected people trapped in flooded buildings. Hence, affected individuals hang different colours of sheets or other materials on windows and roofs while needing some kind of assistance. Specific need is represented by the colour code identifiable by the first responders as follows:

- White – urgent evacuation.
- Blue – water and food.
- Red – medical assistance.

There were in total 19 different coloured sheetings (white, blue, and red) hung on windows of 12 different buildings located in the area of 26 ha. The total number of available windows was 353 on different sites and levels of the buildings. This simulation constituted so-called "ground truth" for the trial purposes. Moreover, there was an assumption that there could not be more than one sheet hung in a window. In addition, the "ground truth" was neither shared with the practitioners trialling the baseline, nor the innovation line.

Both the practitioners working in the baseline (analysing 2D and movie) and in the innovation line (analysing 3D and orthophotomap) were tasked the same: (1) to identify whether there is (or is not) a sheeting in a window, which meant that some individuals require assistance in the identified place, and (2) to recognise the colour of the identified sheeting that codifies a certain need for assistance.

The baseline and the innovation line groups constituted independent variables, presented in a qualitative nominal dichotomy scale, while the dependent variables, defining the quality of the proposed innovation line, including the trialled solution implementation, were represented by efficiency and efficacy of the decision-making process measured both in the innovation line and, for comparison purposes, in the baseline. In order to define the efficiency and efficacy values for the trialled lines, the key performance indicators (KPIs) in quantitative ratio scale were formulated. They were adapted from the classification model evaluation method and reflected:

- Decision efficiency:
  - KPI 1 – time required to make decision on number, locations, and colours of sheetings.
  - KPI 2 – usefulness of the trialled lines measured as the F-score that is an indicator of binary classification of the tested model (baseline and innovation line), triggered by the decision priority criterion (economy versus safety) taken into account.
- Decision efficacy:
  - KPI 3 – *sensitivity*, also known as recall (the ability of the baseline and innovation line to identify locations of the real needs for assistance).
  - KPI 4 – *specificity* (the ability of the baseline and innovation line to identify locations where assistance is not required).
  - KPI 5 – *precision*, also named positive predictive value, or *PPV* (percentage of locations identified as those that require assistance).
  - KPI 6 – *negative predictive value*, or *NPV* (percentage of locations recognised as not requiring assistance).
  - KPI 7 – *accuracy* (percentage of locations correctly identified as those that require assistance).
  - KPI 8 – *likelihood ratio for a positive test*, or *LR+* (probability of truly positive identifications of a need location [real need location identified] divided by the probability of a falsely positive identification of a need location [a need location identified; however, it is not really there]).
  - KPI 9 – *likelihood ratio for a negative test*, or *LR−* (probability of falsely negative identifications of a need location [real need location is not identified] divided by the probability of truly negative identifications of a need

location [locations not requiring any assistance are falsely identified as they do require it]).

In order to measure and calculate the KPIs, primary data were generated, collected, and recorded during the trial. All the KPIs address the crisis management dimension of the trial since they reflect the quality of the decision-making process in the baseline and innovation line.

Moreover, KPIs 2–9 were transferred, adapted, and newly defined for the purpose of the trial from medical research (Realdi et al., 2008; Gregorius & Śpiewak, 2013; Simon, 2015). In medicine science, there are several parameters (KPIs) which allow the evaluation of performance of testing methods, based on quantitative surveys (Krieg, 1975; Price, 2000; Wians, 2009). Transferring definitions of these KPIs into safety and security research is basically feasible in order to evaluate the quality of any new procedure/solution/protocol used for the purpose of a disaster response, such as by comparison of the efficacy of the existing model (baseline) and newly proposed and tested classification model (innovation line). As such, there are some potential options to be taken into consideration while elaborating on the findings of an aerial reconnaissance. For example, a need might be correctly identified, meaning proper indication of a sheeting presence in a window. Another option might be that the need stays not identified, or an incorrect location is appointed (no need exists in a location while disaster managers falsely appoint the window – for example, as a result of a sun reflection) (Zwęgliński, 2020). Such potential options are elaborated in Table 10.1.

The collected primary data were further used as an input to the algorithms presented in the following list in order to quantify and calculate the value of efficiency and efficacy for both the baselines and innovation line trialled.

TABLE 10.1 Recognition Possibilities of Assistance Needs

Identification Type of Need	*Occurrence*	
	A Need Does Exist	*A Need Does Not Exist*
A need is identified.	True positive (TP)	False positive (FP)
A need is *not* identified.	False negative (FN)	True negative (TN)

TP – the existence of a sheet in window was recognised; FP – the sheet does not exist in a window, but it was recognised; FN – the existence of the sheet in window was not recognised; TN – the sheet does not exist in a window, and it was truly recognised that there was no sheet in the window.

There is a classification table including potential options for the identification or misidentification of sheetings hanged out on the buildings.

For efficiency:

- F-score (KPI 2) enables statistical analysis of binary classification of the trialled lines, taking into consideration the relation between precision (PPV) and sensitivity (recall). For the purpose of the trial $F_{0.5}$- and F_2-scores were analysed in order to figure out the difference between the baseline and the innovation line. The chosen F-scores refer to the importance (weight) given to the following decision criteria: costs of the emergency response generated by the decision taken in the baseline and innovation line (economical aspect) versus responding to the utmost number of individuals who need assistance (safety aspect of the flood-affected individuals). Since the $F_{0.5}$-score gives more weight on precision (PPV) and less on sensitivity, as such it puts more attention on minimising false positives (FP) (1). Hence, $F_{0.5}$-score counts more for evaluation of trialled lines mostly from an economic perspective. Literally, it means that when precision (PPV) is higher, so less false-positive locations are identified, then the first responders are not deployed to falsely identified locations in reality. Therefore, the emergency service is not wasting efforts and finances for responding to false alerts (FP), the same as giving priority to economic criteria.

$$F_{0,5} - score = \frac{\left(1+0,5^2\right) \cdot Precision \cdot Sensitivity}{0,5^2 \cdot Precision + Sensitivity} \cdot 100\% \tag{1}$$

F_2-score, reverse to $F_{0.5}$-score, gives more weight on sensitivity (recall) and less on precision (PPV) (2). Thus, it puts more attention on minimising false negatives and, as such, focuses on decreasing the number of unidentified locations of the flood-affected individuals (FN) – see the following sensitivity (recall) algorithm. Therefore, in F_2-score the key attention is put to maximise the identification of truly existing needs of the affected population. Thus, it justifies an interpretation that similar to $F_{0.5}$-score, F_2-score depicts as well the value of the baseline and the innovation line for comparison purposes. However, in this case, the key priority criterion taken into account in the comparison is safety of the flood-related victims, not the overall cost of the operation.

$$F_2 - score = \frac{5 \cdot Precision \cdot Sensitivity}{4 \cdot Precision + Sensitivity} \cdot 100\% \tag{2}$$

Due to the precedent aim of a disaster response, F_2-score is more important for evaluation of the baseline and the innovation line by comparison since it is more relevant to emergency services' profile of activity. On the other hand, the $F_{0.5}$-score cannot be completely neglected since the emergency response costs are valid as well.

As such, the F-scores parameter is relevant for evaluation of the baseline and the innovation line efficiency.

For efficacy:

- *Sensitivity* (KPI 3) is the ability of the baseline and the innovation line to truly identify the individuals who need assistance. High sensitivity tests carry a higher risk of false-positive results (misidentification of a place as needing assistance while there is not such a signal depicted in a window), while a negative result of such a test is very likely to indicate that the subject does not need assistance (3). The high sensitivity of a line (baseline or innovation line) means the line is particularly useful in ruling out who does not need assistance.

$$Sensitivity = \frac{TP}{TP + FN} \cdot 100\% \tag{3}$$

- *Specificity* (KPI 4) is the lines' ability to genuinely identify individuals who do not need assistance. As a high specificity test is usually associated with a higher risk of false-negative results, a positive test result is likely to indicate the presence of needed assistance (4). Therefore, a high specificity of a line means the line is particularly useful in confirming the assistance needed.

$$Specificity = \frac{TN}{TN + FP} \cdot 100\% \tag{4}$$

- *Precision*, also known as positive predictive value (PPV) – KPI 5 – is the percentage of individuals who genuinely need assistance among all those who are identified by a line (baseline or innovation line) as required assistance (including false-positive misidentifications) (5).

$$Precision, PPV = \frac{TP}{TP + FP} \cdot 100\% \tag{5}$$

- *Negative predictive value (NPV)* – KPI 6 – is the percentage of individuals who genuinely do not need assistance among all those who are identified by a given line (baseline or innovation line) as not requiring assistance (including false-negative misidentifications) (6).

$$NPV = \frac{TN}{TN + FN} \cdot 100\% \tag{6}$$

- *Accuracy* – KPI 7 – is the percentage of individuals correctly qualified (7) by the baseline or the innovation line.

$$Accuracy = \frac{TP + TN}{TP + FP + FN + TN} \cdot 100\% \tag{7}$$

- *Likelihood ratio for a positive test* – KPI 8 – is the probability of genuine identification of the individuals in need of assistance compared to the probability of misidentification of the places as needing assistance while it is not needed in reality (8).

$$LR+ = \frac{Sensitivity}{1 - Specificity} \cdot 100\% \tag{8}$$

- *Likelihood ratio for a negative test* – KPI 9 – is the probability of not identifying places where the assistance is truly needed compared to the probability of proper identification of places where assistance is not needed (9).

$$LR = \frac{1 - Sensitivity}{Specificity} \cdot 100\% \tag{9}$$

Besides collecting the quantitative data, there was also a qualitative approach implemented in the trial. A questionnaire-based survey was conducted to complete the results of the trial (Zwęgliński, 2020) with the disaster managers' perception of the solution and trial dimensions (Römer et al., 2016). The answers were collected in the five ratings on the Likert scale, from strongly disagree (coded −2) to strongly agree (coded +2) (Jamieson, 2004; McLeod, 2021). This process was designed and executed to assess the reliability of the data generated and collected during the trial and, in consequence, the reliability of the results achieved in the crisis management dimension (Eyerman, 2018; Feher Waltz et al., 2010). Finally, the trial involved disaster management practitioners who trialled the baseline and the innovation line and observers who were tasked to monitor the entire process in order to identify potential biases impacting the evaluation.

Results

Efficiency

The results presented by Zwęgliński (Zwęgliński, 2020) show that the average timings of the overall process of decision, starting from data collection, through data processing and data analysing, and finally leading to the decision (Meier, 2020), achieved in the baseline and the innovation line runs are noticeable different (KPI 1). The decision-making process takes more time in the innovation line (75 min.) than in the baseline (53 min.). This delay in the innovation line is heavily burdened by the data processing time, including data conversion and transfer in order to generate 3D model and orthophotomap. This element strictly depends on the size of drone flight pictures to be transferred and might be significantly reduced if the solution had been used in the disaster site that was not the case in the trial or by better internet bandwidth than the one available at the training ground where the trial took place. By that the processing time, mainly transferring the data to the machine producing

3D model and orthophotomap, would definitely decrease. However, the overall time difference between the baseline and the innovation line is still satisfactory considering getting an extra solution (3D + orthophotomap) for flood reconnaissance purposes. All in all, an important finding is that pure decision-making time, excluding data transfer but including analytical effort, occurred to be nine minutes shorter in average for the innovation line working groups (working with 3D and orthophotomap) than it appeared to be in the baseline groups (working with 2D movie and drone images).

Usefulness, as a component of efficiency (Culbertson, 2018), is a value related to cost-effectiveness and as such aimed at maximising the effect obtained. Usefulness of a tested model (solution) is assessed in relation to reference points of a referenced model – in this case, the baseline run results. It allows comparison between what has been achieved so far (the baseline) and what is expected or will be generated when the drone flight protocol is improved (the innovation line). For this purpose, analysing the F-score results appears to be a very good technique to figure out how better or worse is the innovation line with respect to the priority criteria (in the surveyed case related to economic aspects of the disaster response versus safety of the affected population) taken into account during decision-making process. The results are presented in Table 10.2.

TABLE 10.2 F-score Results and Differences for the Baseline and the Innovation Line

Type of Need	*Type of Teams*	*F-score Result*		*Difference*	
		$F_{0.5}$	F_2	$F_{0.5}$ *priority – economic aspect of the disaster response*	F_2 *priority – safety of the affected individuals*
Evacuation (white sheeting)	Innovation line	44.72%	61.33%	3.29%	6.32%
	Baseline	48.01%	67.65%		
Water and food (blue sheeting)	Innovation line	43.36%	58.77%	13.39%	27.40%
	Baseline	56.74%	86.16%		
Medical assistance (red sheeting)	Innovation line	59.63%	92.90%	2.87%	7.10%
	Baseline	62.50%	100.00%		
Total	Innovation line	48.45%	68.86%	6.62%	13.85%
	Baseline	55.07%	82.71%		

There are the values of the F-score calculations for the baseline and the innovation line runs and the differences in the values. The results are based on the data collected during the trial.

Efficacy

Neither the baseline nor the innovation line working groups pointed to all of the "ground truth" locations and colours correctly in their assessments. The main difference is that the groups working with baseline avoided pointing out faulty locations. The innovation line teams made both types of mistakes, overlooking and/or pointing to faulty locations.

The data collected in needs identification resulting in the efficacy analysis of the baseline and innovation line runs are presented respectively in Table 10.3 and Table 10.4. KPIs were calculated taking into account average values for the number of sheeting of different colours (white, blue, red) correctly and incorrectly identified by the baseline working pairs (got 2D movie and pictures) and the innovation line pairs (got 3D model and orthophotomap). Data was generated from the same drone flight.

As shown in Table 10.3 the teams working in the baseline correctly identified all "medical assistance" needs (red sheets), while the "evacuation" needs (white sheets) were the most overlooked by them. All in all, the baseline teams have omitted a need in some cases; however, they have never pointed to faulty location, misleading the needs in reconnaissance.

Table 10.4 presents results of the other KPIs:

1. Average *sensitivity* (KPI 3) (3) of the three baseline runs (79.5%) – each pair one run – was higher than for the five innovation line runs (64.1%). It means that the ability to identify individuals who are in need of assistance appears to be better for the baseline. However, it is worth noting that the differences between the baseline and the innovation line results were slightly lower for "evacuation" and "medical assistance" than for "water and food" needs. More

TABLE 10.3 Results of the Needs Identification – Arithmetic Average from the Trialled Teams

Type of Need	*Type of Teams*	*Ground Truth*	*TP*	*FP*	*FN*	*TN*
Evacuation	Innovation line	7	4.2	2.0	3.3	343.5
(white sheet)	Baseline	8	4.7	0.0	2.8	345.5
Water and Food	Innovation line	6	3.2	0.0	2.8	347.0
(blue sheet)	Baseline	6	5.0	0.0	1.0	347.0
Medical assistance	Innovation line	4	4.2	0.4	0.4	348.0
(red sheet)	Baseline	5	5.0	0.0	0.0	348.0
Total	Innovation line	17	11.6	2.4	6.5	332.5
	Baseline	19	14.7	0.0	3.8	334.5

There are arithmetic average values for true-positive, false-positive, true-negative, and false-negative recognitions presented in comparison to the ground truth of the trial scenario. The table includes data for the baseline and innovation line runs.

TABLE 10.4 Efficacy Evaluation for the Baseline and the Innovation Line

Type of Need	*Type of Teams*	*KPI 3 Sensitivity*	*KPI 4 Specificity*	*KPI 5 PPV*	*KPI 6 NPV*	*KPI 7 Accuracy*	*KPI 8 LR+*	*KPI 9 LR−*
Evacuation (white sheetings)	Innovation line	56.0%	99.4%	67.7%	99.0%	98.5%	96.74	0.44
	Baseline	62.7%	100.0%	100.0%	99.2%	99.2%	∞	0.37
Water and food (blue sheetings)	Innovation line	53.3%	100.0%	100.0%	99.2%	99.2%	∞	0.47
	Baseline	83.3%	100.0%	100.0%	99.7%	99.7%	∞	0.17
Medical assistance (red sheetings)	Innovation line	91.3%	99.9%	91.3%	99.9%	99.8%	795.26	0.09
	Baseline	100.0%	100.0%	100.0%	100.0%	100.0%	∞	0.00
Total	Innovation line	64.1%	99.3%	82.9%	98.1%	97.5%	89.43	0.36
	Baseline	79.5%	100.0%	100.0%	98.9%	98.9%	∞	0.21

There are presented the results of the key performance indicators formulated for the solution evaluation purposes. These are sensitivity, specificity, PPV, NPV, accuracy, LR+, and LR−, which have been calculated in percentages for the baseline and the innovation line runs, respectively, to the flood needs to be identified by the trial participants.

than that, it is to be noted that for both the baseline and the innovation line, the highest sensitivity was measured for "medical assistance" while the lowest for "evacuation" needs.

2. Average *specificity* (KPI 4) (4) for the baseline runs (100.0%) was slightly higher than for the innovation line (99.3%). It means that the ability to genuinely identify individuals who do not need assistance is again better for the baseline than for the innovation line; however, the difference is much smaller in comparison to sensitivity. Another aspect is that the differences between the baseline and the innovation line were slightly higher for "evacuation" needs, while for "water and food" and "medical assistance" needs, the two lines were almost equal.
3. Average *PPV* (KPI 5) (5) for the baseline teams (100.0%) was higher than for the innovation line (82.9%). It means that the efficacy of identification of individuals who genuinely need assistance among all of those who are recognised as requiring assistance is also better for the baseline than for the innovation line. The difference for "evacuation" needs between the baseline teams and the innovation line teams was the highest, while it was mediocre for "medical assistance" needs and actually no difference was measured for "water and food" needs.
4. Average *NPV* (KPI 6) (6) for the baseline (98.9%) and the innovation line (98.1%) was almost the same. It means that the efficacy of identification of

individuals who genuinely do not need assistance among all those who are identified as not requiring assistance is almost the same for the baseline and the innovation line runs. Differences between both lines are similar for all types of needs: "evacuation," "water and food," and "medical assistance" needs.

5. Average *accuracy* (KPI 7) (7) for the baseline (98.9%) and the innovation line (97.5%) was almost the same, and also the results are close to the NPV. It means that the ability to truly identify the individuals in need of assistance (windows with sheets) as well as those who do not need it (windows with no sheet on) is almost the same for the baseline and the innovation line. Taking into account that accuracy seems to be the most important KPI from the perspective of disaster management decision-makers (due to the need for proper identification of places where assistance is required and where it is not), this result is of high value.
6. Interpretation of average *LR+* (KPI 8) (8) and *LR−* (KPI 9) (9) for the baseline and innovation line runs is difficult to interpret due to the fact that these KPIs have no maximum value. Additionally, in the trial setup, LR+ reached infinity for the baseline runs, which makes an interpretation even more difficult.
7. Among KPIs 3–7, taking into account all the results generated by the baseline and the innovation line runs the highest average value was reached by the baseline for *specificity* and *PPV* (100.0%), while the lowest average value was measured for *sensitivity* (64.1%) in the innovation line.

Discussion

Reliability of the Results

As Zwęgliński has already shown (Zwęgliński, 2020), the practitioners trialling the solution and the observers were questioned for their opinions on the perception of the solution performance in the given trial environment. This survey was conducted in order to identify potential biases that might heavily influence the reliability of the data collected in the crisis management dimension. Hence, the practitioners were asked to elaborate on their experiences with the solution (13 out of 16 disaster managers participated in this questionnaire), while the observers were questioned on the observations they made during the solution performance (six observers). The data related to the solution dimension concerned the following aspects and were collected by questionnaires. The respondents had to provide their opinion on the following statement – the solution facilitates any of the following:

- Comprehensive overview of the situation.
- Decision-making process.

The data was collected in five ratings on the Likert scale from strongly disagreeing (coded −2) to strongly agreeing (coded +2) with the given statement (Jamieson, 2004; McLeod, 2021). The results of the survey showed that practitioners positively perceived the solution as the one that facilitates a comprehensive overview of the disaster (average score +1.16) and the decision-making process (+0.90). There was not any single disagreement (opinion below 0) towards formulated arguments.

Further, as Zwęgliński reported (Zwęgliński, 2020) the respondents also evaluated the testing conditions in the trial (trial dimension). This survey was conducted as a kind of data quality assessment (Stufflebeam & Zhang, 2017) in order to monitor if the final results of the trial are not heavily biased – such as by shortcomings in simulation of the disaster management reality – and, as such, the impact reliability of the overall findings from the trial. There were six criteria describing the trial dimension: sufficient number of practitioners to trial the solution, the practitioners' proficiency in trialling the solution, sufficient training on the solution before the trial, clarity of the trial technical setup, clarity of the trial organisational setup, and clarity of the scenario. Again, the respondents were asked for their opinions by questionnaires. All the criteria were evaluated as positive (with an average score from +1.17 to +1.34), which gives a premise to assume that the trial findings are sufficiently reliable for the surveyed trial case. However, it has to be noted that the lowest result among these criteria has been revealed for the criterion "sufficient training on the solution before the trial" that might have had certain impact on the data generated in the innovation line. On the other hand, the highest ratings were given to the criteria: clarity of the organisational setup and clarity of the scenario. It indicates that neither the disaster managers trialling the solution, nor the observers of the trial identified any major problem, which might have heavily biased the achieved trial results besides the issue with the insufficient training on the solution for the practitioners.

Validity of the Results

Decision Efficiency

The efficiency of the overall decision process in the innovation line is significantly burdened by data processing time required to produce 3D models and orthophotomaps. It has rather a negative impact on the decision-making process, especially in the specific and dynamic disaster response conditions. This impact could be effectively mitigated by physical provision of the solution (a computer generating 3D and orthophotomap) to the disaster scene, which is easily doable due to the relatively small dimensions of the tool, and with this measure limiting the data transfer time via the Internet. However, it still might not be enough to achieve better results than in the baseline from the time-wise perspective. Therefore, another way of efficiency improvement should be elaborated. One of them might be development of data processing capabilities, which will make the time of 3D model and orthophotomap production significantly shorter.

Nevertheless, the pure decision-making time, including only the analytical work of the teams, appears to be a positive result of the innovation line. Analysing the 3D model and the orthophotomap in the innovation line was shorter than in the baseline, providing in some cases comparable results.

To conclude, the efficiency of the overall process of the aerial reconnaissance of flood-affected population's needs is mainly burdened by technological limitations of the bandwidth used for data transfer. However, the positive finding is that the efficiency of the critical phase dedicated to clear analytical work is higher for the innovation line (shorter time needed) than for the baseline in terms of time consumption.

Concerning the usefulness of the innovation line, the results prove that for both criteria elaborated with the F-score (economic aspect of disaster response and the aspect safety of the affected population) the innovation line performs worse than the baseline. However, the approach, based on the decision priority given to the costs of the disaster response, provides less arguments to definitively state that the baseline is better. This assertion is proved by the F-scores. Differences between the baseline and the innovation line are smaller for $F_{0.5}$ (economical priority) than F_2 (safety of the affected population priority).

All in all, based on this particular trial, the decision on implementation of the proposed solution is not sufficiently justified in respect to the KPIs' results translating to the solution efficiency in the trialled setup.

Decision Efficacy

As has been already described by Zwęgliński (Zwęgliński, 2020), for the particular trial focused on the flood disaster response, neither the teams working with 3D model and orthophotomap nor the teams working with 2D movie and drone image, indicated all the locations of sheets and their colours properly in 100%. It proves there is still some room for improvement in both evaluated lines of conducting reconnaissance of disaster needs.

Taking into account the evaluation of the efficacy of the trialled innovation line, the following discussion points are to be formulated:

- Efficacy KPIs revealed in the course of trialling the innovation line were lower or equal to those measured for the baseline. However, this finding might have been biased to some extent by insufficient training of the disaster managers on 3D and orthophotomap products as tools used for aerial reconnaissance. This limitation arises due to the trial time constraints.
- Assessment of *sensitivity* (KPI 3) leads to the conclusion that if the identification of individuals who are truly in need of assistance has the highest priority, the red colour of sheeting is the most efficient colour as a signal mark (highest rate of correctly identified sheets among three considered colours).
- Assessment of *specificity* (KPI 4) allows for the conclusion that if the identification of individuals who do not need assistance has the highest priority, white colour of sheeting placed in a window as a signal mark might cause misidentifications problems in aerial reconnaissance due to the fact that most of the windows' frames are simply white. It might mislead the analysts of the aerial reconnaissance products, and the sun's reflection might do as well.

- Assessment of *PPV* (KPI 5) might be concluded that if the identification of individuals who genuinely need assistance ("ground truth" cases) among all those who are identified as requiring assistance (TP + FP) has the highest priority, blue colour of sheeting should be used as a signal mark.
- Assessment of *NPV* (KPI 6) leads to the conclusion that if the verification of individuals who do not need assistance has the highest priority, both the innovation line and the baseline proved similar efficacy. Moreover, the colour of sheeting placed in windows almost does not influence this efficacy.
- Assessment of *accuracy* (KPI 7) provides evidence that if the correct identification of the needs location (including correct exclusion of locations where need is not required – no sheeting hanged) has the highest priority, both the innovation line and the baseline perform on a similar level of efficacy. Moreover, similar to the NPV, the colour of sheeting does not influence this efficacy in a significant manner.
- Interpretation of average *LR+* (KPI 8) and *LR−* (KPI 9) leads to the conclusion that *LR+* and *LR−* might be less useful in the efficacy evaluation of the innovation line.

Conclusions

The trial demonstrated that the trial guidance methodology might be successfully implemented for checking potential added value of new solutions for disaster management purposes. One of strengths of the approach presented in this chapter is its potential to be repeated for further verification of already achieved results during future trials. It makes a lot of sense in comparison checks between solutions appearing in the market that claim to possess similar functionalities, being better than traditional tools (the baseline) and, as such, being in the scope of interest of disaster managers and first responders. The new knowledge generated by such trials constitutes a solid base for critical decisions on investing into procurements and implementation of new solutions and protocols to particular disaster management functions.

The results of the trial showed that the proposed solution (innovation line) is not necessarily better than the traditional way of doing aerial reconnaissance (baseline) during massive floods. However, it leaves a lot of room for further search on other solutions that might do so. As the disaster management practitioners are still interested in improving the disaster management function which the solution is expected to serve, then additional trials strictly implementing the same method on the same solution, or other similar solutions, are dearly recommended in order to further increase the reliability of the finding.

Due to dynamically changing safety and security environments, searching for and trialling new solutions in disaster management is a must. This process may be significantly improved by usage of the trial guidance methodology in organising trials that implement firm and evidence-based evaluation methods.

Acknowledgements

This paper has received funding from financial resources for science in 2018–2023 granted by the Ministry of Science and Higher Education of the Republic of Poland, grant no. 3851/7.PR/17/2018/2.

References

BBK (German Federal Office of Civil Protection and Disaster Assistance) Services for Modern Civil Protection. (2017). BBK: Bonn, Germany.

Craig, S. B., & Hannum, K. (2006). Experimental and quasi-experimental evaluation. In K. Hannum, J. W. Martineau, & C. Reinelt (Eds.), *The Handbook of Leadership Development Evaluation* (pp. 19–47). Greensboro: Jossey-Bass and Centre for Creative Leadership.

Culbertson, R. A. (2018). The ethics of efficiency. *Healthcare Executive*, January/February 2018, 50–52. www.ache.org/-/media/ache/about-ache/ethics/ethics-he-resources/jf18_ethics_reprint.pdf.

DRIVER+. (2021a). *Website*. www.driver-project.eu. Accessed 15 November 2021.

DRIVER+. (2021b). *Summary of Trial Poland*. www.driver-project.eu/events-old/trials/trial-1/. Accessed 15 November 2021.

Eyerman, J., Crispino, G., Zamarro, A., & Durscher, R. (2018). *Drone Efficacy Study (DES): Evaluating the Impact of Drones for Locating Lost Persons in Search and Rescue Events* (pp. 15–21). Belgium: DJI and European Emergency Number Association.

Feher Waltz, C., Strickland, O. L., & Lenz, E. R. (2010). *Measurement in Nursing and Health Research* (pp. 49–87). New York: Springer.

Feltynowski, M., & Zawistowski, M. (2018a). Opportunities related to the use of unmanned systems in emergency services. *BiTP*, 51(3), 126–136. https://doi.org/10.12845/bitp.51.3.2018.9.

Feltynowski, M., & Zawistowski, M. (2018b). Threats related to the use of unmanned systems in emergency services. *BiTP*, 51(3), 138–149. https://doi.org/10.12845/bitp.51.3.2018.10.

Foks-Ryznar, A., Ryzenko, J., Trzebińska, K., Tymińska, J., Wrzosek, E., & Zwęgliński, T. (2019). Trial as a pragmatic and systematic approach for assessing new solutions in crisis management and rescue operations. *Proceedings of SPIE 11176, Photonics Applications in Astronomy, Communications, Industry, and High-Energy Physics Experiments 2019*. https://spie.org/Publications/Proceedings/Paper/10.1117/12.2537925. https://doi.org/10.1117/12.2537925.

Fonio, C., & Widera, A. (Eds.). (2020). *Trial Guidance Methodology Handbook. DRIVER+ (Driving Innovation in Crisis Management for European Resilience)*, Brussels, 2020. https://tgm.ercis.org/.

Gilman, D. (2014). Unmanned aerial vehicles in humanitarian response. *OCHA Policy and Studies Series 010*. https://www.unocha.org/sites/unocha/files/Unmanned%20Aerial%20Vehicles%20in%20Humanitarian%20Response%20OCHA%20July%202014.pdf.

Gregorius, A., & Śpiewak, R. (2013). The principles of rational selection of diagnostic tests in allergology. *Alergia Astma Immunologia 2013*, 18(4), 221–230.

Hendriks, M., Vullings, E., van Campen, S., & Hameete, P. (2019). An interoperability framework for trials and exercises. *Proceedings of the 30th International Forum for the Military and Civil Simulation, Training and Education Community (ITEC)*, 13–16 May 2019. Stockholm, Sweden.

Hodgson, J. C., Mott, R., Baylis, S. M., Pham, T. T., Wotherspoon, S., Kilpatrick, A. D., Segaran, R. R., Reid, I., Terauds, A., & Koh, L. P. (2018). Drones count wildlife more accurately and precisely than humans. *Methods in Ecology and Evolution*, 9, 1160–1167.

Jamieson, S. (2004). Likert scales: How to (ab) use them. *Medical Education*, 38(12), 1217–1218.

Krieg, A. F. (1975). Why are clinical laboratory tests performed. *JAMA*, 233, 76–78.

Kurz, F., Rosenbaum, D., Leitloff, J., Maynberg, O., & Reinartz, P. (2011). A real-time camera system for disaster and traffic monitoring. *Proceedings of the International Conference on Sensors and Models in Photogrammetry and Remote Sensing 2011*, 18–19 May 2011, Tehran, Iran, 1–6.

Marturano, F., Ciparisse, J. F., Chierici, A., d'Errico, F., Di Giovanni, D., Fumian, F., Rossi, R., Martellucci, L., Gaudio, P., & Maliz, A. (2020). Enhancing radiation detection by drones through numerical fluid dynamics simulations. *Sensors*, 20, 1770. https://doi.org/10.3390/s20061770.

McLeod, S. (2021). A. Likert scale. *Simply Psychology*. www.simplypsychology.org/likert-scale.html. Accessed 15 November 2021.

Meier, P. (2015). Digital humanitarians. In *How Big Data is Changing the Face of Humanitarian Response* (pp. 73–92). London: Routledge.

Meier, P. (2020). Humanitarian drone/UAV missions: Towards best practices. *We Robotics: 6*. https://docs.google.com/document/d/1xyjOBVe4cz7F-Ed1lGzeT4ZPmC2tJ0JiDUyzWo1Ci9U/edit#. Accessed 15 November 2021.

Meyer, M. (2015). The new killer drones: Understanding the strategic implications of next-generation unmanned combat aerial vehicles. *International Affairs*, 91(4), 765–780.

Oddo, P. C., & Bolten, J. D. (2019). The value of near real-time earth observations for improved flood disaster response. *Frontiers in Environmental Science*, 7, 127.

Price, C. P. (2000). Evidence-based laboratory medicine: Supporting decision-making. *Clinical Chemistry*, 46, 1041–1050.

Realdi, G., Previato, L., & Vitturi, N. (2008). Selection of diagnostic tests for clinical decision making and translation to a problem oriented medical record. *Clinica Chimica Acta*, 393(1), 37–43. https://doi.org/10.1016/j.cca.2008.03.024.

Römer, H., Kiefl, R., Henkel, F., Wenxi, C., Nippold, R., Kurz, F., & Kippnich, U. (2016). Using airborne remote sensing to increase situational awareness in civil protection and humanitarian relief – The importance of user involvement. *Proceedings of the International Archives of the Photogrammetry, Remote Sensing and Spatial Information Sciences*, Vol. XLI–B8, 12–19 July 2016, Prague, Czech Republic, 1363–1370.

Ryzenko, J., & Smolarkiewicz, M. M. (2010). Space-enabled information environment for crisis management. Scenario-based analysis and evaluation in an operational environment. *Acta Astronautica*, 66(2010), 33–39. https://doi.org/10.1016/j.actaastro.2009.03.088.

Schröter, E., Kiefl, R., Neidhardt, E., Gurczik, G., Dalaff, C., & Lechner, K. (2020). Trialling innovative technologies in crisis management – "Airborne and terrestrial situational awareness" as support tool in flood response. *Applied Science*, 10, 3743. https://doi.org/10.3390/app10113743.

Simon, R. (2015). Sensitivity, specificity, PPV, and NPV for predictive biomarkers. *Journal of the National Cancer Institute*, 107(8). ISSN 0027-8874. https://doi.org/10.1093/jnci/djv153.

Soesilo, D., Meier, P., Lessard-Fontaine, A., Du Plessis, J., Stuhlberger, C., & Fabbroni, V. (2017). Drones in Humanitarian Action. A guide to the use of airborne systems in humanitarian crises. *Report: 23–34*. https://irevolution.files.wordpress.com/2011/07/drones-in-humanitarian-actionemail.pdf. Accessed 15 November 2020.

Stufflebeam, D. L., & Zhang, G. (2017). *The CIPP Evaluation Model. How to Evaluate for Improvement and Accountability* (pp. 20–56). New York: The Guilford Press.

Verykokou, S., Doulamis, A., Athanasiou, G., Ioannidis, C., & Amditis, A. (2016). UAV-based 3D modelling of disaster scenes for Urban search and rescue: 106–111. *IEEE International Conference on Imaging Systems and Techniques (IST)*, Chania. https://doi.org/10.1109/IST.2016.7738206.

Wians, F. H. (2009). Clinical laboratory tests: Which, why, and what do the results mean? *Lab Medicine*, 40, 105–113.

Zwęgliński, T. (2019). Metodologiczne podstawy teorii ewaluacji ćwiczeń w zakresie zarządzania kryzysowego i ochrony ludności. In B. Wiśniewski & T. Zwęgliński (Eds.), *Racjonalizacja zarządzania jednolitymi formacjami umundurowanymi odpowiedzialnymi za bezpieczeństwo wewnętrzne* (pp. 252–277). Warsaw, Poland: Szkoła Główna Służby Pożarniczej.

Zwęgliński, T. (2020). The use of drones in disaster aerial needs reconnaissance and damage assessment – Three-dimensional modeling and orthophoto map study. *Sustainability*, 12(15), 6080. https://doi.org/10.3390/su12156080.

Zwęgliński, T., Maksimenko, A., & Smolarkiewicz, M. (2019). Exercising in a radioactive environment – a case study from CBRNE exercise in Chernobyl Exclusion Zone. *Safety & Fire Technology*, 54(2), 160–166. https://doi.org/10.12845/sft.54.2.2019.12.

Zwęgliński, T., & Morgado, C. (2018). Analytical approach to cooperation of the State Fire Service with the Police in Poland in the context of brief comparison with Portuguese reality. *Internal Security*, 10(1), 161–182. https://doi.org/10.5604/01.3001.0012.7498.

11

A MULTIMODAL REMOTE SENSING SYSTEM FOR IMPROVED DECISION-MAKING IN EARTHQUAKE RESPONSE

Elisa Schröter, Gunnar Schwoch, Christian Niermann, Veronika Gstaiger, and Nina Merkle

Introduction

Earthquakes are a natural hazard that pose a variety of challenges. While the origins of earthquakes are widely known, the exact locations where damage occurs and its extent can vary and are therefore unpredictable. Despite numerous approaches to modelling earthquakes and their effects, uncertainty remains as to when the next quake will occur. Earthquakes strike suddenly, so the population cannot be warned quickly and those affected can be taken by surprise at any time of day, night, or year. In addition to the earthquake itself, rockfalls, landslides and avalanches also occur, especially in mountainous regions, and lead to further damage and impairments. Severe earthquakes leave behind destroyed buildings with trapped survivors, injured and dead people, damage to the transport infrastructure, destroyed communication infrastructure, and depending on the area of damage, cut off supplies to the population. Local rescue workers are often affected themselves and aftershocks can further complicate search and rescue operations. It is therefore vital to get a quick and comprehensive overview of the affected area after the event, in order to determine:

- Where exactly damage occurred.
- What kind of help might be needed (e.g. search and rescue teams, experts for chemical and toxic accidents, medical supplies, water and food, mobile shelters).
- By which route the aid can be brought to the damaged areas. (Are the access routes usable? What alternative routes might exist?)
- Which areas need to be prioritised.

Images of the damaged areas, which can be provided by remote sensing technologies, offer an intuitive way to quickly detect and locate damage. In the case of

DOI: 10.4324/9781003256977-18

small-scale damage events with intact access, rapid damage evaluation by a small drone with an optical camera system, as available in many rescue operations, is an option that is being used more and more frequently. However, large-scale, remote, and difficult-to-access regions, such as in mountainous areas, cannot be covered and accessed quickly enough. Here, area-wide images are needed, such as those provided by satellites. This technology has also been used operationally for years, such as in the Copernicus Management Service (2021), and satellite data is evaluated and made available by experts around the clock in the event of a disaster. However, evaluations of satellite data also have their limits: firstly, their ground resolution of up to about 30 cm makes it difficult to reliably detect and classify damage, and secondly, cloud-free weather conditions are necessary for suitable optical satellite images. Since the latter is often difficult in many regions, it might be necessary to wait several days for new images: valuable time that is not always available to those affected. Matching reference data from before the event is a great additional help, but unfortunately, they are rare. In return, emergency responders often are limited to geographic information systems or mapping platforms that incorporate outdated imagery (satellite or airborne) only, which does not depict the disaster situation, and map selective reports or single-frame imagery to build situational awareness (baseline).

One possible alternative for obtaining high-resolution images of the damaged region over a wide area and at a flexible time is to use airborne optical camera systems, such as those developed and operated by DLR. The special feature of the DLR system presented in this chapter is the bundling of competencies and scientific developments from a wide range of disciplines in an overall solution. The latest technologies for image acquisition and processing from the field of remote sensing are combined with an innovative solution for flexible flight planning and control to an autonomous, real-time imaging system. At the analysis level, artificial intelligence (AI) algorithms for image interpretation are networked with routing methods from traffic research. Finally, the resulting data is post-processed and handed over via user-friendly information products and applications at end-use level.

To test and evaluate actual benefits in practice but also potential obstacles to using the system proposed by DLR, our team participated in a full-scale trial of the DRIVER+ project organised by the Austrian Red Cross and Austrian Institute of Technology in 2019 that focused on a high-impact earthquake event and its cascading effects (trial AT). This trial was designed to overcome gaps reported by practitioners in similar crisis scenarios by integrating multiple new solutions into crisis response. The relevant selection for the presented crisis scenario in the Austrian trial has been made from a list of identified gaps for the whole DRIVER+ project (Chapter 4) and targeted the management of spontaneous volunteers, management of psychological stress of those affected, and usage of emerging heterogeneous and real-time data sources to support decision-making. The hypothesis made and assessed by the trial team has been that DLR's multimodal system will improve decision support through near real-time provision of aerial imaging products and its integration with other information sources in an interactive GIS environment (innovation line) (Chapter 6). In addition, the application potential of 3D mapping for on-site and off-site teams of response organisations in international relief operations during earthquakes and cascading disasters was investigated as a test case of the Bavarian Red Cross during the IRONORE exercise, which was conducted in parallel to the trial. It incorporated the final part of the systems processing chain.

Due to the simultaneous involvement of our sub-systems in both a trial and an exercise, the different experiences made in the process, in terms of preparation and evaluation, will be discussed in the final part of this chapter.

Overview of the Solution

The solution consists of three main components. The first component is DLR's 3K camera system, which is used to acquire and process aerial images. The system is mounted on DLR's Dornier 228-101 research aircraft, which can also be used as a remote-controlled drone. With the help of a special control option, flight paths can also be defined or changed at short notice in order to let the aircraft fly to certain areas and take aerial images according to user requests. The third component of the solution is image analysis and visualisation. Additional information is obtained from aerial and satellite images using the latest image evaluation methods and AI algorithms. To process this information according to user requirements, visualise it in a comprehensible way according to user profiles, and transfer it into user-specific formats, such as maps or 3D visualisations, an individualised service is the final link in the process chain. Each of the individual components is explained in more detail in the following figure.

Figure 11.1 shows DLR's solution from flight planning, data acquisition, transmission, and processing to product generation and delivery.

DLR's 3K Aerial Camera System

The main component of DLR's solution is the 3K camera system for recording and directly evaluating aerial images.

With the 3K optical camera system, DLR has developed a cost-effective alternative to metric aerial camera systems with constant elements of internal orientation (Kurz et al., 2012). The overall system is made up of various hardware and software components. The actual camera system forms the airborne core and consists of three commercially available Canon EOS 1Ds Mark III SLR cameras, as well as a real-time GPS/INS unit. The cameras are rotatable and can achieve a maximum field of view of up to 104°. Depending on the application, the desired flight altitude and ground resolution, the system can also be equipped with 35 mm, 50 mm, or 100 mm lenses. During the flight campaign for the conducted trial, 100 mm lenses were used and delivered ground resolutions of 14 cm at a flight altitude of 2,000 m above ground.

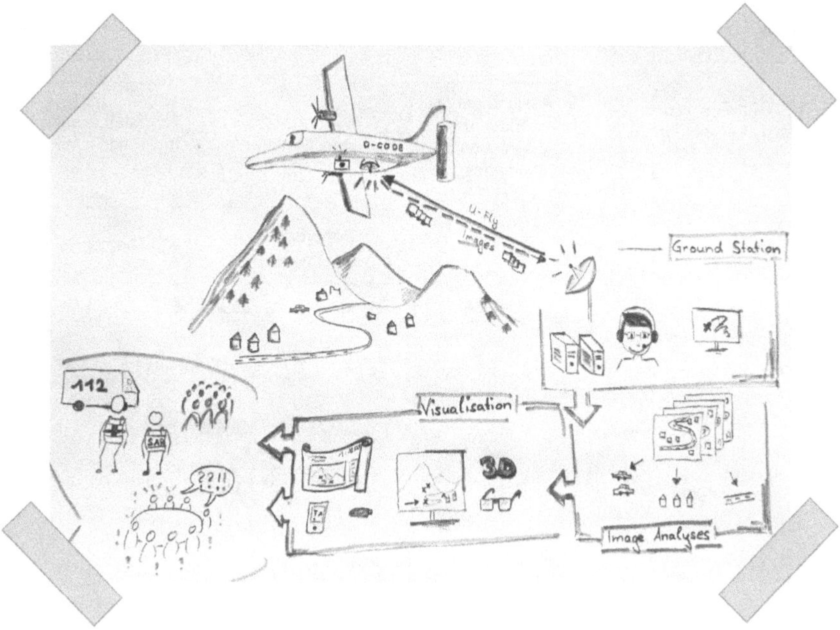

FIGURE 11.1 Overview of the process chain components and workflow.

In addition to the setting options for the acquisition angle and focal lengths, it is also possible to configure different acquisition modes. Thus, image recordings up to 2 Hz can be set and recordings can also be triggered in burst mode, in which a sequence of several recordings with a short interval is followed by a longer pause of several seconds until the next burst. This mode is mainly used for the extraction of the speed of moving objects like cars.

In order to be able to process the recorded image data immediately, a computer infrastructure consisting of three industrial PCs is also on board. Here, the images are georeferenced with high precision using the position and altitude information of the GPS/INS unit and orthorectified with the help of a stored terrain model. Subsequently, further optional evaluation procedures can be implemented. For example, an up-to-date traffic evaluation can be carried out based on the aerial images and information on the number of vehicles, their driving direction and speed can be derived (Leitloff et al., 2014). Furthermore, a fast version of semi-global matching (d'Angelo & Kurz, 2019) is implemented in order to create a 2.5D (also referred to as 3D in the following) surface model of the disaster site from the captured aerial images during the flight. As stated in Chapter 10, the added value of 3D data in disaster management is often limited by the fact that the calculation of a 3D model takes a lot of time. Our method bridges this gap by on board calculations in near real-time.

An illustration of the workflow and required input data of DLR's 3K camera system is depicted in Figure 11.2.

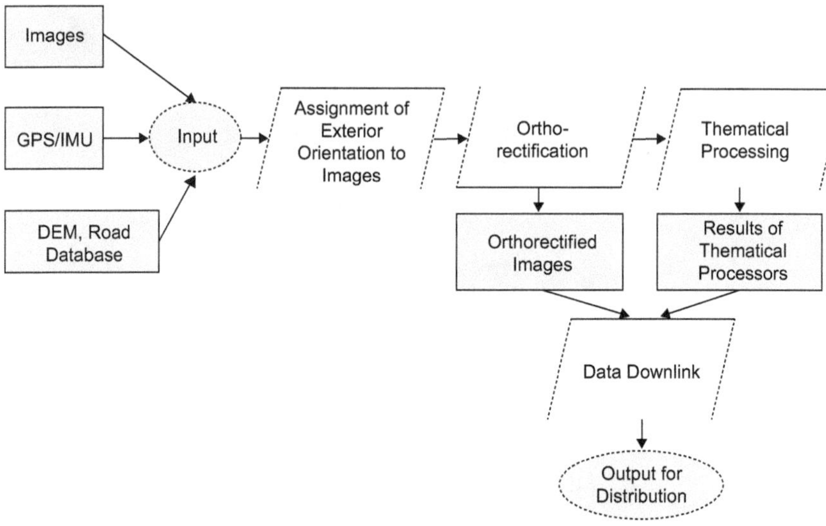

FIGURE 11.2 Data and workflow of DLR's 3K camera system.

The orthorectified aerial images, together with the traffic data and 3D information, can finally be transmitted to a mobile ground station via a microwave data link. For this purpose, there are two antennas on board that can transmit the data at a transmission rate of 11 Mbit/s over a distance of up to 70 km. At the ground station, the data is finally received, assembled into image mosaics, and transformed into the desired data formats that are needed for transmission or evaluation.

The 3K camera system has been used for years for mapping missions as part of science projects. The aerial images are used to develop state-of-the-art image processing and analysis methods and were also used to record the situation of the 2013 floods in Germany. The 3K system can be implemented on the following platforms of the DLR research fleet: Cessna 208B Grand Caravan (D-FDLR), Dornier 228-212 (D-CFFU), and Dornier 228-101 (D-CODE). Among those mentioned, the D-CODE platform was used as a remotely piloted system for the trial.

Remotely Piloted Aircraft Systems as Flight Carrier

Unmanned aerial vehicles or remotely piloted aircraft systems (RPAS) have become an essential part of many areas of industry and are used in a wide range of applications: small drones as a hobby, research and individual missions to monitor infrastructure and fields or to search for missing people or animals, and even military purposes, such as surveillance, with large RPAS. The level of automation varies just as much as the size of the aircraft: direct control via a remote control in line of sight to satellite-based guidance at a great distance from the remote pilot to the aircraft.

Apart from RPAS applications in industry, the use of RPAS in crisis management and disaster control is a broad field of research. In particular, RPAS can make an important contribution to the rapid assessment of a crisis area – which may have limited access – without putting people in danger, or to the large-scale mapping of an area. The long flight duration, which can significantly exceed the performance of manned aircraft, is of particular importance for these missions.

The Dornier 228-101 D-CODE operated by DLR is used as a flight test carrier. The high payload of the Dornier in combination with the extensive modifications to the aircraft result in an optimal carrier for various research tasks. This makes it possible to extensively test and optimise even early research equipment in flight tests before miniaturisation.

With its modern avionics and data link equipment, the D-CODE can be operated as an optionally piloted vehicle. This means that it can be controlled from the ground via a control station just like an RPAS, as well as directly from the conventional cockpit by test pilots just like a regular manned aircraft. This provides the opportunity to explore the use of RPAS in different areas while maintaining a high degree of safety and compliance with legal regulations. Nevertheless, as the system consists of a ground control station and data link, it offers the functionality to operate an aircraft of this size even without pilots on board.

Depending on the predicted purpose of use, it can be decided whether the use of a large fixed-wing RPAS is necessary and justified or if the flexibility and more cost-efficient use of a smaller drone is considered sufficient. In the specific case described here, the use of a large RPAS is preferred, as the dimensions of the disaster area could not be covered nearly as quickly with small, slower drones. However, the additional use of a smaller drone could also meet the desire for different perspectives, such as to identify damage to infrastructure which is not clearly visible from a top-down view.

The use of RPAS in crisis management not only reduces the number of personnel required but also significantly reduces the danger to life and health. In the event of a disaster, for example, in the case of accidents involving the release of radioactive emissions, images and sensor data can be obtained without endangering pilots. The operational purposes are various but require both a safe and defined chain of command supported by assistance systems, as well as flexible operation from the ground with direct feedback to mission control. The usage of such a system in the scope of a trial as presented here is a very efficient way of a deployment of the system in a relevant environment, to test new and adopted features and to derive the benefit such a system can bring by assessing practitioners' qualitative feedback. While technology-related components can also be tested in a laboratory environment, the main purpose of such a system, such as the added value to decision support through the rapid integration of aerial image products, requires an operational setup, as is facilitated by the trial. Finally, practitioners themselves can benefit from the deployment of the system in a trial by being able to understand the workflow and to perceive the added value in depth, both of which can help in the planning of future collaboration or investments in such a technology.

During the trial in Austria, two separate and independent data links were used. For the data link for flight control and mission transmission and monitoring, a data link with less bandwidth but higher range was suitable, in this case a full-duplex high-frequency data link in the L/S band with 1 to 4 GHz. Tracking of the signal was assisted by displays presenting the signal strength.

The use of an aircraft as an RPAS, which therefore goes well beyond the usual commercial dimensions, naturally requires appropriate planning. In the preparation for the trial in Austria, four points of interest (POIs) that could particularly benefit from aerial images in near real-time were selected from the total of 30 POIs defined for the trial (Schwoch & Niermann, 2021). The selected trial POIs included collapses of traffic infrastructure, blockades due to landslide debris, and the assessment of hard to reach or currently inaccessible areas. It was also considered that certain framework conditions must be fulfilled by the deployed aircraft. For example, unlike a helicopter, the D-CODE cannot fly tight turns or even hover at any speed, but it can fly over large areas much faster and thus map them quickly. The flight paths created in combination with the ground control station were transmitted to the D-CODE as a waypoint-based planned mission.

The on-board flight management system converts this mission into a trajectory that can be flown. This trajectory is then sent as a reply via the data link to the ground control station and displayed there. This way, the ground operator can check whether the desired mission objectives can potentially be achieved, even if, from an aeronautical point of view (for example, wind influences or slightly changed flight altitude or speed), a different trajectory results from the waypoint-based flight plan than that predicted by the ground system.

The use of an aircraft such as the Dornier 228-101 as an optionally piloted vehicle, acting as an RPAS, comes with special challenges, as mentioned in the beginning. In particular, mission and flight planning must be considered accordingly. At DLR, the in-house development U-FLY is used for this purpose.

Mission Planning and Control via Ground Station "U-FLY"

The DLR ground control station U-FLY is configurable to different types of missions and aircraft, ranging from small-scale missions with vertical take-off and landing drones to the deployment of larger optionally piloted vehicles, such as the Dornier 228-101. Built-in features, such as collision detection, telemetry range indicators, and flight planning for instrument flight rules operations, enable simple mission planning. Missions are waypoint-based and can be planned, stored, loaded, and activated using U-FLY. The following possible actions can be performed in planning stage:

- Plan departure and arrival using stored and available published information.
- Add waypoints by location manually.
- Create rectangular areas of interest.
- Assign search and scan patterns to areas of interest.

- Define altitude constraints at or between waypoints or for areas of interest.
- Define speed constraints at or between waypoints.
- Define turn behaviour for fixed-wing aircraft (i.e. fly-over or fly-by).

If payload for remote sensing capabilities (e.g. the described 3K camera system) is configured, the automatic calculation for search and scan patterns (Geister et al., 2017) considers the aperture of the sensor, the type of pattern, and the planned altitude at the area of interest. The width of flight strips, in this case the distance between two overflights of an area of interest, is calculated in order to achieve gap-less coverage. To achieve higher resolution of the resulting imagery, a low altitude might be preferred. Low-altitude flights result in smaller flight strips (e.g. for the Creeping Lines pattern; see Figure 11.3) and thus smaller turn radiuses. If the flight envelope requires, procedural turns outside the area of interest are inserted.

When a mission is activated, the mission plan is sent from U-FLY to the onboard system, using the available data link. The aircraft then follows a 4D trajectory, calculated by the flight management system onboard the aircraft. While the aircraft is airborne, flight axes can be controlled by commanding selected altitude, speed, vertical speed, and/or heading. A holding of a defined duration can be issued with U-FLY anytime.

Compared to satellite-based products, the advantage of the described aerial system is that the image products can be scheduled more flexibly and demand-driven regarding the temporal resolution, in contrast to imagery acquired by orbiting

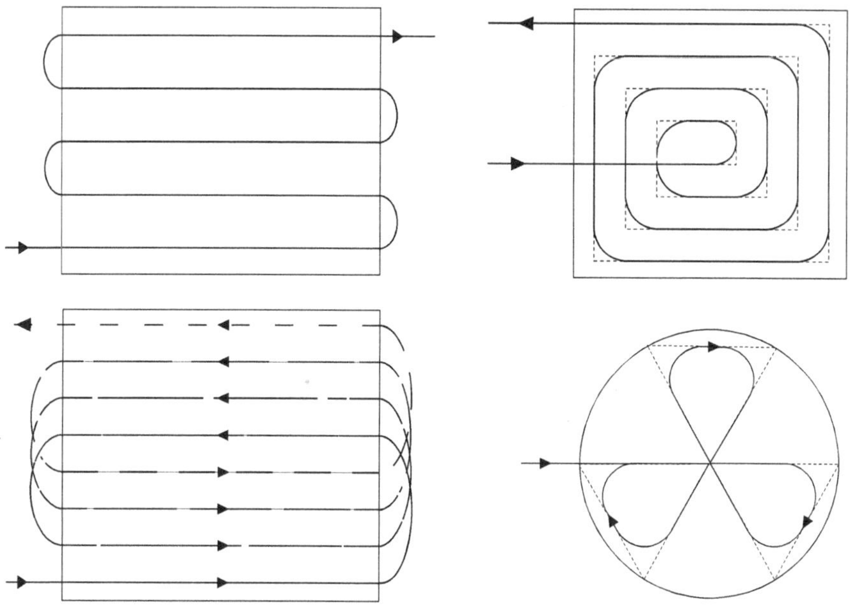

FIGURE 11.3 Scan and Search patterns. Top row, from left: Creeping Lines, Expanding Square; Bottom row, from left: Zamboni, Sector Search.

satellites of which the acquisition window is limited by the next overpass. Depending on the deployment time of the aerial system at the disaster site, this may be associated with faster recording capabilities, and further the possibility of a higher temporal density of repeated imaging for change detection, if needed. The capability of our imaging system further opens up new possibilities for the user to react more flexibly on the situation and adapt mission planning accordingly in order to completely map the areas most relevant to the disaster managers.

From Image Data to Operational Information

To support disaster managers in gathering relevant information from remotely sensed data, the Center for Satellite-Based Crisis Information (ZKI), as an institution of the German Remote Sensing Data Center (DFD) at DLR, constitutes a link between the scientific development of new remote sensing methods and its operational application in disaster management. For years, ZKI has been applying the latest algorithms and newly emerging data sources in an on-demand service to generate up-to-date situation information from remotely sensed data (Lechner & Gähler, 2017). It is capable of processing earth observation data from various platforms (satellite or small to large-scale airborne) to assist in different phases of the disaster management cycle and according to the respective context and data situation. ZKI further merges analysis results with additional geo-data. In close consultation with the users, customised information products are produced by transforming the acquired data into easily accessible and user-friendly formats. With this, ZKI completes the processing chain with its various value-adding services.

To draw conclusions from newly acquired images of the disaster site, it is very important to have information on the situation right before the event. To this end, ZKI runs a rapid search on the availability of reference data in the form of very high-resolution satellite data or aerial imagery provided by local authorities. Its experts assess the data situation and make an individual decision as to which image material is suitable as a reference according to its timeliness and characteristics for the respective disaster event. The most appropriate reference images are obtained from the provider and are processed and disseminated according to the user's requirements – for example, as GIS packages or web services for integration in their own data infrastructures. Further, compiled map products (digital or printed) including additional geo-data to highlight specific infrastructure and labels can be provided to facilitate orientation and interpretation for users without special knowledge in remote sensing (Gähler, 2016). Depending on the requirements and availability, first reference products can be provided already within four to eight hours. Consequently, this process is not yet part of the chain but can take place in parallel to the preparations of the flight campaign and travel of the aircraft to the disaster area already so that data is ideally available as soon as the aircraft arrives on-site. The availability of reference data is a crucial aspect to ease the analysis of the situation, both visually and through automatic processes.

Under the real-world conditions of the trial, ZKI could provide reference maps based on aerial imagery from the federal state of Styria (GIS-Steiermark) with a spatial resolution of 20 cm and a timeliness of about two years at the time of the event for comparison. It demonstrates once again common challenges faced in data acquisition. Oftentimes, due to restrictions of the available data, timeliness of the pre-disaster imagery and spatial resolution or spectral characteristics have to be weighed against one another. The best compromise for comparison with the post-disaster images has to be found depending on the use-case and the analysis method chosen. In this case, the spatial resolution was weighted higher than timeliness by an anticipated only minor change in the building development and transportation infrastructure in this area and the need to enhance conditions for automated analysis by choosing comparable spatial characteristics. Besides the acquisition of reference images, additional geodata to highlight points of interest especially from the traffic and critical infrastructure via additional map layers was obtained and processed to digital, interactive map products and paper maps (Merkle et al., 2020).

DLR's system was chosen by the trial team to assess the potential of real-time images in a set of other real-time data sources, such as crowdsourced situation reports, for decision-making on distribution of incoming national and international assistance in such situations. Consequently, the setup directly linked the acquisition of post-disaster image data in near real-time to third-party software (innovation line of the trial). This introduced a change to DLR's original process chain and limited the process chain to image data provision without the service of additional value-adding through the ZKI module in the context of the trial.

Extended Operational Capabilities

Even though the solution could demonstrate its core feature by providing near real-time image data, the scope of the trial did not allow it to assess the full spectrum of options, including additional value-adding services and latest product developments. To take advantage of the possibility of testing newly emerging methods and products from the process chain in a realistic operational scenario, DLR teamed up with one of its frequent ZKI users, the Bavarian Red Cross (BRK). BRK has been a participant of the full-scale exercise IRONORE that was connected to the DRIVER+ trial and carried out in parallel.

During the IRONORE exercise, BRK practised assisting ongoing rescue operations with their rescue teams within the framework of an international aid mission. It included a quick reaction team responding to CBRN events that has been tasked with the decontamination of injured persons and supply of technical equipment for hazard containment and neutralisation, as well as a special team (named SEG-GILT) for heavy ground, infrastructure, logistic, and transport that provides assessments with drones and amphibian vehicles. The scenarios conducted during IRONORE included a simulated bus accident, a rail tanker crash, a chemical spill accident, and a landslide. The exercise was specifically tailored towards training the particular challenges arising from the alpine environment it was conducted in. As a frequent user of ZKI satellite products, BRK was interested in testing new products provided with the aerial system by DLR under real-world conditions during the IRONORE exercise, especially with respect to its 3D capabilities.

Rapid 3D Mapping

In the case of large-area natural disasters, local people are dependent on the support of rescue forces from outside, such as the BRK. These, in turn, are often unfamiliar with the area and have to orient and organise themselves in an area with incomplete information. Due to these circumstances, a major interest of the BRK team during IRONORE has been the effective communication and information exchange between the team on-site and their integrated command centre based in Schweinfurt, Germany.

Up-to-date aerial photographs, as provided by the DLR system, can already be of great support for an initial assessment of the situation in the damaged area. Beyond that, through technological advances in remote sensing, new opportunities arise constantly for the provision of crisis information. Recently, the addition of 3D information and 3D visualisations created new possibilities to provide emergency forces with even more detailed spatial information on the disaster event and to digitally visualise the affected area as best as possible. Instead of working only with updated 2D maps using the acquired post-disaster imagery, the BRK team was provided with reference and up-to-date 3D situation maps by ZKI during the IRONORE mission to explore its utility value and potential usage barriers.

Just as the 2D references, the ZKI team processed 3D reference maps from digital elevation models and semantic 3D building models obtained from authoritative data sources (GIS-Steiermark) in a rapid mapping activity for an initial use. The 3D reference maps were intended to provide first responders with a first impression of the terrain and on-site conditions already while on their way to the deployment site. During the image acquisition of the trial, special flight patterns with large image overlaps were flown with the RPAS and used for 3D calculations. This formed the basis for a rapid update of the digital 3D reference maps to 3D situation maps of the disaster site (Figure 11.4).

All map products were provided in 3D PDF format to the end-users to achieve independence from special software and hardware requirements and thus enabling fully offline work in case of disrupted communication on-site. These were linked to the 2D representations to optionally provide additional orientation and further information (Merkle et al., 2020).

Innovative Analysis Methods

In addition to providing orientation and coordination assistance for non-local emergency forces, specific evaluations of up-to-date 3D data can also make a valuable contribution to emergency response activities. Landslides are one of the frequent side effects of earthquakes in regions with strong relief. They may

FIGURE 11.4 3D situation map of the disaster site as a result of the aerial mapping process chain.

require additional recovery and clearing work that needs to be planned and coordinated. At this point, evaluations of 3D data can support by identifying landslides and calculating the volume of the moved material with the help of an automated before-and-after comparison. In the context of the full-scale exercise IRONORE, blasting operations were carried out in the area of the Erzberg in Austria, thus deliberately triggering landslides for demonstration purposes. Therefore, not only were aerial images of the area of interest, but also current surface information was required. To demonstrate the potential of change analysis between subsequent flights, the entire terrain was mapped during a total of three overflights before and after the blasting on September 10, 11, and 13, 2019, paying attention to large image overlap in order to derive additional height information from the aerial images. Surface models with ground resolutions of 30 cm were calculated and processed after the aerial surveys (d'Angelo & Kurz, 2019). The data was used to create 3D map visualisations and evaluated separately in the context of change analysis. The results not only reflect the terrain in highly accurate quality, but also clearly show in which areas material has been removed and added. In total, the landslide was calculated over 520 m^3 (Merkle et al., 2020). Its visualisation is shown in Figure 11.5. An automated procedure for 3D change analysis can provide a variety of valuable information and is currently being implemented and tested for operational use for the 3K camera system (Gstaiger et al., 2018).

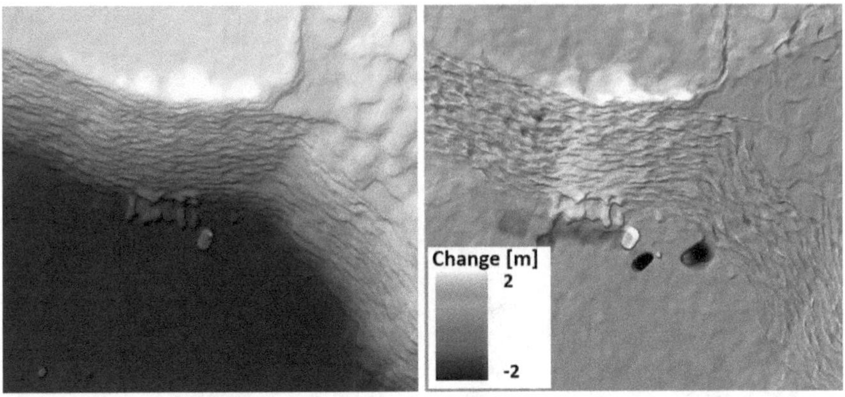

FIGURE 11.5 Digital surface model of September 11, 2019 (left), and change between September 10 and 11, 2019 (right), after the landslide.

In addition to calculating the change in volume, it is relevant whether there are possibly buried people in the affected areas. Vehicles often provide an indication of people in the immediate surroundings. In the area of the landslide at Erzberg, Austria, vehicles were parked and eventually partially buried. In widespread damage areas, a manual search for vehicles or their remains in aerial or satellite images would be very time-consuming, so an AI-based approach for the automated detection of vehicles was used here (Azimi et al., 2018) and detected vehicles and vehicle parts very reliably (Merkle et al., 2020).

However, it is not only the search for vehicles that can be supported by modern AI-based methods, thus saving valuable time in image analysis. With regard to infrastructure, procedures exist for the recognition of roads, for example, to supplement existing databases or show changes (Henry et al., 2021). An automated detection of buildings and classification of their condition/damage was demonstrated in Yuan et al., 2021, using satellite and aerial data from Mozambique and Lebanon. If buildings have been destroyed or damaged, assembly points with temporary shelters are often formed. These can also be identified and the type and number can be used to estimate how much food and drinking water is needed to supply the population (Azimi et al., 2021).

Once damaged areas have been identified, they can be combined with additional data sources, such as population density or information on critical infrastructure in order to support prioritisation. Then, planning begins on how these areas could be accessed. DLR has developed and tested a tool from the field of traffic research that enables special routing for emergency services. If information on traffic jams, road damage, destroyed bridges, or blocked access routes is extracted from the aerial images and integrated into this system, alternative routes are automatically calculated to avoid these disruptions (Bieker-Walz et al., 2019). This approach has already been successfully integrated into crisis management procedures for an evaluation in the event of flooding (Chapter 5). The user feedback received during this trial has yielded encouraging results that integrating this traffic management module into the process chain provides great value for crisis management authorities especially in the immediate aftermaths of the event (Schröter et al., 2020). As the earthquake scenario also dealt with hard-to-reach rural areas and disruptions due to collapses of traffic infrastructure and blockades from landslide debris (Chapter 6), we believe that the positive results obtained on the contextual routing feature of our system in the context of flooding are further transferable to earthquake scenarios as trialled in Austria, as both lead to serious traffic disturbance and the need for alternative routing. Incorporating up-to-date information about crisis-related traffic disruptions into the routing algorithm speeds up access to relief efforts by bypassing inaccessible sections and providing suggestions on alternative routes to villages in need of assistance, thus avoiding unsuccessful attempts to reach these areas that would result in long detours for first responders.

Preparation and Evaluation of Trial and Exercise

After the successful application to participate in the trial, the DLR team was required to join multiple pre-trial meetings. Those were dedicated to discuss a reasonable implementation of the new technology into existing working procedures of the participating practitioner organisations and potentially required adaptations to the proposed solution. Further, the data exchange with other newly introduced technologies of the innovation line has been discussed. The resulting setup has been tested on various levels of detail within those meetings in preparation of the trial.

According to the trial guidance methodology (Chapter 2), the trial team has assessed the effect of the proposed aerial system on disaster management operations on multiple dimensions. Data collection focused on task performance, usability, and interaction between solutions. The perspectives of practitioners, observers, and technical performance were captured via checklists, semi-structured questionnaires, debriefing/focus group discussions, and testbed logs and synthesised on the following aspects:

- How is the information processed?
- What is the quality of information?
- How is the information perceived?

The systematically collected data were analysed to answer research questions related to previously defined objectives for each solution. Regarding the real-time imaging system, the focus has been on evaluating the ability to improve situational awareness and decision-making at incident command level. Each objective was defined and evaluated in the light of the predefined gaps observed in the current crisis management processes.

As opposed to the trial, the preparation and conduction of the IRONORE exercise was comparable to an operational deployment. The BRK team made contact with the ZKI service team to discuss the expected disaster situation at its deployment site and related information needs. It received consultancy concerning the available data and product options in line with the technical circumstances at hand. In the scope of this session, two user groups have declared interest in receiving up-to-date map information from the aerial data acquisition: the response teams on-site and the team of dispatchers in the operations control centre of the BRK. For the exercise, a fairly new but already operational 3D product was chosen. The 3D product was provided within a rapid mapping activity with actual data from the training site acquired by the team of DLR participating in the trial event.

During the exercise, the team of ZKI was available remotely for queries and technical support through its on-call service but did not witness the event on-site. The scope of the preparations and the information exchange prior to product delivery and during the exercise was comparable to an operational deployment. To evaluate the exercise, a feedback form was issued, which is also used in regular operations. Further, an unstructured debriefing session was organised following the exercise to discuss individual feedback with a focus on the user experience with the products and further information needs.

Contrasting both activities, the effort spent on preparing and evaluating the trial was considerably higher than the exercise. However, this also allows for more detailed insights from different perspectives and, thus, a more comprehensive and objective assessment, which one may not get in the context of an exercise.

Results

During the trial, the multimodal remote sensing system proposed by DLR was assessed on the potential improvement of situational awareness from real-time aerial image data provision. Practitioners rated the perceived effectiveness of the system in the crisis-management-oriented dimension by using five indicators (Figure 11.6) closely related to sufficiency, quality and accuracy of the information provided by the system with respect to situation assessment. The opinions were evaluated on a five-point Likert scale from −2 to 2, where −2 represents "Strongly disagree," −1 "Disagree," 0 "Neutral," +1 "Agree," and +2 "Strongly agree."

The assessment was carried out and evaluated separately for each block in which the technology was used – namely, block 2 with a focus on the initial (overall) situation assessment and block 4 on a more detailed situation assessment of a chemical spill due to aftershocks of the earthquake event.

The results revealed that practitioners were generally satisfied with overall performance of the solution in the mentioned aspect – four from five indicators were rated above +1.0 in both scenario blocks (2 and 4) in which the solution was used. The highest rates were assigned to the indicators related to the usefulness of the provided information for further decision-making (1.6 in block 4) and situation assessment (1.33 in block 2). The lowest rating was assigned to the indicator related to time required for reviewing image map products (0.33 for block 2) but showed a

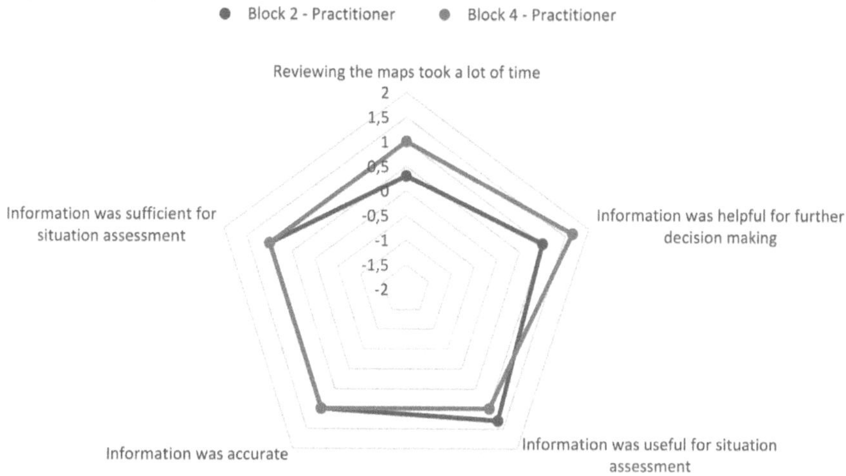

FIGURE 11.6 Practitioner rating of DLR's remote sensing system on its effectiveness for crisis management operations

clear increase towards block 4 (1.0 for block 4) that points to different challenging experiences depending on the demands of the assessed situation.

In the experiences shared by practitioners and observers during the first impression debriefings, observers and practitioners confirmed the high potential of the solution and outlined that the aerial photography products are a valuable source of information for situation assessment and planning activities. Emergency responders were convinced by the quality and provision time of the imagery. Overall, the system was found to support situation assessment for decision-making, especially if reference data is available and thus a comparison can be made between the situation before and after the disaster.

However, by integrating aerial images in a common operational picture, the trial procedure was missing out on the value-adding capabilities of the system, which is not yet fully developed for operation. The feedback discussions have brought up that interpretation of the nadir images is deemed a rather challenging task, especially for emergency managers with little or no training, who require greater assistance with this task. This is confirmed by the records of observations, which show that requested objects were only easily identified by the practitioners in one of two scenarios the solution contributed to. With this, the initial time gain due to data acquisition was significantly compromised by the difficulty of image interpretation for non-experts in earth observation.

In contrast to this comprehensive analysis based on different modalities of data collection, the feedback to the IRONORE activities was based on individual feedback received from users of the BRK only. The 3D visualisations were highly appreciated by the users and found to increase safeness of teams operating on-site in difficult terrain and to support communication between field operations and control centre (Merkle et al., 2020). Recapitulating the usage scenarios revealed that the 3D maps have been mostly used in particular by the integrated control centre to accompany the response teams on-site from remote. By this means, they could provide them with additional support with information, like potentially critical areas on the way.

Potential for Further Development

To overcome the challenges of image interpretation and provide additional information in complex situations (such as still in place but possibly substantially damaged structures), the trial practitioners raised in the discussion sessions that oblique views could be a crucial and supportive resource. Since the offered rapid surface modelling workflow is limited to 2.5D, the resulting map products do not include any oblique view information. On the other hand, field units are increasingly equipped with diverse robotic technology, such as compact drones with consumer-grade camera payload on board, and could contribute detailed data acquisitions from the disaster site. Accordingly, a gradual and multimodal procedure within a shared 3D environment would be worth considering to enable an efficient and application-oriented 3D representation of the disaster site.

Such a comprehensive 3D-mapping approach could thus consist of several levels: the provision of small-scale 2.5D elevation base data from satellite remote sensing for initial assessment on general terrain features, up-to-date and high-resolution surface models based on aerial stereo imaging of the disaster site for rapid visualisation and analysis of the current situation, and large-scale 3D models acquired by consumer-grade drones for high-detail inspection and documentation of specific points of interest, such as affected critical infrastructure. This approach, however, may require some extra training of drone pilots in emergency response on photogrammetric data collection, as many pilots in emergency response are predominantly familiar with search and rescue or reconnaissance flights with fewer requirements on a structured data collection. Among others, the ZKI team therefore supports first responders with demand-oriented training not only on the usage of 2D and 3D geodata for crisis mapping but also data acquisition using their own drone equipment and staff to prepare for this integrated approach (DLR/ZKI, 2021).

However, as results from another trial revealed, highly detailed 3D acquisition using commercially available drones and associated software solutions is still a relatively time-consuming task and may influence the overall effectiveness of such operations. Due to the dependence on the intended use-case, this should always be placed in the specific context and assessed individually.

Within the framework of the exercise, the individual experiences and use-cases shared by the end-users formed the basis for an internal discussion about possible further developments of the products: For the described use-case of utilising the 3D map environment for improved collaboration with remote colleagues, the ability to directly upload geotagged image data to communicate more detailed and up-to-the-minute situational information from on-site teams was identified as a potential enhancement for off-site support. Further, the AI-based road detection as described by Henry et al. (2021), in combination with an automated 3D analysis of the terrain, could provide additional support to additional modules of the introduced process chain, that also incorporates a traffic management platform (Schröter et al., 2020).

Discussion and Conclusions

Based on the experience from testing and assessing the proposed process chain in two trials, and partly in a full-scale exercise, we conclude that the results obtained from trialling allow for a very comprehensive and detailed assessment of the solutions operational value and potential for further development. In particular, the intensive collaboration in the trial preparation phase offers special opportunities for co-creation between practitioners and solution providers towards a user-defined goal in order to further develop and directly validate the technological adaptations to specific user requirements and thus also rethink and adapt the solution, if needed. Further, a trial provides a beneficial environment for cooperation with new user groups, in which opportunities and development goals are still being explored. The trial process therefore holds great potential for innovation but requires flexibility and openness to change on both sides in return. However, the effort spent on preparations has been comparably high in the scope of this research project and is particularly worthwhile when the collaborative potential is fully exploited.

In contrast to the trial approach, an exercise pursues the main goal of training relief workers' skills in a specific crisis context and on building routines in already existing procedures and inter-agency collaboration. In a technological cooperation with end-users, the exercise setup can be used to strengthen existing cooperation and to gain additional user experience and confidence in handling provided technology. However, an exercise usually does not leave scope for an iterative adaptation of a new technology to the individual context. Thus, it requires ready-to-use tools with a technical readiness for operational use and sufficient usage experience on the practitioners' side to be deployed successfully – for example, the close and longstanding cooperation between BRK and ZKI facilitates such an interaction, which is to test new products in real-world but uncritical conditions in the course of an exercise. The evaluation of usage scenarios is further focused on verbal feedback rather than evidence-based observations, which leads to a more subjective perspective and limits the means of analysis for the technology provider in the aftermath.

Nonetheless, from a solution provider's perspective, freely expressed user feedback through first impression reviews was perceived the most insightful of all data collected, as also found in the trial. This user experience is an important input to identify the need for adaptation and create additional ideas for follow-up developments of the technology, which has also been made available in the exercise setup. However, the latter implies longer development cycles, compared to its implementation within the scope of a trial, as changes can only be applied and evaluated in the following exercise.

In conclusion, both trial and exercise have proved that, due to the unique combination of the diverse fields, DLR's interdisciplinary process chain can make a valuable contribution to effective crisis management and to support situation assessment after an earthquake. On the one hand, the developments of the disciplines correspond to the latest state of research, and this unique combination of disciplines is forward-looking. As a research institution, however, DLR enables experimental developments up to a certain level of operationalisation. For full operational use, the technologies must be transferred to partners from industry or public services that can provide the required level of personnel and technical expenditure on standby. For this reason, the solution can be used experimentally but not fully operationally at this stage.

In the future, the DLR team will continue to further develop its components in close cooperation with the end-users. Processes will be optimised and automated, and products will be adapted according to new user requirements derived from practical, experimental use and follow-up exchange with end-users. Trial AT and the international disaster exercise IRONORE are good evidence that developments of this kind are helpful and beneficial for disaster response. The numerous deployments of the systems in a wide range of scenarios continue to prove their value and will proceed in the long term.

References

Azimi, S., Kiefl, R., Gstaiger, V., Bahmanyar, R., Merkle, N. M., Henry, C., Rosenbaum, D., & Kurz, F. (2021). Automatic object segmentation to support crisis management of large-scale events. *The International Archives of the Photogrammetry, Remote Sensing and Spatial Information Sciences*, 433–440. https://doi.org/10.5194/isprs-archives-XLIII-B2-2021-433-2021.

Azimi, S., Vig, E., Bahmanyar, R., Korner, M., & Reinartz, P. (2018). Towards multi-class object detection in unconstrained remote sensing imagery. *Asian Conference on Computer Vision (ACCV)*, 2–6 December 2018, Perth, Western Australia. http://doi.org/10.1007/978-3-030-20893-6_10.

Bieker-Walz, L., Flötteröd, Y.-P., & Sohr, A. (2019). Modelling of emergency vehicles route choice with use of trajectory data. In *MT-ITS 2019 – 6th International Conference on Models and Technologies for Intelligent Transportation Systems*. MT-ITS, 5–7 June 2019, Cracow, Poland. https://doi.org/10.1109/MTITS.2019.8883323.

Copernicus Emergency Management Service. (n.d.). https://emergency.copernicus.eu/. Accessed 8 November 2021.

d'Angelo, P., & Kurz, F. (2019). Aircraft based real-time bundle adjustment and digital surface model generation. *ISPRS Geospatial Week*, 10–14 June 2019, Enschede, Netherlands, 1643–1647. https://doi.org/10.5194/isprs-archives-XLII-2-W13-1643-2019.

DLR/ZKI. (2021). *Exercise and UAV flight campaign in Ahrtal as part of the AIFER project*. www.dlr.de/eoc/en/desktopdefault.aspx/tabid-17471/27737_read-74744/. Accessed 15 December 2021.

Gähler, M. (2016). Remote sensing for natural or man-made disasters and environmental changes. *Remote Sensing of Environment InTech*, 309–338. https://doi.org/10.5772/62183.

Geister, D., Schwoch, G., & Becker, H. (2017). Flight testing of optimal RPAS Scan Patterns. *Journal of Guidance, Control, and Dynamics*, 54(5), 1–17. American Institute of Aeronautics and Astronautics (AIAA). https://doi.org/10.2514/1.C034337.

Gstaiger, V., Tian, J., Kiefl, R., & Kurz, F. (2018). 2D vs. 3D change detection using aerial imagery to support crisis management of large-scale events. *Remote Sensing*, 10(12), 1–19. Multidisciplinary Digital Publishing Institute (MDPI). https://doi.org/10.3390/rs10122054.

Henry, C., Hellekes, J., Merkle, N., Azimi, S., & Kurz, F. (2021). Citywide estimation of parking space using aerial imagery and OSM data fusion with deep learning and fine-grained annotation. *The International Archives of the Photogrammetry, Remote Sensing and Spatial Information Sciences*, XLIII, 479–485. http://doi.org/10.5194/isprs-archives-XLIII-B2-2021-479-2021.

Kurz, F., Tuermer, S., Meynberg, O., Rosenbaum, D., Runge, H., Reinartz, P., & Leitloff, J. (2012). Low-cost optical camera system for real-time mapping applications. *Photogrammetrie Fernerkundung Geoinformation*, 2012(2), 159–176. E. Schweizerbartsche Verlagsbuchhandlung. https://doi.org/10.1127/1432-8364/2012/0109.

Lechner, K., & Gähler, M. (2017). Earth observation based crisis information, emergency mapping services and recent operational developments. *Special Issue of the International Journal of Information Technology and Management. Inderscience Publishers. 4th International Conference on Information and Communication Technologies for Disaster Management*, 11–13 December 2017, Münster, Germany.

Leitloff, J., Rosenbaum, D., Kurz, F., Meynberg, O., & Reinartz, P. (2014). An operational system for estimating road traffic information from aerial images. *Remote Sensing*, 6(11), 11315–11341. Multidisciplinary Digital Publishing Institute (MDPI). https://doi.org/10.3390/rs61111315.

Merkle, N., Gstaiger, V., Schröter, E., d'Angelo, P., Azimi, S., Kippnich, U., Barthel, C., & Kurz, F. (2020). Real-time aerial imagery for crisis management: Lessons learned from an European civil protection exercise. *The International Archives of the Photogrammetry, Remote Sensing and Spatial Information Sciences*, XLIII, 1243–1249. https://doi.org/10.5194/isprs-archives-XLIII-B3-2020-1243-2020.

Schröter, E., Kiefl, R., Neidhardt, E., Gurczik, G., Dalaff, C., & Lechner, K. (2020). Trialing innovative technologies in crisis management – "Airborne and terrestrial situational awareness" as support tool in flood response. *Applied Sciences*, 10(11), 3743. Multidisciplinary Digital Publishing Institute (MDPI). https://doi.org/10.3390/app10113743.

Schwoch, G., & Niermann, C. A. (2021). *Luftbilderfassung im Rahmen einer Katastrophenschutzübung*. DLRK – Deutscher Luft- und Raumfahrtkongress, 31 August–2 September 2021, Germany. https://doi.org/10.25967/550089.

Yuan, X., Azimi, S., Henry, C., Gstaiger, V., Codastefano, M., Manalili, M., Cairo, S., Modugno, S., Wieland, M., Schneibel, A., & Merkle, N. (2021). Automated building segmentation and damage assessment from satellite images for disaster relief. *The International Archives of the Photogrammetry, Remote Sensing and Spatial Information Sciences*, XLIII, 741–748. https://doi.org/10.5194/isprs-archives-XLIII-B3-2021-741-2021.

PART 4

Towards a Paradigm Shift in Assessing Innovative CM Solutions

Part 4 revolves around the application of the DRIVER+ outputs, specifically the TGM, in four important EC-funded projects: Resiloc, ResponseDrone, Stamina, and IN-PREP and STRATEGY. The chapters provide insights into the scalability and the adaptation potential of the TGM, but they also highlight challenges and reflections that can be of interest to researchers and crisis managers alike who would be interested in benefiting from the finding of the DRIVER+ project. This part shows the DRIVER+ paved the ground to a paradigm shift in assessing innovation in CM and speaks to future and promising opportunities in the field.

DOI: 10.4324/9781003256977-19

12

TGM APPLICATION IN A HORIZON PROJECT

Jonathan Hall

Introduction

This chapter describes the approach taken to apply the DRIVER+ trial guidance methodology to a real case of field trial design based upon the requirements of a European project funded under SU-DRS01-2018-2019-2020 – human factors and social, societal, and organisational aspects for disaster-resilient societies. The project was called RESILOC (Resilient Europe and Societies by Innovating Local Communities) and ran from 2019 to the end of 2022. In its few pages, we endeavour to describe the trials, challenges, and successes.

Resilience is defined by the United Nations as "the ability to resist, absorb and accommodate to the effects of a hazard, in a timely and efficient manner." Thus, resilient communities are those in which their citizens, environment, businesses, and infrastructures have the capacity to withstand, adapt, and recover in a timely manner from any kind of hazards they face, either planned or unplanned. In recent years, efforts have been spent to tackle resilience, and there is, still, a long path forward in defining an EU valid and sound approach to the problem.

RESILOC aimed to study and implement a framework of studies, methods, and software instruments that combined the physical with the less tangible aspects associated with human behaviour.

The studies and methods served to create two software instruments:

1. The RESILOC inventory – a comprehensive, live structure for collecting, classifying, and using information on cities and local communities, implemented as a Software as a Service (SaaS).
2. The RESILOC cloud-based platform – for assessing and calculating the resilience indicators of a city or a community and for developing localised strategies and verifying their impacts on the resilience of the community. The cloud platform, a combination of SaaS and PaaS, includes the inventory as its repository.

DOI: 10.4324/9781003256977-20

In doing this, the project sought to make use of built solutions in four field trials in four diverse communities, stakeholders with very different infrastructures and demographics for each of the localities. The trials themselves needed to be underpinned by a robust academic methodological approach which led to using the TGM as timing of this work coincided with the end of the DRIVER+ project in 2020. The first lesson from this is that a project without a defined methodology will willingly accept a proposal to use one already established and carrying the credentials of a previous successful project.

The design was split into two distinct components: establishment of a methodology and creation of a resulting implementation plan.

The methodology was based upon the outcomes of the DRIVER+ project trial guidance methodology (TGM), the components of which were individually analysed, consulted upon, and adjusted to suit the purposes of RESILOC. The process and outcomes of each deliberation are outlined in this chapter.

A resulting implementation plan was created using a panel of noted "trial experts" followed by further extensive consultations with project partners and representatives from the host communities.

The following provides a practical analysis of the processes adopted whilst outlining some of the challenges encountered with such an approach. It will highlight where the process is thought to have worked well and where the authors consider that it could have been improved upon. The challenges and successes are described in some detail, the findings of which have also been shared with broader European crisis management and research partners through the support of the DRIVER+ Network of Centres of Expertise.

It is important to note that this text captures the trial design (methodology) process alone and describes it at a moment in time which pre-dates the delivery/implementation phase. The trials themselves and how they were delivered are presented in other public reports, most notably RESILOC D5.5: Field Trial Execution and Guidelines. The trial evaluation phase is also covered in a report; RESILOC D5.6: Field Trial Validation and a final report – RESILOC D7.2 – which presents evidence gathered from the RESILOC trials.

One of the strongest outcomes of the project was considered to be its description of the process for adapting the trial guidance methodology, originally created to trial innovative practical crisis management tools, into something that could support a different type of trial. It also highlighted the importance of embedding such a process at the earliest stage possible of any project (ideally at the proposal stage) and of getting buy-in from other project partners.

Overall Approach

Establishing a single high-level objective was essential to enable the trials to be designed while the products to be trialled were still undergoing development and many of the mechanisms required to deliver them, such as the proposed local resilience teams (LRTs), were in a pre-formative condition. The objective was stated simply as "the trials will test, verify and improve the project outputs." In a single sentence, this provided focus and three clear elements of the trials to be developed – testing, verifying, and improving.

Throughout the design phase, project partners were consulted extensively through a structured and project-managed process. A group of key consultees was agreed with partners at an early stage and a core team of end-user representatives was established to enable a detailed dialogue to take place ahead of many aspects of the project being finalised.

These two aspects alone (unknown product and the formation of user groups) made the process complex, but starting with a proven methodology such as the TGM and then adjusting it to the specific needs of the project at least provided a framework by which work could commence. Furthermore, employing professional designers with extensive experience of trial and exercise delivery and therefore capable of understanding the desired outcome proved invaluable.

An early study of the project grant agreement (the statement of what the European Commission thought they were funding) led to an immediate review of the structure that would be required to discharge it. The language used was far from clear. For example, the terms "pilots," "exercises," "trials," and "field trials" were used rather indiscriminately and interchangeably despite each having different connotations and meanings to different partners. In addition, there was mixed language around the terms "tabletop" and "full-scale trials," which were also used interchangeably at times. These issues needed working through with partners to achieve a shared understanding not just of what was meant but also what was possible within the confines and resources of a limited project.

Creating an agreed phraseology and glossary is not as easy as one might think when dealing with diverse partners from different sectors and multiple activities. At this point, it was considered helpful to consult with other projects studying disaster-resilient society (DRS) subjects and following similar goals. A parallel activity saw the creation of a "DRS cluster" to provide an environment where matters affecting multiple projects could be discussed. A proposal was made to this group to jointly create a glossary of terms against which each project could rely. Conflicts such as multiple definitions of the terms "resilience" and "field trials" were discussed and multiple sources referenced to provide consensus where possible. The DRIVER+ glossary of terms was influential but not as reliable as one might have hoped due to "soft conflicts" with common usage of terms in other contexts such as the European Civil Protection Mechanism (UCPM) and the United Nations Search & Rescue Guidelines (INSARAG). As a note for future trial designers, consistent language in a proposal might be considered indicative of a clearly understood and applied design process. Conversely, multiple conflicting and undefined terms may be indicative of unclear preparation.

Even the question of what was meant by the title "trial design" needed examination to establish just how far it would go towards enabling execution. As a process of evolution over the first two years of the project, enough components were established with sufficient agreement of partners to ensure that the task could be completed in plenty of time to allow effective execution.

Challenges

Two recurring challenges became apparent throughout the process of applying the TGM in the RESILOC project.

The first challenge was the necessity for engagement of all parties from the early agreement of a methodology through to it resulting in a workable design. Partners will generally be responsive to consultations and requests but not always sufficiently engaged sufficiently to participate as fully as preferred.

The second challenge was the linear nature of projects (a milestone approach) can hinder full engagement of delivery partners in the design process. For example, a key partner may not always be "stood up" or resourced to support all aspects of design for which they can add benefit.

Use of the Trial Guidance Methodology (TGM) as a Starting Point

The TGM states that a trial has an objective and needs to be structured. It also implies a co-creative approach and an open mind. Workshops and tools are essential, as several iterations (especially for preparation) are usually needed. Trials are evolving processes: they grow "in the making," like a handcrafted artefact.

Time should be devoted to adjusting the design. Key decisions must be taken in agreement with different stakeholders that need to be identified. The success of a trial then clearly depends on its design: A robust design will lead you to find appropriate answers to your needs. This trial guidance methodology provides step-by-step guidelines, a list of roles and responsibilities, and tools and methods to perform a trial through a clear, structured, and co-creative approach.

The RESILOC project helpfully agreed use of the TGM at an early stage and, in the absence of an alternative methodology, committed to adhere to its principles. This was a valid decision with limited alternatives, but at the time the decision was taken, there was a limited understanding among those taking the decision of its implications and likely requirements.

It was recognised at an early stage that application of the TGM in its entirety was considerably beyond the capacity of many projects for a number of reasons:

Resourcing: The trial component and budget of a project is likely to be insufficient to facilitate implementation of the TGM in its entirety.

Focus: The intent and focus of the RESILOC project was to create new strategic assessment capability, not to deliver trials. The trials were merely one component of the project with the main aims of extending use of the tools and verifying them in a locality and scenario context.

Training/capability: To have deployed the trial tool in its entirety (even in moderation) may require a training and support regime not available to most projects. This was determined by four members of the trial design team attending training and evaluating its potential impact upon the rest of the project.

Stage of development of solutions: The TGM assumes the availability of a range of solutions to evaluate their performance against assessed gaps. In many projects, a single solution is under development but, in a research context, may not be at a fully developed stage of development at conclusion of the design phase. Later determination of detailed solutions may render analysis of them through trials more complex.

Nature of the tools / products to be trialled: The TGM was developed for creating trials of tactical and/or operational tools. Projects looking at tools for the more strategic level of decision-making (predominantly in preparation for a described scenario rather than during it) may have difficulty flexing the methodology sufficiently.

Alignment of Trial Design, Execution, and Evaluation Activities

European project proposals involving product development often described a series of "work packages" that describe a linear progression from design through to execution and then to evaluation/validation. This cannot be achieved without the full engagement of all parties in each aspect of evolution or disconnects between design and delivery are likely. It will be of benefit to appoint and commission an execution team at an early stage and especially before conclusion of the design phase.

Practical application led to a better understanding that trial management from design to delivery and validation needs to be a wholly iterative process requiring the planned and structured involvement of all relevant parties from the outset.

Management of Partner and User Expectations

Trials rely on an understanding of "what is to be trialled" – described in the TGM as "solution selection." At the start of many projects, however, there may be limited substance to this, and subsequently, it may be far from clear what the solution will be. Without this, it is hard to describe at what phase of a crisis the solution will be used or even at what level within an organisational it might be deployed (strategic, operational, or tactical). Will it support operational decision-making, dynamic and constantly changing in the face of an evolving scenario, or will it be a more strategic product supporting the preparation phases of a crisis or disasters?

The answer to this question may, of course, be unclear, especially where concepts such as perception of threat, the use of on-the-ground sensors and social media analysis are included. Projects may have to run for some considerable time before such an issue can be coherently addressed.

From the commencement of the RESILOC project, it was understood that trial design was to take place in an extremely challenging environment requiring the agreement for a series of trials in communities whilst the project and solutions were still maturing and evolving. Only some way into the project did it start to produce answers to basic questions which, to the practitioner and user, felt obvious prerequisites of understanding to be able to design and prepare for any trial. This may sound critical of the project, but the opposite is true! Research and innovation are never easy or linear, so whilst the involvement of practitioners is essential from an early stage, the expectations of ultimate users requires very careful management and communication to avoid disenfranchising them while a project becomes coherent for them.

Incorporating a Formal Trial Methodology into a Project

First, we considered structural aspects of the project design which created challenges for trial design:

A project designed with community engagement and trials combined into a single activity might find application of the methodology difficult. The two components (community focus and trial design) involve very different groups of partners: community engagement naturally focuses on the localities involved in the trials, whereas trial design relies more on DRR partners and solution developers (systems, process, and IT experts).

Although community (or user) engagement is a core component of successful trial design, it will not necessarily be the primary driver in a project that is largely focused on theoretical threat or behavioural analysis leading to strategic solution development. Design should arguably focus more on the knowledge and understanding of the communities it will work with from the outset and use *their needs* as the basis for solution development rather than allowing the two workstreams to separate. Trial design will not be the only process disrupted by such an approach, but it may be one of the earliest activities to manifest such a disconnect and to demonstrate resulting problems.

If strategic solutions are to be created based upon a *global approach* to disaster risk reduction and proxy inclusion (the use of commonly used datasets) and then tested within a diverse but limited number of communities, the problems of combining trial design and community engagement risks the trials becoming the "battleground" for the resolution of fundamental (and legitimate) research tension. This might of course be helpful for the project overall whilst complicating the trial design process.

Fundamental to the design of any trial process is the methodology that it will adopt. It is this approach that will provide the answer to many questions raised by those involved. It determines and records the answers to many of the why as well as the how questions. A trial design must include a robust methodology, not just a practical description of implementation. A simple sentence such as "Implement an agreed methodology" in a proposal will mask considerable scope for misunderstanding and problems further down the line. It may not only leave fundamental questions of process unanswered, but it may also leave their resolution un-resourced.

An early agreement among sponsors and partners for this is vital, but it has to be accepted that this will sometimes need to be achieved without a good understanding of what it will involve and the detail that will be required. That will follow as the project evolves. This would imply that an experienced trial design partner is a requirement for well-founded proposals.

Successes

The process of trial design is itself a substantial enabler to a complex project. The process of preparations requires partners to address fundamental questions of project design and objectives that are not always well vocalised outside of the trial process.

As an example, the process of capturing the trial objective in a single simple statement is beneficial to assist technology-focused partners "translate" their workings to end-users who may not be so well informed on the impact that is desired.

The trial design process may well result in a natural alliance of those that wish to deliver the most benefit for their stakeholders. In the experience of the authors, it was noted that the same organisations repeatedly engaged more than others, and as a result, their own stakeholders were increasingly deriving tangible and additional unplanned benefit from the design process in the form of a growing understanding of what the project was trying to achieve.

The Place for Broader Networking in Trial Design and Management

Conceived by the same DRIVER+ project that developed the TGM, the Crisis Management Innovation Network Europe (CMINE) was created to bring together practitioners, researchers, policy-makers, and industry into a single network environment where they could communicate and influence the direction of future resilience innovation, whatever their speciality or level of expertise.

When the project closed in 2020, under an agreement with the Research Executive Agency of the European Commission, the custodianship of CMINE passed to an associated organisation, the Resilience Advisors Network, as part of a sustainability and impact strategy designed to strengthen and support its embryonic community. Since then, the network has grown exponentially with many crisis management professionals engaging with it. The community now carries an ambition considerably beyond its conception as, under its umbrella, members meet in a neutral and informal but structured and moderated professional environment to discuss such matters as new product innovation and realistic trial methodologies.

Membership suggests that there is clearly demand for such a network, but CMINE does not seem to be following the same path as many other practitioner forums. In this section, we consider why that might be and why its model may prove helpful for further trial design and management professionals.

The Internet is often described as a "network of networks." It is a connection of multiple computer networks around the world into one entity or interphase. It is not one big computer but, rather, numerous networked computers connected together. CMINE enables such a concept for the trial design and crisis networking world – a network of networks, the connection of groups around the world into one entity, so to speak. Not one big organisation but rather numerous networked individuals, groups, and organisations connected together.

The structure of recent European research initiatives – FP7, Horizon 2020, and the latest, Horizon Europe – is the EU's Ninth Framework Programme for research and innovation. It is the largest programme of its kind anywhere in the world, with a total budget of around €95 billion. To make a substantive difference to European resilience thinking and to influence trials in this arena, it is necessary to be involved.

Involvement can be complex though. Terminology and process alone create considerable barriers to entry. Language, nomenclature, acronyms – all with the potential to put the expert off practising their skills where they are most needed.

What Is CMINE?

At its most simple, CMINE is a professional networking website. It supports the interests of those involved in all aspects of crisis management and civil protection by encouraging innovation in crisis and emergency prevention and resolution. It provides an environment where practitioners, policy-makers, researchers, and industry can connect and discuss ideas relating to process and product innovation as well as trialling and validation in all of its forms, in an informal and neutral environment.

To make itself accessible to the broadest range of experts functioning within the crisis management arena, the site is organised with a communal area for all members and specific groups/forums to address specific interests and specialities. As an example, a single research project or theme may have its own area, as well as a shared area where it interacts with colleagues engaged in related themes. There are specific areas for those interested in specific aspects of the DRIVER+ legacy and its ongoing work such as trial design, the register of crisis management tools/products in development (known as the portfolio of solutions), and training. Finally, there are forums dedicated to specific professional subjects, such as wildfire, flooding, and CBRN-E.

The network provides a number of other added-value services, such as rolling newsfeeds coming out of the main emergency and humanitarian organisations – UNDRR, OCHA, and the Union Civil Protection Mechanism, among others.

Supported by a fully functional event management facility, it also facilitates events for discussing, promoting and administrating trials conferences and meetings by handling basic logistics such as registrations an ticketing.

Cross-Sectorial Resilience Networking with Practitioners and Policy-Makers

CMINE has increasingly become a natural home for aligned H2020 and Horizon Europe projects wishing to share knowledge and streamline processes wherever they can. In discrete groups of like-minded professionals, some innovative and collaborative steps have already been taken in the areas of ethics, etymology, and standardisation.

Ironically, almost every new funded project proposes developing a new platform or tool for a specific interest group or function and spends months building capability appropriate to this interest. These can be very good offering promise and ambition, but they frequently struggle to create a sustainable legacy strong enough to become embedded within the sectors they target. It is rare to see a platform created within a project survive in any useful or meaningful way beyond its end date. By providing a "hub" for communities of practice rather than by being one in its own right, CMINE seems to be providing a different solution.

The Future of Trial Guidance Networking

CMINE does not actually aspire to be a network in its own right: It seeks rather to facilitate the coming together of many of those networks already in place or in the process of being created:

- National networks such as that operating successfully in the UK, Germany, and Spain.
- Thematic networks for specialist subjects, such as community resilience, CBRN, or wildfire.
- Project-based networks, such as the Community for European Research and Innovation for Security (CERIS), and sub-networks, such as that for disaster-resilient societies (the DRS01 cluster of projects).

These have a sustainable and natural home in an environment where they can both function for their niche interest groups whilst also breaking out into the larger and more inclusive network of networks, where everyone can interact as and when appropriate. In this manner, the opportunity exists to gain partners in the mutual space where practitioners and trial designers who wish for sustainability/longevity have the ability to survive and grow in an ever-changing environment of crisis management, disaster risk reduction, and security enhancement.

The recent emergence and realisation of the European Commission's rescEU Union Civil Protection Knowledge Network and CERIS's increasing presence and support for research in the field of disaster resilience and security and the emerging science pillars of the commission and its many agencies lead to more complexity in networking as everyone needs somewhere to converge and to meet with, otherwise potentially excluded parties involved with ground-breaking innovation. The industry, practitioners, and academics are constantly pushing the boundaries and a structured but informal network of networks (such as CMINE) may add an additional dimension.

13

APPLYING THE TRIAL GUIDANCE METHODOLOGY TO EVALUATE RESPONDRONE

A Situation Awareness Platform for First Responders

Max Friedrich, Joonas Lieb, Richard van Oorschot, and Alexander Scharnweber

Introduction

In recent years, natural disasters are causing an increasing threat to humanity. In 2018, globally 61.7 million people were affected by extreme weather events, such as earthquakes, floods, landslide, or wildfires (UNISDR, 2019) Moreover, man-made disasters and harmful accidents occur frequently in globalised society. Responding to any of these types of incidents are different emergency organisations (e.g. local/federal police, firefighting agencies, emergency medic, etc.) with their specific capabilities and expertise.

One key element to prevent severe causalities in any disaster or to administer first aid support to victims of accidents are fast and efficient response times. Hence, the need for innovative and sophisticated technological means to enable first response organisations to quickly assess and react to any disaster situation becomes evident. Emergency response teams are required to quickly gain a thorough understanding of the disaster or accident. In order to quickly obtain situation awareness during a disaster, the use of unmanned aerial vehicles (UAVs) in first response missions is emerging as one key enabler (Drury et al., 2006; Doherty & Rudol, 2007). However, integrating UAVs into first response missions and especially coordinating the missions of multiple UAVs simultaneously while assuring safety is challenging.

In order to address the challenges related to quickly gaining a thorough understanding of the disaster with the help of UAVs, the ResponDrone project aims to develop a situation awareness platform for first response missions. Highly automated UAVs provide critical information and communication services to first responders in emergency and crisis situations.

DOI: 10.4324/9781003256977-21

The ResponDrone Platform

The aim of the ResponDrone platform is to improve and accelerate situation assessment, information sharing, decision-making, and operations management. This is accomplished by providing information stemming from sensors mounted on multiple UAVs through a web-based system to any stakeholder involved in responding to the crisis situation. To ensure that first response organisations can make use of the platform, it needs to be integrated into the organisational structures and the operational workflow of the first responders. This is a challenging task since every organisation has their own organisational structures and workflows that do not necessarily match the organisational structures and workflow patterns a new solution, such as the ResponDrone platform, requires. In order to address this challenge and support the integration of the ResponDrone system into today's first response operations, its design, development, and evaluation have been closely accompanied by first response organisations.

Studies that have been carried out during the ResponDrone project (ResponDrone, 2019) and the continuous consultation with first response organisations showed that the different organisations require different features and solution adaptations in order to best make use of the platform. For this reason, we designed the ResponDrone platform in a modular manner with separate components working together. We developed a number of components that can be viewed as core components (that are necessary) and components that could be considered plug-ins. The possibility of adding plug-in components with the features a specific first response organisation requires makes the platform scalable and adaptable. Figure 13.1 shows the concept of the ResponDrone platform.

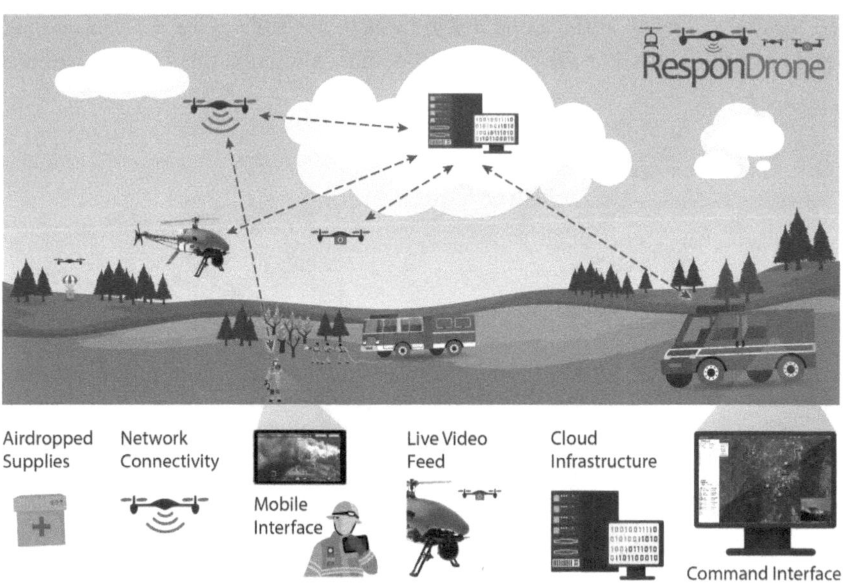

FIGURE 13.1 Overview of the ResponDrone concept.

Multiple highly automated UAVs with various payloads (e.g. camera, communications relay, or package drop functionality) can be commanded via a user interface, either via an on-site command centre web application or a mobile app on a smartphone (for more details, see the section "Multi-Mission Control System and User Interfaces"). The information is shared in real-time via a cloud, allowing all involved stakeholders access to the same information at all times. The following chapters will describe the components we developed.

Unmanned Aerial Vehicles and the Ground Control Station

The UAVs we used in the ResponDrone project are built by Alpha Unmanned Systems. The UAVs from Alpha's catalogue used in ResponDrone are the Alpha 800 (A800) and the Alpha 900 (A900; see Figure 13.2), both of which are gasoline powered, leading to extended flight time when compared to electric UAVs. These UAV helicopters share some important features: Both were designed to comply with both civil and military standards and can operate in demanding conditions of wind, rain, and other weather scenarios, certified for IP65 or higher in their critical systems.

The UAV's Ground Control Stations (GCS) and tracking antennas have also been designed for adverse conditions, being capable of offering operational ranges up to 50 km with live-video transmission. The GCS is called G-Case Duo (see Figure 13.2), which is a rugged computer with a high-brightness dual screen, capable of working off-grid for hours. It has an expansion bay for any additional equipment that the mission may require.

Alpha's UAVs are controlled by Vector autopilots on board the aircraft and Vision Air Software installed in the computer in the G-Case. One of the most interesting features of the software is the capability of commanding multiple UAV helicopters

FIGURE 13.2 Alpha 800 (upper left), Alpha 900 (lower left), and the G-Case ground control station (right).

from a single GCS. This reduces the personnel needed at the emergency site while increasing the operational capacity of the involved entities by enabling multiple, parallel UAV missions at once.

Multi-Mission Control System and User Interfaces

In order to manage and control missions and make sure all stakeholders have the correct information, the platform needs an appropriate user interface. Therefore, the ResponDrone platform has a web-based multi-mission control system with two user interfaces, which are the On-Site Command Center (OSCC) web application and the first responder mobile application (see Figure 13.3). These applications are the user interfaces for the different first responder roles, allowing them to coordinate missions and have access to all the data that is gathered. The purpose of the web-based platform is to provide first responders with the capability to manage the disaster response missions including (but not limited to) task assignments, the creation of incident reports, and the planning and supervision UAV missions.

Traffic and Mission Management System (TMM)

In the ResponDrone platform, the central component responsible for coordination and planning of the UAV missions is the Traffic & Mission Management component (TMM). The TMM is a software component which serves as a gateway between mission requests originating from the end-user via the OSCC's user interface and the UAV's GCS. As such, it is responsible for accepting mission requests, finding safe, deconflicted trajectories to ensure that the UAVs do not collide, and executing these trajectories (i.e. sending them to the UAV and monitoring their performance). Furthermore, the TMM is responsible for feeding UAV telemetry data back into the ResponDrone system, which provides the necessary live information about UAV status to all stakeholders and enables first responders to maintain situation awareness.

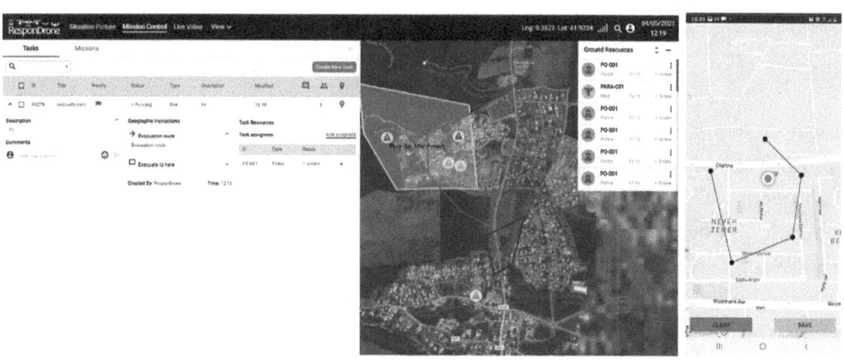

FIGURE 13.3 The On-Site Command Center (OSCC), the desktop interface of the ResponDrone platform (left), and the first responder mobile app (right).

The TMM uses a state-of-the-art 4D conflict detection and resolution algorithm developed at DLR for deconflicting UAV trajectories, called NDMap (Kuenz et al., 2017). The NDMap software module allows for the identification of conflicts between 4D trajectories of different UAVs. The 4D trajectories are trajectories of UAVs, where the fourth dimension represents time: each point on a 4D trajectory denotes that at a given time, a given position in 3D space will be occupied by the UAV. Polygons are, in a similar fashion, defined primarily in two dimensions, and, in addition, contain intervals for altitude and time (in other words, the altitude and time of existence of a polygon can be bounded). Furthermore, the NDMap is able to consider 4D no-fly zones (i.e. temporally adjustable no-fly zones) for the planning of the UAV's trajectories.

Communications Relay

The ResponDrone platform also provides a communications relaying service that uses the platform's UAVs to transport INESC TEC's Payload Communications Relay (PCR). The PCRs form a private aerial wireless network infrastructure that provides on-demand wireless communications to first responders and victims in emergency scenarios. This network allows first responders to access the Internet, communicate among themselves, use the ResponDrone mobile app running on their smartphones, and relay data from and to the command centres or the cloud. Being a private network, it works independently of the state of existing communications infrastructures. Thus, in addition to offering wireless communications, the UAVs are able to extend the coverage and capacity provided by remote cellular base stations or fixed Wi-Fi access points, taking advantage of the line-of-sight component induced by the UAV's altitude to avoid obstacles to radio signal propagation. The network traffic generated by the ground users connected to the PCR is intelligently forwarded to the Internet either through the UAV's air-ground data link or through an onboard cellular connection to a remote base station.

The communications relaying service can be requested on-demand by an on-field commander to meet the dynamic communications needs of the scenario, including the provision of communications coverage in a given area of interest, or even follow selected first responders (moving on foot or in vehicles) to provide them with permanent communications. These requests are processed by the airborne network controller (ANC), which may be running locally on-site or in the cloud. The ANC calculates the positions of the UAVs and the radio-frequency parameters of the PCRs to achieve the mission's objectives. The UAVs can be repositioned over time, taking into account the dynamics of the disaster, the constraints imposed by the scenario, and the behaviour of people, while ensuring their safety. Given the long flight times of the UAVs that are used, the coverage can be consistently provided for hours.

The PCR hardware weighs approximately 600 g, which enables it to be easily integrated onto a wide range of UAVs. Moreover, experimental tests showed that the PCR can provide communications to users on the ground up to 1 km away from the UAV and enable users to stream video up to 500 m away. Figure 13.4

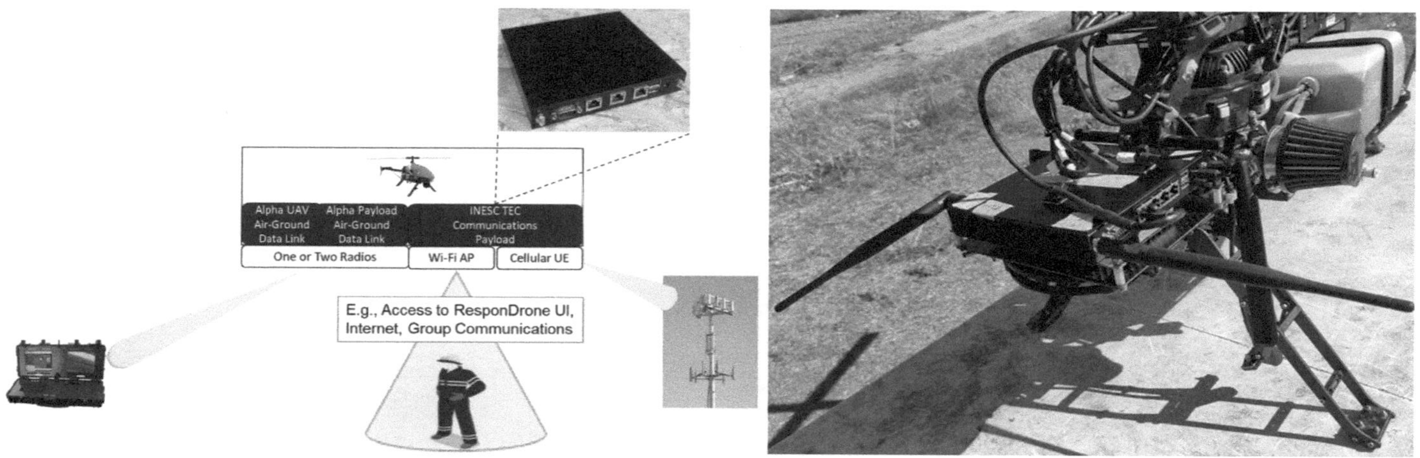

FIGURE 13.4 The communications system of the ResponDrone platform providing on-demand Wi-Fi access to first responders.

shows the concept of the communications relaying service and the hardware component, mounted on a UAV.

Video Streaming System

The type of camera gimbal used for the ResponDrone missions is an intelligent dual sensor payload developed by THALES. This payload has the particularity of having both electro-optical (EO) and infrared (IR) cameras coupled with an artificial intelligence-enabled computing device embedded into the same enclosure. This hardware architecture enables the usage of an on-board real-time video analytics system, leveraged by machine learning algorithms in order to, for example, automatically track specific entities of interest, such as victims, first response units, fires, or vehicles in real-time.

The intelligent dual camera system is composed of two main elements: (1) an enclosure containing both EO and IR cameras, along with the embedded computing device, and (2) the gimbal. The first one is responsible for acquiring, processing, and streaming video feeds and processed data – that is, run the video analytics algorithms for detecting and geo-locating a victim or a fire, for example. The gimbal is responsible for pointing the camera payload towards the direction requested by the user using the first responder mobile application or the OSCC user interface (see the section "Multi-Mission Control System and User Interfaces") by means of manual or fully automatic control schemes.

Decision Support System

The ResponDrone Decision Support System (DSS) is a service that produces insights to the first responders, based on the analysis of crisis and environmental data.

DSS is based on CEA List's ExpressIF software (CEA, 2022) that allows representing expertise. Expert knowledge is modelled in the form of rules that are close to natural language. Thanks to different logic like Boolean logic and fuzzy logic (Iphar et al., 2021), knowledge can be formulated in certain or vague terms as appropriate. ExpressIF is then able to apply this knowledge on different cases. The inference engine is able to execute more than one million rules per second. Within the scope of the ResponDrone project, different use-cases for first responders were developed. This knowledge is editable through a web interface in order to customize the behaviour of the DSS.

Firstly, DSS helps prioritising the buildings to evacuate or protect: ExpressIF can connect to geographical information systems to retrieve all buildings and monuments in a given region and assign them risk and vulnerability scores, according to the fire line, wind direction and speed, and several other factors. Secondly, ExpressIF is able to reason on digital terrain models. This new functionality allows the DSS estimations for the fire spreading according to the slopes, the wind, and several other factors. This estimation is then used to assess the risk and vulnerability scores.

Finally, we developed an estimation of the water dropout points when an aerial fleet is necessary to fight a fire.

Applying the Trial Guidance Methodology

From the very beginning of the project in May 2019, ResponDrone has always been an end-user-centred project. Through this end-user-driven, iterative design and development approach, it was ensured that the end-users actively participate in the design, development, and evaluation of the platform. For example, an initial evaluation of the OSCC and first responder mobile app user interfaces was performed with the end-user partners using mock-ups of the two systems (Friedrich et al., 2021). Altogether, nine end-user organisations are part of the consortium, whom the technical partners can continuously consult through a dedicated end-user panel and the very first work package was dedicated to study current end-user procedures and needs. Additionally, a design thinking workshop was held over three days together with the system developers and the end-user partners. The aim of the workshop was to gather ideas for solutions that were both operationally useful and technically feasible (for further reading, please see Friedrich et al., 2020). Figure 13.5 shows some impressions from the workshop.

After the initial design was finished and the technical development of the platform was progressing, we started to plan the final system demonstration. The original plan foresaw a demonstration of the ResponDrone platform within the scope of a regular field exercise. Without doubt, the integration of the platform into such

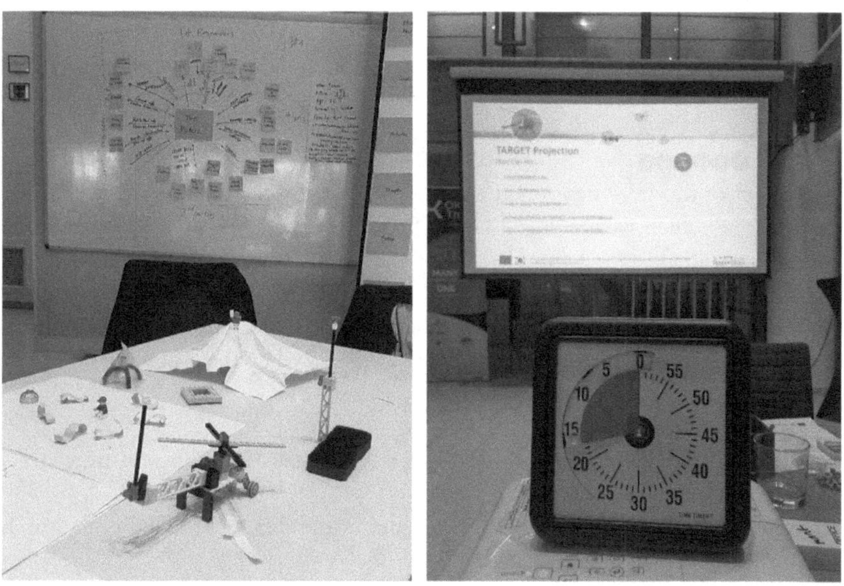

FIGURE 13.5 Impressions from the design thinking workshop.

an exercise would have been a great opportunity to demonstrate the platform and show its unique capabilities to external parties. However, due to the dynamic and unforeseeable nature of an exercise, it would have been extremely difficult to gather the data needed to properly evaluate the platform and show its benefits compared to current first response operations. Therefore, we decided to look for approaches and methods suitable for testing and evaluating the platform. In the course of this, we quickly identified the Trial Guidance Methodology (TGM) as the most suitable approach for ResponDrone, as it provides a comprehensive framework for planning and executing trials with the aim to evaluate novel socio-technical system solutions in a standardised fashion. Although we were already halfway through the project, we decided to adopt the TGM for the evaluation of the ResponDrone platform. For this purpose, we modified the respective work package contents and task structures according to the three main phases of the TGM cycle: (1) preparation, (2) execution, and (3) evaluation. By doing so, we made sure to follow the TGM as closely as possible. The following sections describe how we integrated the different phases of the TGM into our workflow and present the core results of each phase. It is worth noting that at the time of writing this chapter, the trial run has not yet taken place. Therefore, we will describe the envisioned results for the phases that have not yet been conducted – namely, the trial run and the evaluation phases.

Results of Applying the TGM

This section presents the results that we obtained by going through each step of the TGM. The methods we used and the key findings are outlined for each step separately.

TGM Phase 1: Preparation

Step Zero

The results of the aforementioned end-user studies and the design thinking workshop, during which we specifically identified the current gaps and the operational context, provided a good foundation for the whole step zero of the TGM. We categorised the results in the categories "End-user expectations," "Potential for improvement in current operations," and "Needs." The end-user expectations towards the ResponDrone platform can be summarised by the following three key points:

- Increase speed and accuracy of disaster assessment supporting on-time decision-making.
- Quick identification of scale and kinetics of the disaster and ability to search for living organisms.
- Improved technological capability for sharing information among all involved stakeholders.

Based on this information, the findings from the end-user studies and the results of the design thinking workshop, we derived the potential for improvement in current operations. These results can be summarised as a lack of the following:

- A comprehensive overview of the disaster situation.
- An overview of available resources, units in the field, and positions of victims.
- The availability of communication means.

In a similar manner as for the identification of potential improvements, specific needs of first responders during a disaster situation, such as a forest fire, including search and rescue missions, were derived from the findings of the end-user studies and the results of the design thinking workshop. These are the need for the following:

- Real-time, relevant, and reliable information on the situation in the disaster area to support on-time decision-making.
- Positions and health conditions of victims, first response units deployed in the field, available resources, and potential hazards.
- Technological capabilities for sharing information among all involved stakeholders.
- Easy and safe system deployment and operation.

With regard to the trial context, the design thinking workshop also provided valuable input. The first responders mentioned that during a disaster situation, it is crucial to quickly obtain a comprehensive overview of the situation. This would be especially difficult in unmanageable and large regions with limited means of communication (due to disrupted communications infrastructure) and potentially hazardous conditions, such as bush fires or earthquakes.

Six-Step Approach

The following six-step approach to plan the trial was conducted via three online workshops with all involved partners. The workshops were conducted online because the COVID-19 pandemic unfortunately prevented us from meeting physically in person. During the workshops, we gathered the views of each partner regarding their objectives for the trial run, including specific research questions they would like to have answered. In order to facilitate the discussions and gather all relevant data, we used the online platform Concept Board (Ondreicsik, 2022) during the work sessions of the workshops. Using the virtual concept boards, each partner developed data collection plans and evaluation approaches for their individual research questions. The workshops were supervised and mediated by the work package leaders.

From the material that each partner developed, we finalised each phase (except for the solution selection phase since our goal was to test the ResponDrone platform) of the six-step approach by taking care that as many of the partner objectives as possible were met and their ideas included. This was not always trivial since the objectives among the different end-user organisations differ (to some extent) and may perhaps also differ from the objectives a technical partner has. Whereas the technical partners would, for instance, be more interested in testing and demonstrating the systems they had developed, the end-user partners were especially interested in how well the ResponDrone platform supports them during their work and how its performance compares to their currently used systems.

Scenario Formulation

Due to the differing objectives and our ambition to cover as many objectives as possible, we decided to go for a twofold approach during the scenario formulation phase, meaning that the activities will be split into a demonstration and a trial run. This meant that we first created scripted scenarios for specifically demonstrating the capabilities of each component. The purpose of these scenarios was to show how we envision the ResponDrone platform to work during a firefighting and a search and rescue mission in areas with disrupted/destroyed communications infrastructure. Second, for the trial run, we created unscripted scenarios to evaluate the performance of the ResponDrone platform compared to the equipment and procedures first response organisations currently use.

In order to show the benefits of the ResponDrone platform over currently existing systems that first responders use, the performances of these systems need to be compared. As mentioned earlier, at the time of writing this chapter, the trial run has not yet taken place. Therefore, we will switch to future tense for the remainder of this paragraph. The experimental design for the evaluation trials as part of the trial run will include two conditions: (1) "Currently used equipment" and (2) "ResponDrone platform." During the evaluation trials, the participating first response organisation will be confronted with two similar, unscripted scenarios, both of which will feature a search and rescue mission. In both scenarios, the first responders will be asked to find an unknown number of missing persons and bring them to safety, first using their own equipment and procedures (in our case involving the use of one manually flown UAV) and then using the ResponDrone platform (involving multiple automated UAVs). During each scenario, we will collect the data as defined in the created data collection plans and will compare the data between the two scenarios. Table 13.1 shows an example for how the trial objectives were broken down into research questions, data collections plans, and evaluation approaches.

TGM Phase 2: Execution

After successful completion of the six-step approach and thus the preparation phase, we started to plan for the dry runs and the actual trial run.

TABLE 13.1 Example Results of the Different Steps of the TGM Preparation Phase

TGM Phase	*Example Result*
Trial objective	Show that the ResponDrone platform improves situation awareness compared to current first response operations.
Research questions	Does ResponDrone lead to a more accurate perception of the mission and disaster situation (i.e. position and status of victims, first response units deployed in field, available resources, and potential hazards) when compared to current operations?
	Is the subjectively perceived situation awareness better with the ResponDrone platform compared to current operations?
Data collection plans	In predefined time intervals, ask the first responders to draw the positions of hazards, victims, resources, and so on, on the map (SAGAT method; Endsley, 1988).
	After the trial, ask participants to rate their subjectively perceived situation awareness.
Evaluation approaches	Measure the distance between the actual and the reproduced positions on the map (SAGAT method; Endsley, 1988).
	Situation awareness questionnaires: SAGAT (Endsley, 1988) and SASHA (SJU, 2022)

Dry Runs

The dry runs were initially scheduled for spring 2021 and the trial run for fall 2021. Ideally, we would have liked to hold two dry runs, one dry run focusing on the correct technical integration and functionality of each sub-system and one end-user focused dry run, during which the end-users would have the opportunity to test out the system and practice for the actual trial run. However, due to the COVID-19 restrictions, we were limited in our ability to travel and therefore unable to perform the second dry run. The first dry run was successfully completed in October 2021. All the individual components were tested and complete missions (starting from requesting a UAV mission via the first responder mobile application until the UAV reached the desired location) were flown. During the first dry run, we identified a number of technical issues we had not been able to detect before and that needed to be fixed and tested before the trial run.

Trial Run

Travel limitations and restrictions regarding in-person meetings due to the COVID-19 pandemic unfortunately also forced us to adapt the final trial run. The original plan foresaw the active participation of all end-user organisations of the project consortium as either observers (to gather data) or actual trial participants using the platform in various crisis scenarios. However, the situation forced us to reduce the number of participants and limit participation to the minimum amount of people necessary to conduct the trial run. These include representatives of the technical partners and altogether four representatives of one end-user organisation.

The trial run will consist of two main trials. The first trial will be a capability demonstration, during which we show how the ResponDrone platform is meant to be operated and demonstrate its unique advantages. The platform will be operated by the four participating first responders, and they will conduct two scripted scenarios, one featuring a wildfire and the other one featuring a search and rescue mission. The second part of the trial will be the actual evaluation of the platform compared to currently used equipment. As described in the previous chapter, the first responders will conduct two similar search and rescue missions, first with their own currently used equipment and procedures and then with the ResponDrone platform. The partner acting as evaluation coordinator (i.e. DLR) will send two representatives to gather the data needed for the evaluation.

In order to give the rest of the consortium and external, interested parties the opportunity to take part in the trial run, we will record the most important parts of the trials on video and present the videos at dedicated online workshops after the trial run. In addition to presenting the videos, the partners will also be able to operate a simulated version of the ResponDrone platform.

TGM Phase 3: Evaluation

For evaluating the data gathered during the trial, we classified the measures into six different evaluation topics. Table 13.2 presents the topics and respective measures for evaluating the platform. Some measures may only be collected during the trial with the ResponDrone platform and some measures during both conditions: "Currently used equipment" and "ResponDrone platform." The measures collected in both conditions will be the ones used to compare the performance of the ResponDrone platform to the currently used equipment. The other measures serve as further evaluation measures for the ResponDrone platform. As presented in Table 13.2, we plan to conduct structured interviews with the participating first responders. We hope to collect informative data on the platform and potential suggestions for improvement during these interviews. The results will be used for dissemination and exploitation purposes.

Conclusion

In this chapter, we introduced the ResponDrone platform, which aims to make it easier for first responders to gain an improved situation awareness. One of the main ways we are doing that is by enabling all users to request and set up UAV missions and viewing their output in real-time. Regular users do not need a detailed knowledge of how UAVs fly. Splitting the technical UAV operation off to dedicated UAV pilots and assisting technology means that the impact of the gathered data is significantly increased. One of the main advantages of the way that the ResponDrone

TABLE 13.2 Measures Taken into Consideration for Evaluating the ResponDrone Platform

Evaluation Topic	*Measure*	*Current Equipment*	*ResponDrone*
Situation awareness	SAGAT (Endsley, 1988)	Yes	Yes
	SASHA (SJU, 2022)	Yes	Yes
	Eye tracking at OSCC	No	Yes
	Structured interview	Yes	Yes
Situation assessment	Time to find an item (e.g. victim)	Yes	Yes
	Eye tracking at OSCC	No	Yes
Decision-making	Structured interview	No	Yes
Usability (OSCC and mobile app)	System Usability Scale (SUS; Brooke, 1996)	No	Yes
	Debrief questionnaire and interview	No	Yes
	Eye tracking at OSCC	No	Yes
Operational efficiency	Information flow questionnaires	Yes	Yes
	Structured interview	No	Yes
Technical performance	For example, available Wi-Fi bandwidth	No	Yes

platform has been designed and built comes from its flexibility. It is designed as a true system of systems, where it can be implemented as is but also can be relatively easily adapted to the unique situation of each end-user and their procedures. The main chain of user interfaces, TMM and UAV component can be enriched by a variety of payloads and other plug-in components as the different first response organisations require them. The platform's structure using shared repositories and application programming interfaces (API) means that components can use each other's data easily, incrementally building knowledge and insight.

Furthermore, we presented how we integrated the TGM to design the evaluation trials of the ResponDrone platform despite the restrictions we faced amid the COVID-19 pandemic. We have found that the TGM brought us a way to structure our thoughts and streamline processes, especially at stages where it was sometimes not quite clear how to continue. We did have to adapt the TGM a little bit as our aim was different from its original; ResponDrone adopted the TGM with a running start in the middle of the project and did not follow the methodology from a problem definition per se. Instead, the project had gone through quite a bit of end-user led co-development already, and the TGM came into play in the evaluation phase. We found that the TGM (and especially its well-defined steps) allowed for this, and we can easily adapt the process. Additionally, due to COVID-19, we had to change a number of sessions from in-person to remote workshops, and we would advise others to use in-person as much as possible.

The next step in the development cycle of the ResponDrone platform is to conduct the trial run for demonstrating and evaluating the platform. After the trial run, we will follow the steps of the evaluation phase of the TGM to ensure that the results are properly analysed and disseminated.

Acknowledgements

The authors thank Juan Perrela (Alpha Unmanned Systems), Eli Neeman (IAI), Jean-Philippe Poli (CEA), Annette Temme (DLR), Michael Borkowski, and André Coelho, Eduardo Almeida, Helder Fontes, José Ruela, Manuel Ricardo, and Rui Campos from the INESC TEC team for their valuable contributions to this chapter.

References

Brooke, J. (1996). SUS – A "quick and dirty" usability scale. In P. W. Jordan (Ed.), *Usability Evaluation in Industry* (pp. 189–194). London: Taylor & Francis. https://doi.org/10.1201/9781498710411.

CEA. (2022). *ExpressIF*. https://expressif.cea.fr/.

Doherty, P., & Rudol, P. (2007). A UAV search and rescue scenario with human body detection and geolocalization. In M. A. Orgun & J. Thornton (Eds.), *AI 2007: Advances in Artificial Intelligence. AI 2007. Lecture Notes in Computer Science* (Vol. 4830, pp. 1–13). Heidelberg: Springer.

Drury, J. L., Riek, L., & Rackliffe, N. (2006). A decomposition of UAV-related situation awareness. *Proceeding of the 1st ACM SIGCHI/SIGART Conference on Human-Robot Interaction – HRI '06*. https://doi.org/10.1145/1121241.1121258.

Endsley, M. (1988). Situation awareness global assessment technique (SAGAT). *Proceedings of the IEEE 1988 National Aerospace and Electronics Conference*. https://doi.org/10.1109/naecon.1988.195097.

Friedrich, M., Borkowski, M., & Lieb, J. (2020, October). A multi-UAS platform to accelerate situation assessment in first response missions – Identification of user needs and system requirements using design thinking. *Proceedings of the 2020 AIAA/IEEE 39th Digital Avionics Systems Conference (DASC)*. https://doi.org/10.1109/dasc50938.2020.9256469.

Friedrich, M., Mnatsakanyan, S., Kocharov, D., & Lieb, J. (2021). Respondrone – A multi-UAS platform to support situation assessment and decision making for first responders. *Proceedings of the Advances in Human Factors in Robots, Unmanned Systems and Cybersecurity Conference*, 110–117. https://doi.org/10.1007/978-3-030-79997-7_14.

Iphar, C., Boudet, L., & Poli, J. P. (2021). Topography-based fuzzy assessment of runoff area with 3D spatial relations. *Proceedings of the 2021 IEEE International Conference on Fuzzy Systems (FUZZ-IEEE)*. https://doi.org/10.1109/fuzz45933.2021.9494395.

Kuenz, A., Schwoch, G., Korn, B., Forster, C., Gerz, T., Grewe, V., Matthes, S., Graupl, T., Rippl, M., Linke, F., & Radde, M. (2017). Optimization without limits – The world wide air traffic management project. *Proceedings of the 2017 IEEE/AIAA 36th Digital Avionics Systems Conference (DASC)*. https://doi.org/10.1109/dasc.2017.8102019.

Ondreicsik, A. (2022). Conceptboard – Virtual collaboration workplace mit Whiteboard Funktionen. *Conceptboard*. https://conceptboard.com/de/.

ResponDrone. (2019). *D15.2: Report of field studies*. https://respondroneproject.com/wp-content/uploads/2020/04/ResponDrone_D15.2_Report_Field_Studies.pdf.

SJU. (2022). *Situation Awareness for SHAPE (SASHA)*. https://ext.eurocontrol.int/ehp/?q=node/1609.

UNISDR (United Nations for Disaster Risk Reduction). (2019). Extreme weather events affected 60 million people. *Press Release*, Geneva, Switzerland.

14

STADEM

An Adapted Trial Guidance Methodology (TGM) for Pandemic Management

Michael Middelhoff, Adam Widera, Georg Neubauer, Dražen Ignjatović, and Bernd Hellingrath

Introduction

The development of the STADEM methodology within the STAMINA project is rooted in the underlying trial guidance methodology (TGM) as result of the DRIVER+ project (Driving Innovation for European Resilience) and formulated in the Cen Workshop Agreement on Systematic assessment of innovative solutions for crisis management – trial guidance methodology (Widera et al., 2020). As stated, the methodology is designed to "enable practitioners to systematically assess the added value of an innovative solution, enhance dialogue and co-operation among solution providers and practitioners [and] support the goal to have a more objective assessment in the procurement process" (Widera et al., 2020, p. 5). The TGM is accordingly a systematic toolbox that addresses in the first place the practitioners and their empowerment to steer the development of innovative solutions. Given the primary focus on crisis management (CM), adaptations are needed to implement the methodology in specialised domains.

The STAMINA project (Demonstration of intelligent decision support for pandemic crisis prediction and management within and across European borders) aims to develop an intelligent decision support platform for pandemic management (PM). In contrast to the overall CM perspective within the DRIVER+ project, STAMINA focuses on innovating pandemic management specifically. To draw a sophisticated view on the innovation of pandemic management across Europe, STAMINA hosts a total of 12 national trial events with one additional cross-border trial. Solutions developed within the project need to comply with medical standards and integrate well into the existing environment. To ease these requirements, STAMINA limits the solution space to project internal developments that are driven by initially recorded practitioner needs. While the project addresses pandemic management, it is itself affected by the ongoing COVID-19 pandemic,

DOI: 10.4324/9781003256977-22

which lead to a series of organisational implications, which in turn motivated methodological changes to assure rigorous assessment of the innovation potential of solutions.

Transitioning from the DRIVER+ origin to the STAMINA context, this chapter further discusses changes to the TGM towards a new trial guidance methodology in pandemic management called STADEM (STAMINA Demonstrations Methodology). The most striking differences stem from the closed solution space, which affects the solution selection within the trial preparation, the medical context of pandemics, which has implications on the trial design and test cases, and finally the project setup of 13 individual trials during the COVID-19 pandemic, which limits the possibilities for on-site observations. The next sections will provide a deeper overview on the STAMINA project and the trials, the STADEM methodology and its phases, as well as lessons learnt from the methodology adaptations towards further evolutions of the TGM.

STAMINA Project Overview

STAMINA (Demonstration of intelligent decision support for pandemic crisis prediction and management within and across European borders) is an EU-funded project in the Horizon 2020 "Secure societies – Protection freedom and security of Europe and its citizens" work programme under the topic "Demonstration of novel concepts for the management of pandemic crises," which started in September 2020 until an extended end of February 2023 (formerly August 2022). It has a consortium of 38 partners from research, medical institutions, and industry. The project will enable national agencies and first responders to anticipate pandemic threats and plan daily efforts to enhance health security. The objectives of the project include the creation of real-time web and social media analytics for the detection of possible disease outbreaks, predictive modelling of pandemic outbreaks, and early warning systems. The crisis management part will define the roles of key actors during crisis management and include protection and diagnostic devices for first-line screening. Finally, STAMINA develops a common operational picture as the main interface that enables coordinated response and demonstrates its application within and across EU borders.

STAMINA takes a distinct perspective on crisis management with a strong focus on pandemic scenarios and dedicated practitioner-oriented trials. That makes it challenging to satisfy the practitioner needs while performing scientific rigorous evaluations. The DRIVER+ trial guidance methodology followed a similar objective yet focused a much wider scope and featured an open call for solutions. With the development of STADEM, the need of a more specialised TGM for solution evaluations in pandemic management should be addressed. An influencing factor on the applied methodology is the absence of a common technology testbed. Compared to DRIVER+, STAMINA performs a direct application of solutions to the trial scenarios, where each trial performs individual integration processes. To avoid duplicated work, STAMINA investigates the development of general data integration and harmonisation solutions, which is feasible due to the specific pandemic management context.

The 12 trials are driven by an initial gap analysis performed by the practitioner organisations in the project. Each trial has, therefore, a set of individual objectives with partly overlap due to common decision problems in pandemic management. Yet every trial has a unique scenario according to the national environment. The ongoing COVID-19 pandemic triggered many trials to take this as a leading pandemic case because relevant data is available and current

TABLE 14.1 Overview of STAMINA Trial Objectives

Country	*Objective*
Austria	Early pathogen detection, contact tracing, communication and documentation, visualisation, and decision support.
Belgium/Netherlands	Detection and monitoring of ESBL (extended-spectrum beta-lactamase) resistant genes, infected persons, food, and water sources.
Czech Republic	Communication methodology in public health care for pandemic management, information processing, and situational awareness.
France	Forecasting of West Nile virus (WNV) and measles, modelling spreading of WNV and measles outbreak, and rapid diagnostic testing.
Greece	Interorganisational communication and hospital and intensive care unit (ICU) bed capacity planning for COVID-19 outbreak.
Lithuania	Communication and cooperation between pandemic management response units, modelling, and contact tracing of pandemic outbreaks.
Romania	Regional reserve force for human and logistical resources to manage several emergencies occurring simultaneously.
Slovenia	Hospital bed and ICU bed management, rapid PCR (polymerase chain reaction) testing, and disinfection of ambulances.
Spain	Shared communication for situational awareness, communication strategy with citizens, and harmonised documentation.
Tunisia	Real-time early warning for COVID-19 and WNV detection, rapid testing capabilities, hospital capacity management, and predictive modelling.
Turkey	Forecasting of ICU bed capacities and monitoring of infectious agent dissemination during COVID-19 outbreaks.
United Kingdom	Comparison of modelling and prediction methods for epidemic parameters in COVID-19 outbreaks.

gaps can be observed in real-time. Additional topics, like handling of other pathogens such as the West Nile virus (WNV), hospital capacity management, or communication and cooperation strategies, further extend the scope. Table 14.1 lists the trial events within STAMINA with a short description of the main objectives. Every trial applies one or multiple solutions to address the objectives. An additional cross-border trial will focus on international communication and collaboration in pandemic management and builds on experiences collected during national trials.

STADEM Methodology

The following description of the STADEM methodology is based on the fundamentals of the TGM as described in (Widera et al., 2020) and adaptions achieved during the STAMINA project (Neubauer et al., 2021). STADEM builds a trial out of the same three dimensions, which need to be considered to allow the evaluation of a trial. The solution dimension states the objectives the individual solution tries to reach in the trial context and the functions it offers to do so. The crisis management dimension states the objectives for the improvement in the activities to address the situation and the roles, responsibilities, and processes needed for that. The trial dimension states the objectives for the overall trial and the sessions it is composed of. All three dimensions have dedicated performance measurements to assess their individual influence on the overall achieved outcome:

- Solution dimension:
  - The influence of a solution (its use, functionalities, etc.) on the trial.
  - The added value a solution brings to pandemic management (PM) functions.
  - Practitioner assessment of the solution.
- Crisis management dimension:
  - The influence of the crisis management (roles, responsibilities, etc.) on the trial.
  - The impact that a solution has on the performance of the PM organisation.
- Trial dimension:
  - The influence of the trial organisation (logistics, staff, hardware/software, etc.) on the trial.
  - The changes that the trial organisation brings to the success of the trial.

Identical to the TGM, the STADEM is structured out of three phases: preparation, execution, and evaluation. The goal of a trial is the systematic assessment of an innovative solution's added value to a specific PM context and gap or need. Hence, the evaluation phase is the most important phase and should be considered from the beginning.

Preparation Phase

The preparation phase of STADEM is closely aligned to the original preparation within the TGM. Some adaptations are presented, which are motivated by the specific pandemic management context and driven by the project organisation.

Step Zero

Identify Gaps

The core of a trial are the gaps that should be addressed by conducting a trial. Every gap is based on the underlaying need to improve the current situation. A gap should be reasoned by formulating the need to improve and describe current deficiencies, as this eases the understanding for non-CM professionals. Hence, it is recommended that CM professionals are involved in the identification and formulation of the trial gaps.

Describe the Trial Context / Pandemic Management Map

As the gap depicts a potential for improvement, it is entwined with a bundle of roles, responsibilities, situations, equipment, and the like that describe the situation in which this specific gap occurs. This is done by depicting the trial context. From that onward, it is important to define the format of trials: tabletop, field, or mixed approach. According to the context, some gaps might favour or require a certain kind to realise the situation. Based on this, the location/facility of the trial should be determined. This should be done at the early stage as it impacts the data collection, evaluation, and scenario, as well as the whole execution phase. Within STAMINA, practitioners in pandemic management demonstrated a national pandemic management map, depicting the responsible authorities, response units, and information exchanges in case of a pandemic outbreak. Such a view can greatly support the description of the trial context and foster the common understanding.

Six-Step Approach

The six-step approach is an iterative process in which selected steps are updated in follow-up iterations. It is recommended to do one full run-through of all six steps and then iterate in the most useful way. The process is co-creative, which means that all stakeholders should be involved. The trial team should discuss the gap(s) and underlying need to reach a common agreement and understanding on the solutions

Step 1: Identify the Trial Objective

The objective is defined as the result to be achieved. It addresses the gaps and proposes an, at least partly, improved situation but does not necessarily close the gaps completely. In this step, the gaps and the trial context should be used to identify the main challenge within the current situation. This can best be done by discussing the baseline, which is a depiction of the "gap process." It is necessary to understand the challenge that is raised by the gap, whether it is a technical process, knowledge/skill challenge, or a combination of those. Furthermore, it is important to know what the needs are that lead to the challenge, what is preventing the solution currently and if there are time conditions in addressing it.

Step 2: Formulate Trial Questions

The trial question is a reformulation of the objective and by that has the nature to be answered by the trial. In contrast to the objective, the question is more open to potential solutions that can fulfil the objective and eases the communication. The iterative nature of the process will lead to a question that gets more and more detailed with time.

Step 3: Formulate Data Collection Plan

The aim of a trial is to gather data to analyse how an innovative socio-technical solution can address a gap. This is only possible by comparing the baseline to the new setup that uses a solution. Reference data can be obtained in three ways: using historic data from past events if available, performing a baseline run in the situation as it would happen today, or simulating the baseline, which is the most difficult and data intense option.

Step 4: Formulate Evaluation Approach

Once a decision is made on the type of data needed, it must be detailed which techniques and tools will be used to analyse the set of data to answer the trial questions. The data collection plan is key here, as it gives a clear indication of the evaluation approaches that must be considered. The starting point is the application of the three-dimensional performance measurement approach. The trial dimension offers an overview of the initially defined targets and the actual results.

Step 5: Formulate Scenario

The trial can address multiple gaps, which leads to multiple baselines that are consolidated into the scenario. The baselines in combination already determine the roles, processes, equipment, and situations needed for the trial. In a first step, it is recommend analysing whether those baselines can form one coherent scenario or demand more distinct scenario scenes.

Step 6: Solution Demonstration

While the former steps are basically taken from the TGM, the last step, formally called solution selection, is adapted towards a solution demonstration. Here, the practitioners intuitively connect the identified gaps with a subset of the available STAMINA solution toolset. Given that STAMINA develops and utilises a solely project internal solution space, it is more important that the practitioners become familiar with the developed solutions to further steer the ongoing work in the trial preparation. While the feasibility assessment of solutions in the selection step was originally skipped without substitute, it became clear that this might lead to a large knowledge gap between the end-users and solution providers. A demonstration and detailed engagement with solutions is therefore important.

Execution Phase

Besides the execution of the actual trial, the execution phase also foresees preparation, rehearsals, and tests to assure a smooth event. This will also allow for the possibility to adjust and check that all that is needed for the trial is in place and is functioning properly.

Due to the direct selection of solutions without a detailed feasibility assessment within STAMINA and without a proper solution demonstration, these meetings had to overcome the created knowledge gap. This emphasises the need for a solution demonstration even if solely project internal solutions are applied. Furthermore, the COVID-19 pandemic demanded online meetings, which further hindered the information exchange and required multiple follow-up discussions resulting not in one but a series of trial integration meetings (TIM) for each trial.

The trial run aims at collecting all the data necessary for the evaluation of the solutions by comparing the baseline with the use of the innovative selected solutions. It is recommended to plan sufficient time to perform a final testing of the technical setup and equipment, training of the participants, running the trial sessions, and performing data collection, including debriefings, questionnaires, interviews, and so on. The total duration of the trial run depends on the complexity of the scenario, as well as the number of gaps and solutions to be assessed.

Evaluation Phase

The main objective of the evaluation phase is to make sense of the data collected during the trial. This sensemaking process is key to understand if and to what extent the solutions are relevant and innovative for a CM organisation. Before diving into the analysis, the collected data should be verified. Practically, this means checking whether the actual data matches with the data collection plan. In case data is missing or major deviations are spotted, corrective actions might be necessary. These actions depend on potential problems of the dataset and include, for instance, further interviews with trial participants and/or repairing a broken dataset. A comprehensive analysis should be preceded by the organisation of the dataset according to the three dimensions: trial, CM, and solutions. This is crucial to understand which data refer to which dimension. Moreover, the data collected in each dimension (e.g. data regarding the trial context, data regarding the CM effectiveness, and data focusing on the functionalities of the solutions) should be related to the KPIs and metrics defined in the preparation phase. Given that the STAMINA trials have not been executed yet, experiences and adaptations to the evaluation phase are open to future work.

Lessons Learnt

Adapting the TGM to a specific crisis management domain and applying the STADEM during the trial preparation and first steps into the execution lead to some lessons learnt, which have partly been presented in the adaptions of the methodology described earlier. This section summarises the identified learnings made up to this point of the project's lifetime.

While the TGM targets crisis management in general, specific domains motivate further adaptations of the methodology just like the STAMINA project does for pandemic management. A domain is characterised by its individual problem statements and legacy systems in place used by the practitioners. One example of such legacy tools is the pandemic management map, which is used in some European countries to depict the structure and cooperation between national response units in the preparation or during pandemic outbreaks. The trials can utilise these domain-specific tools to ease the communication between practitioners and solution providers. STAMINA did this by using the pandemic management map as a suggested tool to be used in the description of the trial context. Therefore, TGM adaptations should use domain-specific terminologies and tools.

The original solution selection in the TGM was motivated to open the solution space beyond the project consortium and to include innovations from third-party organisations. DRIVER+ was able to do so because of the project design, and it was reasonable given the open crisis management perspective. Other projects and, hence, adaptations of the TGM might not be able to provide this opportunity, or it might not be reasonable due to strict requirements. The adjusted solution selection in STAMINA showed that even if only project internal solutions are used, a knowledge gap between practitioners and solution providers can occur, hindering the later developments. Instead, the solution space available to practitioners should be at least demonstrated prior to the trial planning.

Having face-to-face meetings during the trial preparation is seen as crucial to foster the common understanding and to engage in the co-creation process between practitioners and solution providers as it is intended by the TGM. At latest, the TIM is supposed to be the first in person meeting among key stakeholders. STAMINA, while being itself focused on pandemics, showed that project reality might be affected by outside conditions like the COVID-19 outbreak. Face-to-face meetings cannot be replaced with online meetings in the same manner, as this highly affects the communication, especially in an international consortium. Additional preparation and follow-up activities are recommended to achieve the same level of quality. If physical meetings need to be held online, they should be planned as a series of meetings to achieve the intended outcomes.

The co-creation process of the TGM is key to align the practitioner needs to the functionality provided by the solutions. This happens during the extensive process starting with the TIM followed by the two dry runs and ending in the trial itself. While the solutions in STAMINA received inputs from an initial gap analysis stating current needs of practitioners, their development and involvement in trials was not predefined. Solutions were still able to develop freely according to their own objectives. Practitioners later selected feasible solutions for their trials, which resulted in some solutions being selected in multiple trials and some not at all. From the project perspective, it is further ensured that every solution is at least trialled once at latest in the cross-border trial. The initial gap analysis and the trial objectives are both rooted in the practitioner needs but do not necessarily match. If a project uses an internal solution space, the solution providers need to receive trial-specific specifications from the practitioners.

Conclusion

This chapter showed the adaptation of the TGM for a specific crisis management domain in the STAMINA project. STAMINA is an EU-funded Horizon 2020 project that demonstrates innovative solutions in pandemic management. Solutions are assessed in 12 national and one additional cross-border trial based on current practitioner needs. The TGM was adapted for pandemic management, which led to the development of STADEM. While the fundamental components of the TGM, like the three trial dimensions – solution, crisis management, and trial – and the phases – preparation, execution, and evaluation – are taken over, detailed steps have been adjusted to meet project requirements. These changes have been tested during the execution of STADEM, which resulted in lessons learnt. The COVID-19 pandemic further affected the project execution and hence motivated recommendations for future TGM implementations.

While STAMINA is at this point an ongoing project and hence not all phases of STADEM have been executed, it was demonstrated that the TGM is adjustable to specific domains within crisis management. Future work will evaluate the execution and evaluation of the trials, which might motivate additional changes to the methodology. Details of the methodology are always subject to change. Yet the concept of the trial dimensions and the trial phases have proven to be effective guidelines in the trial management. Future projects can build on these findings and provide further methodological adaptations.

References

Neubauer, G.; Ignjatovic, D.; Schimak, G.; Widera, A.; Middelhoff, M.; Scheuer, S.; Aumayr, G.; Jaho, E.; Rainer, K. (2021). The benefits of trials for pandemic management. *29th Interdisciplinary Information Management Talks-Pandemics: Impacts, Strategies and Responses*, IDIMT 2021, 197–203.

Widera, A., Hellingrath, B., Bonnamour, M. C., Lund, D., Fonio, C., Stolk, D., Dubost, L., Pasedag, K., Gerevini, M., Zwęgliński, T., & Friedemann, M. (2020). *Systematic Assessment of Innovative Solutions for Crisis Management-trial Guidance Methodology* (CWA 17514: 2020 E). https://www.cencenelec.eu/media/CEN-CENELEC/CWAs/RI/cwa17514_2020.pdf.

15

INTEROPERABILITY AND STANDARDISATION SUPPORTING PREPAREDNESS AND RESPONSE TO DISASTERS

Angelos J. Amditis, Eleftherios Ouzounoglou, Panagiotis Michalis, Fay Misichroni, Leonidas Perlepes, and Evangelos Sdongos

Introduction

European countries nowadays confront the rising spectre of transboundary crises (Boin et al., 2014; Boin & Rhinard, 2008). These are crises with cross national borders and policy boundaries with speed and ease, threatening the continuing functioning of critical infrastructures and the well-being of many citizens. Three challenges (related both to operational and technical interoperability) need urgent attention in order to support an effective EU-coordinated response: (1) member states need to develop shared response planning and training activities, (2) countries need to share information in real-time, (3) civil protection agencies need to coordinate, even in an EU-level, the use of critical resources to ensure a timely and effective response.

At the same time, civil protection agencies and first responders, during their primary mission of preserving society's safety and security, face a multitude of challenges. They often deal with life-threatening situations, hazardous environments, uncharted surroundings, and limited awareness. Therefore, armouring public safety services is critical to enhance their protection and augment their operational capacities. Intelligent, integrated, interconnected, and seamless tools and services are needed that add layers of protection against the dangers and augment their situational awareness, ensuring on the same time the interoperability among different teams and levels of command and supporting also the important for the EU-wide cross-nation collaboration, eliminating barriers related to technology compatibility or even the language related ones.

In this context, the EU has clearly recognised (e.g. in Mandate M/548) the whole field of "crisis management and civil protection" as one of the three priorities for establishing standards in the security sector to facilitate response, effectiveness, efficiency, and cooperation as top priorities and to properly develop

DOI: 10.4324/9781003256977-23

FIGURE 15.1 The risk management cycle and how H2020 IN-PREP and H2020 STRATEGY are positioned to its different phases.

the appropriate standards in order to strengthen response forces and cooperation, stimulate interoperable command and control, and provide common frameworks for education, training, and exercises.

In this chapter, two EU-funded projects are presented, contributing to these challenges, the first providing solutions related to preparedness, training, and response planning (H2020 IN-PREP) and the second (H2020 STRATEGY) focusing on the establishment and validation of a significant number of (pre-)standardisation activities covering the whole cycle of risk management cycle as depicted in Figure 15.1. STRATEGY capitalises on related standardisation gap analysis results of previous projects, such as ResiStand and DRIVER+, as well as on the technological and methodological results of IN-PREP (among others) in order to make the synthesis of proposed standardisation items in which STRATEGY is focusing.

In the following sections, details about the aforementioned projects, including their overall goals and methodological framework and their specific technical developments, evaluation programmes, and connection with DRIVER+ project results, are presented.

The IN-PREP Project

IN-PREP[1] was a three-year project (extended to 42 months and implemented from 2017 to 2021), funded by the European Commission's Horizon 2020 programme (GA 740627), focused on building a system that can better prepare civil protection practitioners (fire brigades, emergency medical services, police and civil protection agencies) at all levels of command, in responding collaboratively to urgent natural and man-made transboundary crises. The IN-PREP consortium was made up of 20 partners from seven countries in the EU, comprising seven crisis management practitioners and technology providers. Guided by feedback from practitioners, a mixed-reality preparedness platform was developed, which is a novel IT-based tool for response planning and scenario building to integrate command and control and information systems, situational awareness modules, and a decision support mechanism. This innovative IT tool allows practitioners to train and prepare for transboundary crises, using new and existing scenarios, in a virtual environment. In addition, IN-PREP created a first-of-its-kind cross-organisational Handbook of Transboundary Preparedness and Response Operations[2] addressing operational and procedural best practices and technical recommendations for further development of relevant systems. The principal aim for both outputs is to improve preparedness with realistic training in realistic scenarios of disasters and causes of crises, focusing, therefore, mainly on the preparedness phase of the risk management cycle shown in Figure 15.1.

By training with IN-PREP, end-users can identify weaknesses in their response planning for transboundary events, and they can adapt their response accordingly. IN-PREP can be used to prepare for the strategic crisis management tasks. So if, for instance, an organisation concludes that its communication capabilities are not sufficient for transboundary crises, IN-PREP can help to strengthen crisis communication capacities by producing a transboundary scenario in which the existing capacities are challenged, tested, and assessed. In short, IN-PREP has been designed to help users in (1) identifying weaknesses in response planning, (2) recognising actors in transboundary networks, (3) becoming familiar with crisis decision-making in a transboundary context, (4) knowing what type of information is needed and how to get it, and (5) practicing communicating with multiple publics and stakeholders.

The overall IN-PREP system has been demonstrated and tested regarding its effectiveness to support response planning and training through a rich training programme of three tabletop exercises (TTX) and three demonstrations (demos) specifically aimed at first responder and civil protection agencies seeking next generation, joint operational training.

The STRATEGY Project

STRATEGY[3] is a European project funded by the Horizon 2020 programme for a period of 36 months. STRATEGY develops a pan-European framework of the pre-standardisation activities for systems, solutions, and procedures, addressing crisis management, validated by sustainable tests and evaluation frameworks, which will improve the crisis management and disaster resilience capabilities, including European Defence Agency (EDA) initiatives in the CBRN-E area. Based on the needs identified from previous EU initiatives (including the IN-PREP project on response planning and training) and the EU priorities, the project addresses the following eight streams within crisis management: (1) search and rescue, (2) critical infrastructure protection, (3) response planning, (4) command and control, (5) early warning systems and rapid damage assessment, (6) CBRN-E, (7) training, and (8) terminology.

STRATEGY streamlines and validates technical and organisational interoperability in a fully transboundary configuration with means of standards through the implementation of use-cases involving EU and national standardisation bodies. Activities focus on selecting and implementing existing, evolving, and new standards within solutions, tools, and procedural guidelines and recommendations. The validation of the pre-standardisation activities of STRATEGY takes places along a set of table-top exercises (TTXs) followed by a full-scale exercise (FSX). The technical environment of DRIVER+ play a vital role in STRATEGY as it is employed in order to adapt, configure, integrate, and test systems and solutions to be used for implementation of the selected standards during the TTX and the FSX.

Interoperable and Standardised Solutions for Crisis Management and Disaster Response

The IN-PREP Architecture and Evaluation Programme and the Integration with DRIVER+ Solutions

The IN-PREP system developments were strongly driven by the feedback collected from end-users, clearly identifying the need for improvements in collaborative response training (including scenario building) and collaborative response planning. Figures 15.2 and 15.3 present the two different configuration views of the IN-PREP MRPP (multi-reality preparedness platform) in order to support these processes.

In this context, the main actors that use or operate the system are as follows:

1. *The exercise trainer*, who is responsible for setting exercise objectives and creating the exercise scenario. The trainer is also able to interact with trainees during a training session by monitoring progress and editing the scenario on the fly.

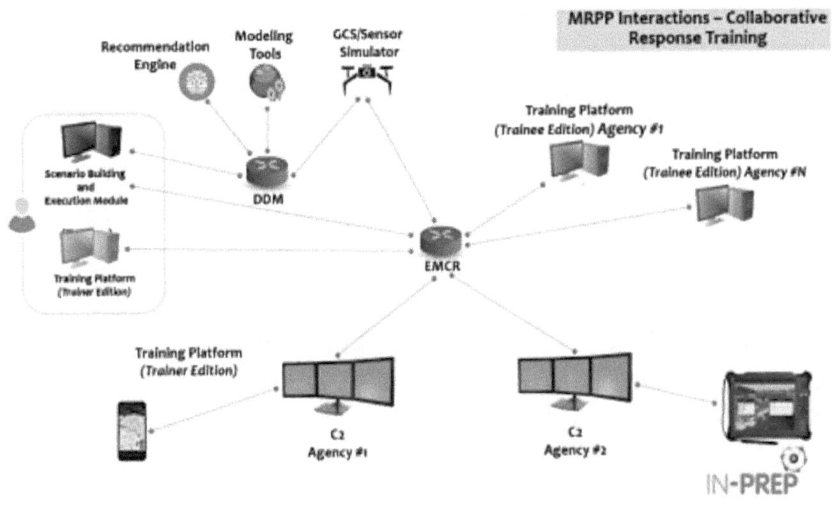

FIGURE 15.2 Collaborative response training using the IN-PREP solution.

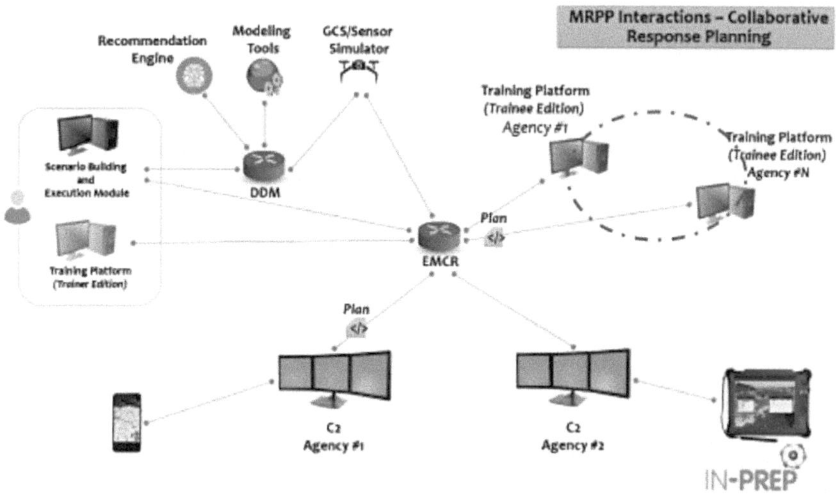

FIGURE 15.3 Collaborative response planning using the IN-PREP solution.

2. *The exercise participants or trainees*, who take their positions behind the IN-PREP command and control and incident management systems and are receiving the digitised scenario in the timeline set by the trainer, prompted to take actions accordingly as they were in a real incident. Notably, the IN-PREP system allows multi-agency collaboration by interconnecting the individual agencies, allowing them to share the operational picture, allocate resources, and communicate to each other.
3. *The evaluators*, who are provided with the digitised output of the entire training session and the automated results extracted by the platform, the latter correlating trainees' actions, communication, and responses with a set of KPIs or objectives.
4. *The planners*, who oversee the digitisation of individual agencies' response plans within the system and the joint plan emerging and presented to the strategic level of command.

The Communication Framework of IN-PREP

The communication framework of IN-PREP is primarily part of the "application layer" of the TCP/IP model. In IN-PREP, there are two modules proposed as a solution that is responsible for this communication, as shown in Figures 15.2 and 15.3:

1. The Emergency Message Content Router (EMCR)
2. The Data Distribution Module (DDM)

All IN-PREP sub-systems bidirectionally communicate with these two modules in order to exchange information in a reliable and efficient way, in real-time. The

EMCR is mainly used for exchanging operational information, such as dispatching, availability of assets, tasking, and notifications from the field, whereas DDM facilitates the interconnection of the training platform with the preparedness layers and decision-making tools. Special attention has been paid to incorporating and complying to information and data exchange standards, ensuring in this way that the IN-PREP system is extendable, scalable, and ready to be connected, with little or no effort, to other legacy or external sub-systems.

Data Distribution Module (DDM) and Technical Collaboration Between DRIVER+ and IN-PREP Projects

The DDM is a communication framework aiming at interconnecting different tools within MRPP. Based on widely accepted standards, the DDM enables real-time interoperable communication, ensuring prompt delivery of information. A data adaptation layer facilitates interoperability between different tools of MRPP in complex scenarios. DDM consists of a set of tools that each one performs a different role related to schema registration (Confluent Schema Registry),[4] message distribution (Apache Kafka),[5] messaging broker coordination (Apache ZooKeeper),[6] storing of geospatial data and modelling results (Geoserver),[7] and storing of files (WebDAV).[8]

As IN-PREP objectives have been supplementary to DRIVER+ project's one, the IN-PREP and DRIVER+ consortiums have launched a communication channel, aiming to find ways to collaborate and exploit the solutions provided by one project in the context of the other. The IN-PREP DDM and the DRIVER+ testbed[9] have been identified as potential collaboration points. The DRIVER+ solution has been extended by IN-PREP, in order to be able to accommodate different kinds of information (such as the retrieval of modelling tools' results).

The two consortiums technical collaboration started, having as its milestone the usage of DDM (enhanced with DRIVER+ testbed component) on IN-PREP TTX3 in the Netherlands focusing on a flood-related scenario. In May 2019, the DRIVER+ project conducted a tabletop trial in the Netherlands (The Hague). In the context of this trial, the DRIVER+ project has also developed the so-called LCMS connector that facilitated the integration of the Dutch nationwide crisis management system (called LCMS)[10] with the DRIVER+ testbed. This connector has been used also in the context of IN-PREP TTX3 in order for the messages (injects) created by IN-PREP tools to be sent to LCMS. This further proves the importance of interoperability and standardisation in crisis management for successful integration of systems.

Kafka Topics and AVRO Schemas

The DDM is extendable by design. New Kafka topics can be easily added in order to enhance the IN-PREP communication framework capabilities. The same applies for the AVRO schemas that can be hosted in the schema registry.

Two topics are available to be used in IN-PREP DDM: the "standard-cap" (adopted from DRIVER+ project) and "task." These two topics are used in order to efficiently manage, classify, and optimise the messages exchanged between the various components consisting the MRPP. The messages' representation format follows well established standards, as EDXL-CAP,[11] EDXL-DE,[12] and OGC SensorThings API.[13]

IN-PREP Trials

The IN-PREP training programme consisted of three tabletop exercises (TTX) and three full-scale exercises (FSX).[14] In IN-PREP, a TTX has been defined as an exercise where key personnel are involved and simulated scenarios are discussed to assess plans, policies, and procedures in order to enhance preparedness of crisis managers while an FSX as a multi-agency exercise with a prepared scenario involving personnel in the field (e.g. joint field office) and personnel in crisis management teams to train based on response plans and enhance preparedness of all personnel that are involved in crisis management response teams. Table 15.1 shows an overview of the IN-PREP exercises.

In the three TTXs, participants passed through the process of dealing with a simulated disaster scenario. Participants review and discuss the actions they would

TABLE 15.1 The IN-PREP Exercises' Theme, Dates, and Places

IN-PREP trial	First TTX: Industrial accident (CBRN)
Date	November 26–30, 2018
Place	Spoleto, Italy
IN-PREP trial	Second TTX: Biological attack involving a cruise ship and urban areas
Date	June 17–19, 2019
Place	Savona, Italy
IN-PREP trial	Third TTX: Massive flood
Date	October 7–10, 2019
Place	Kampen, Netherlands
IN-PREP trial	First FSX: Cross-border terrorist attack
Date	December 4–6, 2019
Place	Belfast, Northern Ireland
IN-PREP trial	Second FSX: Transport incident and CI cascading effects
Date	October 28, 2020
Place	Genova and Savona, Italy (semi-virtual due to the COVID-19 outbreak)
IN-PREP trial	Third FSX: Forest fire with transport accident
Date	February 4, 2021 (virtual due to the COVID-19 outbreak)
Place	Rhodes, Greece

take in a particular emergency, testing their emergency plan in an informal, low-stress environment, nevertheless using a realistic scenario and a timeline. In this context, participants provide solutions to problems as they arise and change while the trainer monitors their response as new situations are injected, creating unexpected challenges as the scenario progresses. Specific objectives has been set for the TTXs that included the familiarisation of end-users with the IN-PREP platform and to provide training exercises for the local officers and the organisations involved in crisis management. The demonstration of how the IN-PREP integrated tools for scenario building and response planning adds value to the response and operating capabilities of the hosting nations and facilitates interorganisational coordination, to test, evaluate, and provide feedback and (additional requirements) on all IN-PREP integrated components. Moreover as the project progressed, goals were set to test the MRPP regarding functionality, effectiveness and usability, flexibility and adaptiveness and the integration capabilities with legacy systems (i.e. LCMS), to incorporate lessons learnt from other EU-projects (DRIVER+) by using their technical testbed, to train the trainees on using the IN-PREP solutions and to test the collaboration and information-sharing between the tactical and strategic team (through LCMS) and collaboration with other agencies through an information-driven platform.

Significant evaluation highlights have been extracted which are presented in detail in the related to knowledge capitalisation publicly available deliverable of the project.[15] These included the importance of defining in detail the trainees that should participate in TTXs and the tools that should be available in each table of the exercise and clearly defining roles and responsibilities for the participants. Moreover, it has been identified that a balance is needed between a scenario which is realistic and meets all the objectives of the host organisation but also includes as many tools as possible, as well as clearly planning for the integration of the local systems. The tools, in general, and how the information is processed should be clearly explained to the trainees before the exercise, so the trainees trust their results (e.g. simulation results from the modelling tools) while the scenario-building tool should be flexible enough to allow adaptations to the scenario during the exercise. In this context, it has been clearly identified the need to learn from other research projects (e.g. DRIVER+) and their methodology to prepare exercises.

The full-scale exercises (FSX)/demonstrations involved full-scale deployment of all levels and ranks of participants from many agencies designed to be high stress and labour intensive involving actors in realistic role-play and providing real-world conditions and role-playing for all participants while scenarios were based on data from previous incidents. As with the TTXs, also FSXs set a specific set of general but also targeted objectives including the demonstration of the IN-PREP tool and its functionality for the development of plans and the training of responders in transboundary crisis as well as to test the command and control structures which are required to support an effective response to crisis and to test the cross-border protocols for major incident activation and the sharing of information. Moreover, special focus has been given on the crisis management at both strategic and tactical level and how IN-PREP tools could be used to facilitate an exercise on this level that are run both virtually and in the field. Regarding the specific IN-PREP tools, the scenario-building tool has been extensively used to create all the injects of the exercise scenario and the command and control application for communicating injects to all the different trainees from different organisation, ensuring that all trainees at the different locations had a realistic overview of the situation while modelling tools supported the emergency managers in their decision (weather-plume-evacuation modelling). Finally, during the FSXs, objective related to the actual final use of the system has been set that were related to testing the civil protection and command and control structures which are required to support an effective collaborative response, the readiness of civil protection agencies to handle excessive needs and the emergency response plans (of fire brigades and police).

Through these FSXs, the consortium was able to extract some significant lessons learnt and evaluation highlights. Firstly, it was made clear that the confidence of the actors to the tools affected their ability to do the role being asked of them; therefore, there is a need for training activities supported by user manuals and training aids so that focus could remain on exercise delivery and for more interoperability training to highlight the flow of information from the operational level to the tactical and strategic levels. It has been however well-noted that the automated circulation of exercise events (through the IN-PREP platform) added an additional level of control and predictability that allowed the training delivery team to pay closer attention to the needs of the students/participants and less on content delivery. Specifically, storing of scenarios and option to "replay" them allows for the comparison of actions taken. In other words, the impact (and effectiveness) of response measures can be compared if several versions of a scenario are played while collaboration between actors was enhanced since the platform allowed for the "tracking" of actions taken by other organisations. However, it has been identified that the use of different terminologies remains a challenge in multi-organisational settings, while the visualisation of command chains and responsibilities would be an additional asset to understand the ways of working of other organisations. Modelling tools were considered useful to enhance the emergency planning and to build the planned response activities on evidence. However, it was also mentioned that it would be important to better understand the models, the way they work and the data they are based on to assess the results they produce.

An interesting aspect regarding the two last IN-PREP FSX was related to the fact that they had to be realised semi-virtual (partial deployment of systems and teams on the field) of even fully virtual due to the COVID-19 outbreak, therefore with very limited physical interaction between the developers and the trainees as well as between the trainer and the trainees and finally among the trainees. The MRPP was proven to play an important role in facilitating an almost entirely remote exercise and provided a basis for envisioning new exercise formats for the future.

Summarising, each exercise provided significant feedback regarding the necessary functionalities of the systems and their integration, as well as supported the identification of the proper preparatory steps needed for ensuring their optimal organisation and end-user engagement. In this context, guidelines such as the trial guidance methodology (TGM) developed in the context of DRIVER+ project should be further supported and adopted towards harmonising the exercise planning (including those related to EC-funded projects), as well as be further standardised by incorporating aspects related to realisation of remote exercises, given the experience gained during the COVID-19 pandemic and the IN-PREP lessons learnt.

Pre-standardisation for Crisis Management: The Use, Extension, and Further Standardisation of DRIVER+ Results in the STRATEGY Project

STRATEGY aims to propose, test, and validate a new pre-standardisation framework that is bringing together main stakeholders in the crisis management area. The project involves industry that is developing products for natural hazards and CBRN-E crisis management, related research organisations, standardisation experts, national standardisation bodies, and first responders. The project employs a realistic environment in order to trigger the proper first responders' and experts' feedback when assessing existing and evolving standards or proposing new ones.

As part of the standardisation processes that are carried out in STRATEGY, the activities involve the production of pre-standards, initiating at least one CEN workshop agreement (CWA) process per stream (eight in total). These also collaborate with workgroups and technical committees of standardisation bodies to support real cases, propose at least four new work item proposals (NWIPs) and draft at least one EN standard.

This environment is generated by using related products and solutions developed by the STRATEGY partners in the frame of previous research projects and will support specific use-cases setup in the frame of eight TTXs and a FSX.

STRATEGY capitalises on related standardisation gap analysis results of previous projects, such as ResiStand and DRIVER+, in order to make a synthesis of proposed standardisation items that are selected to be assessed during the project. For example, the ResiStand Assessment Framework (RAF) is employed to provide an assessment of standardisation initiatives in the domain of disaster resilience and crisis management to ensure that the efforts invested in the production of standardisation deliverables are justified (ResiStand Consortium, 2017).

The outcomes of DRIVER+ project are also put into practice in order to provide a common technical infrastructure for all exercises that are carried out in STRATEGY project by using the testbed technical infrastructure (TTI).[16] The main objectives of utilising DRIVER+ infrastructure for STRATEGY purposes are listed as follows:

1. Set up a common technical infrastructure that can be used for all exercises that will be conducted in STRATEGY to integrate, test, and validate the standardisation items.
2. Provide a common way of working to prepare, execute, and evaluate exercises.

The trial guidance methodology (TGM) (Fonio & Widera, 2020) is employed to facilitate all TTXs and FSXs and to enhance the cooperation framework by providing a common information and simulation space for exchanging messages between end-user solutions and simulations. The TGM provides guidelines on how to assess solutions in non-operational contexts (such as a trial) through a structured approach (CEN, 2020). The TGM directly addresses the context of crisis management and deals with investigating and assessing innovation through a broad set of tools available within the DRIVER+ testbed that are expected to be utilised for the purposes of STRATEGY project. Based on the STRATEGY requirements, the methodology is further advanced to the Exercise Guidance Method (XGM).[17]

CEN Workshop Agreement (CWA) on Standardisation of Digital Specifications for Scenario-Building Activities

The proposed standardisation workshop has resulted from the STRATEGY project and aims to provide specifications for digital scenarios for search and rescue exercises. The digitalisation of exercise scenarios resulted from an analysis of existing standardisation gaps and needs carried out by the project. The main objective is to better support cross-border and cross-platform interoperability and to assist exercise planners in search and rescue related activities. This is due to the fact that the majority of the crisis management exercises are following traditional approaches, such as paper-based scenarios. This is an insufficient practice leading to complications in the design phase of an exercise due to the involvement of numerous actors and the existence of different types of exercises which make the preparation of exercises a particularly challenging issue. The design phase is also considered to be of key importance for the successful operation and completion of an exercise. However, exercise planners have currently limited options in terms of available digital solutions to enhance the preparation of the design phase but also to effectively guide a planner through the process of large-scale exercises in the crisis management field where numerous scenarios and events exist. An opportunity thus exists for the development of a digitised process which can provide a useful tool to exercise planners with regards to the planning and the implementation of search and rescue exercises to enable the effective preparation of large-scale exercises involving multiple agencies and cross-border interoperability. The main scope of this workshop is the development of an interoperable and harmonised process in support of the planners of search and rescue exercises.

The CWA specifies a digital process for the planning of search and rescue exercises and recommendations on the type of digital information exchanged for search and rescue scenarios, the use of data models for exchanging scenario characteristics, and the interoperability characteristics that digital scenario specifications need to have in order to be able to be employed by the majority of scenario-building tools. The activities carried out in the CWA also involved a review of existing scenario managers/builders that have been derived from past EU projects IN-PREP, Heimdall,[18] and DRIVER+. More specifically, the trial management tool (TMT) of

DRIVER+, as open-source software (MIT license) provides an attractive scenario editor and execution manager for testing crisis management systems.[19] In TMT, several scenarios can be created to assess specific aspects of the trialled solutions and execute multiple storylines. During execution of the trial, the trial staff can use the TMT to keep track on activation of these storylines and injects (messages that trigger an action in a simulator, a solution, or a role-player). The aforementioned capabilities of the TMT are planned to be utilised in the CWA in order to demonstrate the incorporation of the proposed specifications into a scenario manager during the TTX but also the FSX of the STRATEGY project.

Conclusions

In this chapter, the activities of two closely interlinked EC-funded projects have been presented, providing solutions spanning from collaborative response planning and training, through scenario building, to the (pre-)standardisation of significant items related to the crisis management domain. In these projects and during their trials preparation and execution, clear conclusions have been extracted regarding the importance of operational and technical interoperability and of the well-defined methodology for designing and executing exercises, necessary for the evaluation and validation of tools and procedures, as well as, as proposed by the STRATEGY project, (pre-)standardisation items. In this context, results of the DRIVER+ related to the testbed supporting data and information interoperability and the TGM providing guidance for exercise planning have already provided significant support to the implementation of these projects. Moreover, the (pre-)standardisation items of STRATEGY include, among others, the definition of an interoperable digital scenario description, accumulating the acquired knowledge in developing scenario-building tools both in IN-PREP and DRIVER+. Both the interlinked nature of the two projects and the successful reuse of DRIVER+ results further highlight the importance of inter-project collaboration, which, as an absolute prerequisite, has the adoption of open and well-defined standards, methods, and tools, ensuring interoperability and fast and efficient integration among the technological solutions being developed for supporting civil protection agencies during their mission of preserving society's safety and security.

Notes

1 www.in-prep.eu, https://cordis.europa.eu/project/id/740627.
2 www.in-prep.eu/handbook.
3 https://strategy-project.eu, https://cordis.europa.eu/project/id/883520.
4 Schema Management – Confluent Platform – Confluent Docs, www.confluent.io/confluent-schema-registry.
5 "Apache Kafka," http://kafka.apache.org.
6 Apache ZooKeeper, https://zookeeper.apache.org.
7 Geoserver, http://geoserver.org.
8 Webdav, www.webdav.org.
9 DRIVER-EU/test-bed, https://github.com/DRIVER-EU/test-bed.

10 LCMS, www.lcms.nl.
11 Common Alerting Protocol Version 1.2, OASIS Standard, 01 July 2010, http://docs.oasis-open.org/emergency/cap/v1.2/CAP-v1.2-os.html.
12 Emergency Data Exchange Language (EDXL) Distribution Element, v. 1.0 OASIS Standard EDXL-DE v1.0, 1 May 2006, https://docs.oasis-open.org/emergency/edxl-de/v1.0/EDXL-DE_Spec_v1.0.doc.
13 OGC SensorThings API Part 2 – Tasking Core http://docs.opengeospatial.org/is/17-079r1/17-079r1.html.
14 www.in-prep.eu/ttx.
15 D7.6 Knowledge Capitalisation Report on Lessons Learnt using IN-PREP in exercises/demos, https://ec.europa.eu/research/participants/documents/downloadPublic?documentIds=080166e5debe3a79&appId=PPGMS
16 www.driver-project.eu/a-pan-european-test-bed-for-crisis-management-capability-building.
17 https://github.com/STRATEGY-EU/XGM.
18 https://heimdall-h2020.eu, https://cordis.europa.eu/project/id/740689.
19 DRIVER-EU/scenario-manager, https://github.com/DRIVER-EU/scenario-manager.

References

Boin, A., & Rhinard, M. (2008). *Managing Transboundary Crises: What Role for the European Union?* (SSRN Scholarly Paper No. 2193998). Social Science Research Network. https://papers.ssrn.com/abstract=2193998.

Boin, A., Rhinard, M., & Ekengren, M. (2014). Managing transboundary crises: The emergence of European union capacity. *Journal of Contingencies and Crisis Management*, 22(3), 131–142. https://doi.org/10.1111/1468-5973.12052.

CEN. (2020). *CEN Workshop Agreement (CWA) 17514:2020 (2020) Systematic Assessment of Innovative Solutions for Crisis Management – Trial Guidance Methodology*. https://elib.dlr.de/137745/1/CWA17514_2020.pdf.

Fonio, C., & Widera, A. (Eds.). (2020). *Trial Guidance Methodology Handbook. DRIVER+ (Driving Innovation in Crisis Management for European Resilience)*, Brussels, 2020. https://tgm.ercis.org/.

ResiStand Consortium. (2017). *D1.3: Assessment Framework for Standardisation Activities, ResiStand H2020 Project*. http://resistand.eu/sites/default/files/resistand/public/content-files/deliverables/ResiStand_D1.3_Assessment-Framework_v10pw30012017_FINAL.pdf.

CONCLUSIONS

Chiara Fonio, Adam Widera, and Tomasz Zwęgliński

The editors of the book, as well as all of the authors, do not have any further doubts on an urgent need for boosting and thorough implementation of innovative solutions to the crisis management world. There have already been many arguments formulated for this statement throughout the book – just to mention the dynamically changing (1) natural environment due to the climate change, (2) political and economic environment due to the recent military conflicts in Europe and worldwide, (3) reality of our everyday life due to the pandemic, and last but not least, (4) world complexity on each and every level of our functioning due to globalisation processes. There are of course many more aspects heavily impacting our lives; however, these four are already more than enough to prove that the future might expose us to new challenges, testing our societal resilience towards upcoming risks. Therefore, being prepared, no matter how trivial it might sound, is still extremely valid, just obligatory.

One might ask whether it is realistic to be prepared in such an uncertain world as it is today. The world that compiles the old and the new types of threats and hazards. The world that provides clear evidence that high-impact, low-probability events, such as pandemics and wars, suddenly materialise in a short period of time, if not in parallel, pushing individuals and entire nations into crises. Are we able to prepare for each and every scenario? Rather not. However, what we can definitely do is work on decreasing the level of uncertainty. It might be achieved by providing crisis managers and, more broadly, protecting entire societies with crisis management systems, with innovative solutions that limit this uncertainty, such as by generating a valid, common, operational picture of a disaster or crisis, facilitating efficient coordination and communication efforts, ensuring effective civil protection assets management, providing reliable and real-time simulations and prognosis, and so on.

DOI: 10.4324/9781003256977-24

The growing sector of research and development delivers a wide spectrum of new solutions that claim to cover many existing gaps and respond to a lot of crisis managers' and first responders' needs. However, matching the solutions' functionalities with the gaps and needs and checking if they really meet end-users' expectations in an optimal manner are still complex challenges to be achieved. Therefore, in this book, we proposed a paradigm shift that basically entails trialling innovative solutions according to a robust methodology and in a realistic and controllable environment offered by the testbed infrastructure.

All elements of the rigorous approach to trials offered by the trial guidance methodology and testbed technical infrastructure, including a set of supporting tools, have been scrupulously explained in Parts 1 and 2 of the book. Then Part 3 elaborated on practical implementation of the DRIVER+ environment into action by providing evidence for the paradigm usability by generating qualitatively and quantitatively measurable results of new solutions' performance in a set of sequential trials that have been proved to be valid and reliable. Finally, Part 4 opened new horizons for the paradigm that has been successfully taken up and adapted by the latest European research and development projects as a catalyst for changing the way in which innovation is conceived.

Thus, the results presented in the book lead to a conclusion that trialling new solutions in an organised and robust manner makes a lot of sense. It basically generates new, methodologically proven, and experience-based knowledge on the solution among such end-users like crisis managers and first responders. Then they might share their opinions and work together with the solution providers on tailoring its functionalities best to their most urgent needs. Moreover, the new knowledge gives crisis managers and first responders a chance to verify and objectivise their initial perception and expectations towards the trialled solution. From the end-user perspective, such knowledge does count before purchasing such solutions and, what is even more important, before introducing it into daily crisis management practice.

As we close this work, let us refer to Dr Mike Ryan, head of the WHO Emergencies Program. During the EU Civil Protection Forum – Think Tank, in the peak of the COVID-19 pandemic on March 16, 2021, he said, "Technology in crisis management is useless as long as it proves to save lives." We hope that you, dear reader, do not have any further doubts on the need of trialling new solutions. If this is the case, let us check them in safe conditions of DRIVER+ before you use them to save lives!

ANNEX

5.1 Glossary

after-action review (AAR) A retrospective session that allows participants (e.g. trainees) to review prior (training) performance. AAR sessions are not intended to be critical evaluations of performance so much as collaborative meetings held to enhance learning. Ideally, AARs should be conducted to focus on what was planned to occur during training, what actually did occur during training, why events unfolded as they did, and what should be modified during subsequent training or performance sessions.

baseline An agreed reference value or set of values which can be derived from past experience and are often used for comparing with ongoing performance data, values, and/or outcomes.

centre of expertise (CoE) A (group of) organisation(s) or institute(s) that offers support in the form of training, technical or methodological support, and/or consultancy services in implementing the trial guidance methodology and/or its supportive tools and methods for organising trials or exercises.

co-creation A process in which different domains work together to create value. Within the TGM, this means that crisis management practitioners, solution providers, and researchers work with one another, share knowledge and experience, and create the trial together.

common information space (CIS) A software tool, part of the testbed technical infrastructure, that enables data exchange between solutions and in between solutions and simulators through this testbed technical infrastructure. This software tool comes with its own architecture, standardised connectors for linking up solutions, and schemas for data exchange.

crisis management (CM) The process by which an organisation deals with a disruptive and unexpected event that threatens to harm the organisation or its stakeholders. The study of crisis management originated with large-scale industrial and environmental disasters in the 1980s. It is considered to be the most important process in public relations.

Crisis Management Innovation Network Europe (CMINE) A platform that facilitates and fosters cooperation and synergies among crisis management professionals. A broad variety of stakeholders, including practitioners, researchers, industry representatives, and/or policy-makers, can exchange knowledge and best practices and initiate cooperation on crisis management topics.

data collection A process for gathering information by different means (including activities such as web monitoring).

data collection plan A document that details the means, persons, and kinds of data to be collected in the trial.

dry run 1 A first rehearsal of a trial, focusing on the technical integration of solutions, reference implementation of the testbed, and scenario validation; it also serves as a readiness review to approve the maturity of technical solutions.

dry run 2 A full-scale rehearsal of a trial without external end-user participation, aimed at detection of technical issues and last second fine-tuning; dry run 2 is organised as a complete mirror of the trial.

evaluation A systematic process that compares the result of measurement to recognised criteria to determine the discrepancies between intended and actual performance. Gaps in performance are inputs into the continual improvement process

evaluation approach The conceptual way of designing and conducting the evaluation that represents the document that details the methods and techniques to collect the data from the data collection plan.

evaluation coordinator The role mainly responsible for preparing and executing the evaluation of the trial. This role can be executed by one person or a team.

evaluation phase The third phase of the TGM; the process of estimating the effectiveness, efficiency, utility, and relevance of a service or facility.

inject A piece of information activated by the trial staff and made available to the participants (i.e. users of the solution). Participants are assumed to take action on each inject they receive just as they would have this been a real crisis (by using the solutions). Injects can, for instance, be supplied via a telephone call played out by a role-player or via a simulator providing a view or other type of information about the state of the simulated crisis.

innovation An implementation of a new or significantly improved product (service) or process or a new marketing or organisational method in business practices, workplace organisation, or external relations.

innovation line An extension of the baseline by adding innovative solutions and embedding them in the baseline-process. Through this, the foreseen changes become visible.

key performance indicator (KPI) A quantifiable measure that an organisation (person or group of people that has its own functions with responsibilities, authorities, and relationships to achieve its objectives) uses to gauge or compare performance (measurable result) in terms of meeting its strategic and operational objectives (result to be achieved).

legacy system A (crisis management) system currently in operational use.

measurement A process to determine a value.

mitigation A limitation of any negative consequence of a particular incident.

need A prerequisite identified as necessary for or desired by the user to achieve an intended outcome, implied or stated.

objective A result to be achieved. An objective can be strategic, tactical, or operational and can relate to different disciplines and can apply at different levels.

objective assessment An assessment that avoids human biases as best as possible by relying on numerical data and predefined observation questionnaires.

observation A method of data collection in which the situation of interest is watched and the relevant facts, actions, and behaviours are recorded.

observer support tool (OST) A software tool, part of the testbed technical infrastructure, that enables observers to create digital observations and send them through this testbed technical infrastructure for storage by the after-action review tool.

participants coordinator The role mainly responsible for organising the participants of the trial and formulating a realistic scenario. This role can be executed by one person or a team.

practitioner A person with knowledge, experience, or ability needed to effectively and timely respond to crises in order to minimise damage to society.

prerequisite A step of the TGM that includes the gap definition and trial context. It is a part of the preparation phase of a trial.

preparation phase The first phase of organising a trial. Contains two sub-phases: step zero and the six-step approach. Must be completed before entering the execution phase.

reference data Data which, by general agreement, may be used as a standard or as a basis for prediction and/or comparison with observed data (IEV 191-14-18).

research ethics The ethics of the planning, conducting, and reporting of research; this pertains in particular to rules and guidelines for the participation and protection of individuals taking part in the research activities.

research question (RQ) A question that a trial sets out to answer. It is used to focus on specific aspects and determine the evaluation approach.

scenario A pre-planned set of storylines that drives an exercise, as well as the stimuli used to achieve exercise project performance objectives. In the context of DRIVER+, scenarios are defined for trials, not for exercises.

six-step approach An iterative, co-creative approach for preparing a trial by following its six specific steps and iterating between them.

socio-technical solution Any solution that does not only come with technological and technical functionalities but by being used it is automatically exposed to the social system of the user. This includes the user's culture, value and belief system, and so on but also factors of the human-machine interaction.

solution is a means that contribute to a crisis management function. A solution is either one or more processes or one or more tools with related procedures.

solution provider An entity providing the solution that will be assessed within a trial.

step zero The definition of the trial objective and the trial context which are used as a starting point for the six-step approach.

technical coordinator The role mainly responsible for managing the solution providers, implementing the testbed technical infrastructure and managing the training on the solutions. This role can be executed by one person or a team.

testbed technical infrastructure (TTI) The software tools and middleware to systematically create an appropriate (life and/or virtual) environment in which the trialling of solutions is carried out. The testbed infrastructure can enable existing facilities to connect and exchange data.

trial An assessment of the performance, qualities, or suitability of solutions for current and emerging needs in such a way that practitioners can execute it following a pragmatic and systematic approach.

trial action plan The main trial planning document, facilitating collaborative planning and supporting execution of the trial. It covers all areas related to the trial organisation and is used to record efforts, circulate decisions, and assess progress.

trial committee The core team organising the trial. This team consists of at least a trial owner, and evaluation coordinator, a participant coordinator, and a technical coordinator but can include more persons with specific responsibilities, like a logistical coordinator and/or safety officers.

trial context The context of use. Users, tasks, equipment (hardware, software, and materials), and the physical and social environments in which a product (a solution) is used.

trial guidance methodology (TGM) A structured approach from designing a trial to evaluating the outcomes and identifying lessons learnt.

trial guidance tool (TGT) A software tool that guides trial design, execution, and evaluation in a step-by-step way (according to the trial guidance methodology), including as much of the necessary information as possible in form of data or references to the portfolio of solutions.

trial integration meeting (TIM) A formal design meeting of trial committee, solution provider(s), and CM practitioners to detail out – based on the selected solution(s), the trial objective, the gap to bridge, and the scenario's concept – how the trial will be executed and what is needed to make this happen.

trial management tool (TMT) A software tool, part of the testbed technical infrastructure, that enables trial staff to create a digital version of the trial scenario in a Gantt-chart-like timetable consisting of different storylines with injects. During the run of this trial, the scenario is played out by the trial staff activating, pausing, or stopping different storylines, thereby controlling when certain injects and use-cases are fired towards the participants.

trial owner The role mainly responsible for managing the trial. This role can be executed by one person or a team.

trial script A document describing the timetable of the trial and providing all necessary practical information, like logistical package and planning, catering information, and a safety plan, to execute the trial event.

5.2 Systematic Literature Review

ID	*Title*	*Author*	*Year*
18	Conventional Medical Education and the History of Simulation in Radiology	Chetlen, Alison L.; Mendiratta-Lala, Mishal; Probyn, Linda; Auffermann, William F.; DeBenedectis, Carolynn M.; Marko, Jamie; Pua, Bradley B.; Sato, Takashi Shawn; Little, Brent P.; Dell, Carol M.; Sarkany, David; Gettle, Lori Mankowski	2015
55	Simulation forward processes of surgical care	Pucher, Philip H.; Darzi, Ara; Aggarwal, Rajesh	2013
65	D-DEMATEL: A new method to identify critical success factors in emergency management	Zhou, Xinyi; Shi, Yangqiuyan; Deng, Xinyang; Deng, Yong	2017
72	A risk assessment tool for improving safety standards and emergency management in Italian onshore wind farms	Astiaso Garcia, Davide; Bruschi, Daniele	2016
76	Project training evaluation: Reshaping boundary objects and assumptions	Lee-Kelley, Liz; Blackman, Deborah	2012
79	Evaluation of Medical Management During a Mass Casualty Incident Exercise: An Objective Assessment Tool to Enhance Direct Observation	Ingrassia, Pier Luigi; Prato, Federico; Geddo, Alessandro; Colombo, Davide; Tengattini, Marco; Calligaro, Sara; La Mura, Fabrizio; Michael Franc, Jeffrey; Della Corte, Francesco	2010
83	Development, initial reliability and validity testing of an observational tool for assessing technical skills of operating room nurses	Sevdalis, Nick; Undre, Shabnam; Henry, Janet; Sydney, Elaine; Koutantji, Mary; Darzi, Ara; Vincent, Charles A.	2009
98	A service oriented architecture for decision support systems in environmental crisis management	Vescoukis, Vassilios; Doulamis, Nikolaos; Karagiorgou, Sofia	2012
123	Striving to be resilient: What concepts, approaches and practices should be incorporated in resilience management guidelines?	Adini, Bruria; Cohen, Odeya; Eide, Aslak Wegner; Nilsson, Susanna; Aharonson-Daniel, Limor; Herrera, Ivonne A.	2017
131	Expanding the use of simulation in open vascular surgical training	Pandey, Vikas A.; Wolfe, John H. N.	2012
150	Emergency transportation network design problem: Identification and evaluation of disaster response routes	Nikoo, Nariman; Babaei, Mohsen; Mohaymany, Afshin Shariat	2017

ID	*Title*	*Author*	*Year*
159	A virtual reality based fire training simulator integrated with fire dynamics data	Cha, Moohyun; Han, Soonhung; Lee, Jaikyung; Choi, Byungil	2012
160	Prepositioning of supplies in preparation for a hurricane under potential destruction of prepositioned supplies	Galindo, Gina; Batta, Rajan	2013
164	Decision support system for emergency management: Road tunnels	Alvear, Daniel; Abreu, Orlando; Cuesta, Arturo; Alonso, Virginia	2013
173	A continuous approximation approach for assessment routing in disaster relief	Huang, Michael; Smilowitz, Karen R.; Balcik, Burcu	2013
190	Post-earthquake emergency assessment of building damage, safety and usability Part 2: Organisation	Anagnostopoulos, S.; Moretti, M.	2008
227	Design of fault simulator	Gabbar, Hossam A.; Sayed, Hanaa E.; Osunleke, Ajiboye S.; Masanobu, Hara	2009
234	Crowd simulation for emergency response using BDI agents based on immersive virtual reality	Shendarkar, Ameya; Vasudevan, Karthik; Lee, Seungho; Son, Young-Jun	2008
236	A nature-inspired decentralized trust model to reduce information unreliability in complex disaster relief operations	Kostoulas, Dionysios; Aldunate, Roberto; Pena Mora, Feniosky; Lakhera, Sanyogita	2008
255	Group value and intention to use – A study of multi-agency disaster management information systems for public safety	Lee, JinKyu; Bharosa, Nitesh; Yang, Jing; Janssen, Marijn; Rao, H. R.	2011
256	Towards a Lightweight Approach for On-site Interaction Evaluation of Safety-critical Mobile Systems	Holl, Konstantin; Nass, Claudia; Villela, Karina; Vieira, Vaninha	2016
263	Flood Emergency Management Using Hydrodynamic Modelling	Liu, Yongzhi; Zhang, Wenting; Cui, Xinmin	2012
273	A general computational recognition primed decision model with multi-agent rescue simulation benchmark	Nowroozi, Alireza; Shiri, Mohammad E.; Aslanian, Angeh; Lucas, Caro	2012
275	Modeling and simulation method of the emergency response systems based on OODA	Huang, Yanyan	2015

ID	*Title*	*Author*	*Year*
284	Developing shared situational awareness for emergency management	Seppänen, Hannes; Mäkelä, Jaana; Luokkala, Pekka; Virrantaus, Kirsi	2013
292	A real-time stochastic evacuation model for road tunnels	Capote, Jorge A.; Alvear, Daniel; Abreu, Orlando; Cuesta, Arturo; Alonso, Virginia	2013
323	Design of formative evacuation plans using agent-based simulation	Zarboutis, Nikos; Marmaras, Nicolas	2007
339	Research on Efficiency of Collaborative Allocation System of Emergency Material Based on Synergetic Theory	Dou, Liangtan; Sun, Ying; She, Lian	2012
350	eStorys: A visual storyboard system supporting back-channel communication for emergencies	Malizia, A.; Bellucci, A.; Diaz, P.; Aedo, I.; Levialdi, S.	2011
371	Detection of undesirable communication patterns in multi-agent systems	Gutiérrez, Celia; García-Magariño, Iván; Fuentes-Fernández, Rubén	2011
393	Scenario-based design: A method for connecting information system design with public health operations and emergency management	Reeder, Blaine; Turner, Anne M.	2011
414	An expert system for an emergency response management in Networked Safe Service Systems	Liu, X.; Li, W.; Tu, Y. L.; Zhang, W. J.	2011
416	A container multimodal transportation scheduling approach based on immune affinity model for emergency relief	Hu, Zhi-Hua	2011
500	Simulation for team training and assessment: case studies of online training with virtual worlds	Heinrichs, W. L. R.; Youngblood, P.; Harter, P. M.; Dev P.	2008
556	Obstetric simulation as a risk control strategy: course design and evaluation	Gardner, R.; Walzer, T. B.; Simon, R.; Raemer, D. B.	2008
808	Managing the inconceivable: participatory assessments of impacts and responses to extreme climate change	Toth, F. L.; Hizsnyik, E.	2008
820	Reconstruction and Exploration of Large-scale Distributed Operations – Multimedia tools for Evaluation of Emergency Management Response	Pilemalm, S.; Andersson, D.; Hallberg, N.	2008

ID	*Title*	*Author*	*Year*
847	An assessment of activity-based modeling and simulation for applications in operational studies, disaster preparedness, and homeland security	Henson, K.; Goulias, K.; Letters, Golledge, R.	2009
936	Defining Team Performance for Simulation-based Training: Methodology, Metrics, and Opportunities for Emergency Medicine	Shapiro, M. J.; Gardner, R.; Jay, G. D.; Lindquist, D. G.; Salisbury, M. L.; Salas, E.	2008
980	Development and evaluation of ontology for intelligent decision support in medical emergency management for mass gatherings	Haghighi, P. D.; Burstein, F.; Zaslavsky, A.; Arbon, P.	2013
1019	Expert system CRIPS: support of situation assessment and decision making	Dellwing, H.; Schmitz, W.	2007
1041	A general methodology for data-based rule building and its application to natural disaster management	Rodríguez, J. T.; Vitoriano, B.; Montero, J.	2012
1165	Utilizing simulation technology for competency skills assessment and a comparison of traditional methods of training to simulation-based training	Tuttle, R. P.; Cohen, M. H.; Augustine, A. J.; Novotny, D. F.; Delgado, E.; Dongilli, T. A.; Lutz, J. W.; DeVita, M. A.	2007
1195	The workpad user interface and methodology: Developing smart and effective mobile applications for emergency operators	Humayoun, S. R.; Catarci, T.; deLeoni, Massimiliano; Marella, A.; Mecella, Massimo; Bortenschlager, M.; Steinmann, R.	2009
1420	Community Assessment for Public Health Emergency Response (CASPER): An Innovative Emergency Management Tool in the United States.	Schnall, Amy; Nakata, Nicole; Talbert, Todd; Bayleyegn, Tesfaye; Martinez, DeAndrea; Wolkin, Amy	2017
1421	Short simulation exercises to improve emergency department nurses' self-efficacy for initial disaster management: Controlled before and after study.	Jonson, Carl-Oscar; Pettersson, Jenny; Rybing, Jonas; Nilsson, Helene; Prytz, Erik	2017
1424	Public Health System Research in Public Health Emergency Preparedness in the United States (2009–2015): Actionable knowledge base.	Savoia, Elena; Lin, Leesa; Bernard, Dottie; Klein, Noah; James, Lyndon P.; Guicciardi, Stefano	2017

ID	*Title*	*Author*	*Year*
1427	Interactive plant simulation modeling for developing an operator training system in a natural gas pressure-regulating station.	Yongseok Lee; Changjun Ko; Hodong Lee; Kyeongwoo Jeon; Seolin Shin; Chonghun Han	2017
1431	Traffic evacuation simulation based on multi-level driving decision model.	Yuan, Shengcheng; Chun, Soon Ae; Spinelli, Bruno; Liu, Yi; Zhang, Hui; Adam, Nabil R.	2017
1437	Identifying and explicating knowledge on method transfer: a sectoral system of innovation approach.	Hvannberg, Ebba	2015
1439	A 3-year Health Care Coalition Experience in Advancing Hospital Evacuation Preparedness.	Lowe, John J.; Hansen, Keith F.; Sanger, Kristine K.; Obaid, Jannah M.	2016
1449	A dynamic model for disaster response considering prioritized demand points.	Rivera-Royero, Daniel; Galindo, Gina; Yie-Pinedo, Ruben	2016
1457	Influencing Factors on Social Media Adoption in County-level Emergency Management Departments.	Schuwerk, Tara J.; Davis, Allison	2013
1465	Testing a methodology to improve organisational learning about crisis communication by public organisations.	Palttala, Paulina; Vos, Marita	2011
1478	Multi-purpose 3-D Real Estate: Understanding the Role of 3-D Technology for Enhancing Resilience	Christensen, Pernille H.; McIlhatton, David; Blair, Neale; Grunninger Bonney, Courtney	2016
1484	A Simulation Tool for Examining the Effect of Communications on Disaster Response in the Oil and Gas Industry.	Singh, Arvind; Adams, Richelle; Dookie, Isa; Kissoon, Shiva	2016
1489	Data Model Development for Fire Related Extreme Events – An Activity Theory and Semiotics Approach	Chen, Rui; Sharman, Raj; Rao, H. Raghav; Upadhyaya, Shambhu J.	2013
1513	Using Twitter in crisis management for organisations bearing different country-of-origin perceptions.	Xu, Jie; Wu, Yiye	2015
1514	High Fidelity Simulation to Evaluate Emergency Management in Urgent Care Centers.	Tabor, Megan; Vaughn, Brooke L.	2017

ID	*Title*	*Author*	*Year*
1520	Challenges in coordination: differences in perception of civil and military organisations by comparing international scientific literature and field experiences.	Pramanik, Roshni	2015
1521	The creation of a training model to support decision-making of emergency management practitioners: A design research study.	McCarthy, Nora; Neville, Karen; Pope, Andrew; Gallagher, Anthony; Nussbaumer, Alexander; Steiner, Christina M.	2016
1524	Towards the development of a decision support system for multi-agency decision-making during cross-border emergencies.	Neville, Karen; O'Riordan, Sheila; Pope, Andrew; Rauner, Marion; Rochford, Maria; Madden, Martina; Sweeney, James; Nussbaumer, Alexander; McCarthy, Nora; O'Brien Cian	2016
1535	Impact of an Education Intervention on Missouri K-12 School Disaster and Biological Event Preparedness.	Rebmann, Terri; Elliott, Michael B.; Artman, Deborah; VanNatta, Matthew; Wakefield, Mary	2016
1543	Team regulation in a simulated medical emergency: An in-depth analysis of cognitive, metacognitive, and affective processes.	Duffy, Melissa C.; Azevedo, Roger; Sun, Ning-Zi; Griscom, Sophia E.; Stead, Victoria; Crelinsten, Linda; Wiseman, Jeffrey; Maniatis, Thomas; Lachapelle, Kevin	2015
1548	Increasing emergency medicine residents' confidence in disaster management: use of an emergency department simulator and an expedited curriculum.	Franc, Jeffrey Michael; Nichols, Darren; Dong, Sandy L.	2012
1557	Educating the Next Generation to Respond to a Bioterrorism Event.	Sobel, Annette L.; Fisher, Beth M.	2012
1560	Context-Specific, Scenario-Based Risk Scales.	Yu, Michael; Lejarraga, Tomás; Gonzalez, Cleotilde	2012
1573	25 Years of MCDA in nuclear emergency management.	Papamichail, K. Nadia; French, Simon	2013
1580	Relationships Between Mental Health Distress and Work-Related Factors Among Prefectural Public Servants Two Months After the Great East Japan Earthquake.	Fukasawa, Maiko; Suzuki, Yuriko; Obara, Akiko; Kim, Yoshiharu	2015
1581	Investing in Disaster Management Capabilities versus Pre-positioning Inventory: A New Approach to Disaster Preparedness	Kunz, Nathan; Reiner, Gerald; Gold, Stefan	2014

ID	*Title*	*Author*	*Year*
1582	Proposing "the burns suite" as a novel simulation tool for advancing the delivery of burns education.	Sadideen, Hazim; Wilson, David; Moiemen, Naiem; Kneebone, Roger	2014
1583	Using Real-Time Decision Tools to Improve Distributed Decision-Making Capabilities in High-Magnitude Crisis Situations.	Moskowitz, Herbert; Drnevich, Paul; Ersoy, Okan; Altinkemer, Kemal; Chaturvedi, Alok	2011
1587	Disaster Preparedness in Philippine Nurses.	Labrague, Leodoro J.; Yboa, Begonia C.; McEnroe-Petitte, Denise M.; Lobrino, Ledwin R.; Brennan, Mary Geronima B.	2016
1592	Examining the Role of Social Media in Effective Crisis Management: The Effects of Crisis Origin, Information Form, and Source on Publics' Crisis Responses.	Jin, Yan; Liu, Brooke Fisher; Austin, Lucinda L.	2014
1593	Mechanisms of Control in Emergent Interorganisational Networks.	Marcum, Christopher Steven; Bevc, Christine A.; Butts, Carter T.	2012
1594	Does message placement influence risk perception and affect?	Lachlan, Kenneth; Spence, Patric R.	2014
1603	Improving Communication in Crisis Management by Evaluating the Relevance of Messages.	Netten, Niels; van Someren, Maarten	2011
1604	Applying Analytical Hierarchy Process to Supplier Selection and Evaluation in the Hospitality Industry: A Multiobjective Approach	Chung, Kaie-Chin	2015
1606	The use of emergency operations centres in local government emergency management.	Sinclair, Helen; Doyle, Emma E. H.; Johnston, David M.; Paton, Douglas	2013
1607	Assessing and improving cross-border chemical incident preparedness and response across Europe.	Stewart-Evans, James; Hall, Lisbeth; Czerczak, Slawomir; Manley, Kevin; Dobney, Alec; Hoffer, Sally; Palaszewska-Tkacz, Anna; Jankowska, Agnieszka	2014
1616	Training and learning for crisis management using a virtual simulation/gaming environment.	Walker, Warren E.; Giddings, Jordan; Armstrong, Stuart	2011
1622	Team Coordination in Escalating Situations: An Empirical Study Using Mid-Fidelity Simulation	Bergstöm, Johan; Dahlström, Nicklas; Henriqson, Eder; Dekker, Sidney	2010

ID	*Title*	*Author*	*Year*
1630	Crisis Management Dilemmas: Differences in Attitudes towards Reactive Crisis Communication Strategies among Future Business Professionals in Croatia	Tipuric, Darko; Skoko, Bozo; Jugo, Damir; Mesin, Marina	2013
1631	Engineering Trust in Complex Automated Systems.	Lyons, Joseph B.; Koltai, Kolina S.; Ho, Nhut T.; Johnson, Walter B.; Smith, David E.; Shively, R. Jay	2016
1632	Psychological Effects of Disaster Relief Activities on Japan Ground Self-Defense Force Personnel Following the 2011 Great East Japan Earthquake.	Dobashi, Kosuke; Nagamine, Masanori; Shigemura, Jun; Tsunoda, Tomoya; Shimizu, Kunio; Yoshino, Aihide; Nomura, Soichiro	2014
1644	Jordanian nurses' perceptions of their preparedness for disaster management.	Al Khalaileh, Murad A.; Bond, Elaine; Alasad, Jafar A.	2012
1648	A priority driven ABC approach to the emergency management of high energy pelvic trauma improves decision making in simulated patient scenarios.	Daurka, Jasvinder S.; Rankin, Iain; Jaggard, M. K. J.; Lewis, Angus	2015
1652	Validity evidence of non-technical skills assessment instruments in simulated anaesthesia crisis management.	Jirativanont, T.; Raksamani, K.; Aroonpruksakul, N.; Apidechakul, P.; Suraseranivongse, S.	2017
1658	The effect of a simulation-based training intervention on the performance of established critical care unit teams.	Frengley, R. W.; Weller, J. M.; Torrie, J.; Dzendrowskyj, P.; Yee, B.; Paul, A. M.; Shulruf, B.; Henderson, K. M.	2011
1659	Interprofessional non-technical skills for surgeons in disaster response: A qualitative study of the Australian perspective.	Willems, Anneliese; Waxman, Buce; Bacon, Andrew K.; Smith, Julian; Peller, Jennifer; Kitto, Simon	2013
1675	Atmospheric dispersion and impact modeling systems: How are they perceived as support tools for nuclear crises management?	Benamrane, Yasmine; Boustras, Georgios	2015
1677	REACT: A paraprofessional training program for first responders-A pilot study.	Marks, Madeline R.; Bowers, Clint; DePesa, Natasha S.; Trachik, Benjamin; Deavers, Frances E.; James, Nicholas T.	2017
1681	Crisis Leadership in an Acute Clinical Setting: Christchurch hospital, New Zealand ICU Experience Following the February 2011 Earthquake.	Zhuravsky, Lev	2015

ID	*Title*	*Author*	*Year*
1682	Building health care system capacity to respond to disasters: successes and challenges of disaster preparedness health care coalitions.	Walsh, Lauren; Craddock, Hillary; Gulley, Kelly; Strauss-Riggs, Kandra; Schor, Kenneth W.	2015
1687	Development and evaluation of an offshore oil and gas Emergency Response Focus Board	Taber, Michael J.; McCabe, John; Klein, Raymond M.; Pelot, Ronald P.	2013
1688	The Emerging Role of Higher Education in Educating and Assessing Future Leaders for Disaster Relief.	Buschlen, Eric; Goffnett, Sean	2013
1697	Assessment of the reliability of the Johns Hopkins/Agency for Healthcare Research and Quality hospital disaster drill evaluation tool.	Kaji, A. H.; Lewis, R. J.	2008
1699	The Rapid Disaster Evaluation System (RaDES): A Plan to Improve Global Disaster Response by Privatizing the Assessment Component.	Iserson, Kenneth V.	2017
1701	The Role of Law in Public Health Preparedness: Opportunities and Challenges.	Jacobson, Peter D.; Wasserman, Jeffrey; Botoseneanu, Anda; Silverstein, Amy; Wu, Helen W.	2012
1704	The effectiveness of a disaster training programme for healthcare workers in Greece.	Bistaraki, A.; WaddinTGTon, K.; Galanis, P.	2011
1705	Knowledge, Experiences and Training Needs of Health Professionals about Disaster Preparedness and Response in Southwest Ethiopia: a cross sectional study.	Berhanu, Negalign; Abrha, Hailay; Ejigu, Yohannes; Woldemichael, Kifle	2016
1714	Culpable leaders, trust, emotional exhaustion, and identification during a crisis.	Kovoor-Misra, Sarah; Gopalakrishnan, Shanthi	2016
1717	Learning crisis resource management: Practicing versus an observational role in simulation training – a randomized controlled Trial.	Lai, Anita; Haligua, Alexis; Dylan Bould, M.; Everett, Tobias; Gale, Mark; Pigford, Ashlee-Ann; Boet, Sylvain	2016
1718	A Socio-Physical Approach to Systemic Risk Reduction in Emergency Response and Preparedness.	Ross, William; Gorod, Alex; Ulieru, Mihaela	2015

ID	*Title*	*Author*	*Year*
1722	How Simple Hypothetical-Choice Experiments Can Be Utilized to Learn Humans' Navigational Escape Decisions in Emergencies.	Haghani, Milad; Sarvi, Majid; Shahhoseini, Zahra; Boltes, Maik	2016
1723	Disaster spread simulation and rescue time optimization in a resource network.	Hu, Zhi-Hua; Sheng, Zhao-Han	2015
1727	A survey of the practice of nurses' skills in Wenchuan earthquake disaster sites: implications for disaster training.	Yin, Huahua; He, Haiyan; Arbon, Paul; Zhu, Jingci	2011
1728	Health worker and policy-maker perspectives on use of intramuscular artesunate for pre-referral and definitive treatment of severe malaria at health posts in Ethiopia.	Kefyalew, Takele; Kebede, Zelalem; Getachew, Dawit; Mukanga, David; Awano, Tessema; Tekalegne, Agonafer; Batisso, Esey; Edossa, Wasihun; Mekonnen, Emebet; Tibenderana, James; Baba, Ebenezer Sheshi; Shumba, Constance; Nankabirwa, Joaniter I.; Hamade, Prudence	2016
1764	Full-scale regional exercises: Closing the gaps in disaster preparedness.	Klima, D. A.; Seiler, S. H.; Peterson, J. B.; Christmas, A. B.; Green, J. M.; Fleming, G.; Thomason, M. H.; Sing, R. F.	2012
1785	Computer-based collaborative training for transportation security and emergency response	Velasquez, Juan D.; Yoon, Sang Won; Nof, Shimon Y.	2010
1803	A Review of Critical Infrastructure Interdependency Simulation and Modelling for the Caribbean.	Dookie, Isa Jeziah; Singh, Arvind; Pooransingh, Akash; Rocke, Sean	2016
1815	Emergency nurses and disaster response: An exploration of South Australian emergency nurses' knowledge and perceptions of their roles in disaster response.	Hammad, Karen S.; Arbon, Paul; Gebbie, Kristine M.	2011
1818	Multievent Crisis Management Using Noncooperative Multistep Games.	Gupta, Upavan; Ranganathan, Nagarajan	2007
1825	Physical and mental health status of soldiers responding to the 2008 Wenchuan earthquake.	Wei Qiang Zhang; Chaojie Liu; Tian Sheng Sun; Jing Zhao; Ju Qiang Han; Yong Hong Yang; Shu Jun Li; Ya Qun Ma	2011

ID	*Title*	*Author*	*Year*
1826	Burn Disaster Response Planning in New York City: Updated Recommendations for Best Practices.	Leahy, Nicole E.; Yurt, Roger W.; Lazar, Eliot J.; Villacara, Alfred A.; Rabbitts, Angela C.; Berger, Laurence; Chan, Carri; Chertoff, Laurence; Conlon, Kathe M.; Cooper, Arthur; Green, Linda V.; Greenstein, Bruce; Lu, Yina; Miller, Susan; Mineo, Frank P.; Pruitt, Darrin; Ribaudo, Daniel S.; Ruhren, Chris; Silber, Steven; Soloff, Lewis	2012
1827	The impact of an online interprofessional course in disaster management competency and attitude towards interprofessional learning.	Atack, L.; Parker, K.; Rocchi, M.; Maher, J.; Dryden, T.	2009
1859	Teaching Critical Management Skills to Senior Nursing Students: Videotaped or Interactive Hands-On Instruction?	Baxter, Pamela; Akhtar-Danesh, Noori; Landeen, Janet; Norman, Geoff	2012
1864	Development, initial reliability and validity testing of an observational tool for assessing technical skills of operating room nurses.	Sevdalis, N.; Undre, S.; Henry, J.; Sydney, E.; Koutantji, M.; Darzi, A.; Vincent, C. A.	2009
1874	Barriers to implementing infection prevention and control guidelines during crises: Experiences of health care professionals.	Timen, A.; Hulscher. M. E.; Rust, L.; van Steenbergen, J. E.; Akkermans, R. P.; Grol, R. P.; van der Meer, J. W.	2010
1880	Leaders as emotional managers: Emotion management in response organisations during a hostage taking in a Swedish prison.	Alvinius, Aida; Boström, Malin Elfgren; Larsson, Gerry	2015
1882	Using pictograms for communication.	Clawson, T. H.; Leafman, J.; Nehrenz, G. M., Sr; Kimmer, S.	2012
1885	The Disaster Preparedness Evaluation Tool: psychometric testing of the Classical Arabic version.	Al Khalaileh, M. A.; Bond, A. E.; Beckstrand, R. L.; Al-Talafha, A.	2010
1888	Impact of crisis resource management simulation-based training for interprofessional and interdisciplinary teams: A systematic review.	Fung, Lillia; Boet, Sylvain; Bould, M. Dylan; Qosa, Haytham; Perrier, Laure; Tricco, Andrea; Tavares, Walter; Reeves, Scott	2015
1893	Manifest leadership styles in a Caribbean cross-sector network.	Cooper, Tracy	2016

ID	*Title*	*Author*	*Year*
1913	The Borsele files: the challenge of acquiring usable data under chaotic circumstances.	Gouman, Rianne; Kempen, Masja; Van der Heijden, Eddy; Wijngaards, Niek; De Vree, Philip; Capello, Toon	2008
1894	A controlled before-and-after evaluation of a mobile crisis partnership between mental health and police services in Nova Scotia.	Kisely, Stephen; Campbell, Leslie Anne; Peddle, Sarah; Hare, Susan; Pyche, Mary; Spicer, Don; Moore, Bill	2010
1927	Using Monte Carlo simulation to refine emergency logistics response models: a case study.	Ruth Banomyong, Apichat Sopadang	2010
1930	Interprofessional team dynamics and information flow management in emergency departments.	Gilardi, Silvia; Guglielmetti, Chiara; Pravettoni, Gabriella	2014
1936	Assessing the reliability and the expected performance of a network under disaster risk.	Günneç, Dilek; Salman, F.	2011
1949	A reassessment and review of the Bam earthquake five years onward: what was done wrong?	Motamedia, M. H. K.; Saghafinia, M.; Bafarani, A. H.; Panahi, F.	2009
1950	Parameter-Based Data Aggregation for Statistical Information Extraction in Wireless Sensor Networks.	Jiang, Hongbo; Jin, Shudong; Wang, Chonggang	2010
1953	Toward the regulation of ubiquitous mobile government: a case study on location-based emergency services in Australia.	Aloudat, Anas; Michael, Katina	2011
1964	A decision support system for debris-flow hazard mitigation in towns based on numerical simulation: a case study at Dongchuan, Yunnan Province.	Fangqiang Wei; Kaiheng Hu; Peng Cui; Qun Guan	2008
1966	Creating order from chaos: part I: triage, initial care, and tactical considerations in mass casualty and disaster response.	Baker, M. S.	2007
1980	An emergency logistics response system for natural disasters.	Tovia, F.	2007
1998	Developing shared situational awareness for emergency management.	Seppänen, Hannes; Mäkelä, Jaana; Luokkala, Pekka; Virrantaus, Kirsi	2013
1999	Dynamic decision support for managing regional resources: Mapping risk in Allegheny County, Pennsylvania.	Chalfant, Brian A.; Comfort, Louise K.	2016

ID	*Title*	*Author*	*Year*
2000	Emergency crowd evacuation modeling and simulation framework with cellular discrete event systems.	Jafer, Shafagh; Lawler, Ryan	2016
2015	Supporting collaborative sense-making in emergency management through geo-visualization.	Wu, Anna; Convertino, Gregorio; Ganoe, Craig; Carroll, John M.; Zhang, Xiaolong (Luke)	2013
2022	An optimization approach for ambulance location and the districting of the response segments on highways	Iannoni, Ana Paula; Morabito, Reinaldo; Saydam, Cem	2009
2026	Two complementary mobile technologies for disaster warning.	Samarajiva, Rohan; Waidyanatha, Nuwan	2009
2030	"G.A.T.E": Gap Analysis for TTX evaluation.	Cacciotti, Ilaria; Di Giovanni, Daniele; Pergolini, Alessandro; Malizia, Andrea; Carestia, Mariachiara; Palombi, Leonardo; Bellecci, Carlo; Gaudio, Pasquale	2016
2037	Big Board: Teleconferencing over maps for shared situational awareness.	Heard, Jefferson; Thakur, Sidharth; Losego, Jessica; Galluppi, Ken	2014
2054	State Mandate Influences on FEMA-Approved Hazard-Mitigation Plans Under the Disaster Management Act of 2000.	Olonilua, Oluponmile	2016
2061	A Procedural Construction Method for Interactive Map Symbols Used for Disasters and Emergency Response.	Peng, Guoqiang; Yue, Songshan; Li, Yuting; Song, Zhiyao; Wen, Yongning	2017
2099	Multi-objective evacuation routing optimization for toxic cloud releases.	Gai, Wen-mei; Deng, Yun-feng; Jiang, Zhong-an; Li, Jing; Du, Yan	2017
2126	Supporting community emergency management planning through a geocollaboration software architecture.	Schafer, Wendy A.; Ganoe, Craig H.; Carroll, John M.	2007
2138	Giving meaning to tweets in emergency situations: a semantic approach for filtering and visualizing social data.	Onorati, Teresa; Díaz, Paloma	2016
2175	Acil durum servislerinin yer seçimi: Analitik Hiyerarşi Yöntemi ve CBS entegrasyonu	Erden, Turan; Coşkun, Mehmet Zeki	2010
2177	Decentralized Coordination in RoboCup Rescue.	Ramchurn, Sarvapali D.; Farinelli, Alessandro; Macarthur, Kathryn S.; Jennings, Nicholas R.	2010

ID	*Title*	*Author*	*Year*
2194	A dynamic decision support system based on geographical information and mobile social networks: A model for tsunami risk mitigation in Padang, Indonesia.	Ai, Fuli; Comfort, Louise K.; Dong, Yongqiang; Znati, Taieb	2016
2199	Interoperable architecture for joint real/virtual training in emergency management using the MPEG-V standard.	Ardila, Laura; Pérez-Llopis, Israel; Esteve, Manuel; Palau, Carlos E.	2015
2207	Early warning and mass evacuation in coastal cities.	Hissel, François; Morel, Gilles; Pescaroli, Gianluca; Graaff, Herman; Felts, Didier; Pietrantoni, Luca	2014
2213	Resource-Poor Settings: Response, Recovery, and Research: Care of the Critically Ill and Injured During Pandemics and Disasters: CHEST Consensus Statement.	Geiling, James; Burkle Jr, Frederick M.; West, T. Eoin; Uyeki, Timothy M.; Amundson, Dennis; Dominguez-Cherit, Guillermo; Gomersall, Charles D.; Lim, Matthew L.; Luyckx, Valerie; Sarani, Babak; Christian, Michael D.; Devereaux, Asha V.; Dichter, Jeffrey R.; Kissoon, Niranjan	2014
2216	Do or die–Strategic decision-making following a shock event.	Bonn, Ingrid; Rundle-Thiele, Sharyn	2007
2220	Application of Traffic Simulation Modeling for Improved Emergency Preparedness Planning.	Sisiopiku, Virginia P.	2007
2233	Training decision-makers in flood response with system dynamics.	Berariu, Romana; Fikar, Christian; Gronalt, Manfred; Hirsch, Patrick	2016
2235	Forming a global monitoring mechanism and a spatiotemporal performance model for geospatial services.	Xia, Jizhe; Yang, Chaowei; Liu, Kai; Li, Zhenlong; Sun, Min; Yu, Manzhu	2015
2241	Agent-oriented modeling and development of a system for crisis management.	García-Magariño, Iván; Gutiérrez Celia	2013
2243	Comparing four operational SAR-based water and flood detection approaches.	Martinis, Sandro; Kuenzer, Claudia; Wendleder, Anna; Huth, Juliane; Twele, André; Roth, Achim; Dech, Stefan	2015
2261	Optimal Path Selection under Emergency Based on the Fuzzy Comprehensive Evaluation Method.	Yunhua Zhu; Xiao Cai	2015
2262	Cross-domain integrating and reasoning spaces for offsite nuclear emergency response.	Xie, Tian; Li, Cong-dong; Wei, Yao-yao; Jiang, Jian-jun; Xie, Rui	2016

ID	*Title*	*Author*	*Year*
2267	Geotagging Twitter Messages in Crisis Management.	Ghahremanlou, Lida; Sherchan, Wanita; Thom, James A.	2015
2270	Modeling and representation for earthquake emergency response knowledge: perspective for working with geo-ontology.	Xu, Jinghai; Nyerges, Timothy L.; Nie, Gaozhong	2014
2272	Emergency Management Decision Making during Severe Weather.	Baumgart, Leigh A.; Bass, Ellen J.; Philips, Brenda; Kloesel, Kevin	2008
2276	Preparing for Emergency Situations.	Asproth, Viveca; Amcoff Nyström, Christina	2010
2277	Towards a Holistic Framework for the Evaluation of Emergency Plans in Indoor Environments.	Serrano, Emilio; Poveda, Geovanny; Garijo, Mercedes	2014
2284	Context-based automatic reconstruction and texturing of 3D urban terrain for quick-response tasks.	Bulatov, Dimitri; Häufel, Gisela; Meidow, Jochen; Pohl, Melanie; Solbrig, Peter; Wernerus, Peter	2014
2311	Simulating individual, group, and crowd behaviors in building egress.	Chu, Mei Ling; Parigi, Paolo; Law, Kincho H.; Latombe, Jean-Claude	2015
2325	Supporting synthesis in geovisualization.	Robinson, Anthony C.	2011
2332	Simulating effects of signage, groups, and crowds on emergent evacuation patterns.	Chu, Mei; Parigi, Paolo; Latombe, Jean-Claude; Law, Kincho	2015
2336	Towards A Framework for Simulation-Based Evaluation of Personal Decision Support Systems for Flood Evacuation.	Knyazkov, Konstantin; Balakhontceva, Marina; Ivanov, Sergey	2014
2356	Time-history simulation of civil architecture earthquake disaster relief-based on the three-dimensional dynamic finite element method.	Liu Bing; Qi Yaoguang; Du Jiyun	2014
2357	The Role of Simulation and Modeling in Disaster Management.	Steward, Duane; Wan, Thomas	2007
2359	Study of efficiency of USAR operations with assistive technologies.	Hamp, Quirin; Gorgis, Omar; Labenda, Patrick; Neumann, Marc; Predki, Thomas; Heckes, Leif; Kleiner, Alexander; Reindl, Leonhard M.	2013
2376	Modeling the emergency evacuation of the high rise building based on the control volume model.	Wu, Guan-Yuan; Huang, Hao-Chang	2015

ID	*Title*	*Author*	*Year*
2378	Building Capacity for Community Disaster Preparedness: A Call for Collaboration Between Public Environmental Health and Emergency Preparedness and Response Programs. (Cover story)	Gamboa-Maldonado, Thelma; Marshak, Helen Hopp; Sinclair, Ryan; Montgomery, Susanne; Dyjack, David T.	2012
2398	Regional Approach to Competency-Based Patient Care Provider Disaster Training: The Center for Health Professional Training and Emergency Response.	Scott, Lancer A.; Smith, Clay; Jones, E. Morgan; Manaker, L. Wade; Seymore, Andrew C.; Fulkerson, Jacob	2013
2401	9. Assessments: Structure, concepts, and methods.	Nordic Societies of Public Health	2014
2406	Modeling IoT-Based Solutions Using Human-Centric Wireless Sensor Networks.	Monares, Álvaro; Ochoa, Sergio F.; Santos, Rodrigo; Orozco, Javier; Meseguer, Roc	2014
2413	Development and Implementation of an Adapted Evacuation Planning Methodology in the Framework of Emergency Management and Disaster Response: A Case Study Using TransCAD.	Andrews, Steven; Wang, Haizhong; Ni, Daiheng; Gao, Song; Collura, John	2010
2419	Self-reported preparedness of New Zealand acute care providers to mass emergencies before the Canterbury Earthquakes: A national survey.	Al-Shaqsi, Sultan; Gauld, Robin; McBride, David; Al-Kashmiri, Ammar; Al-Harthy, Abdullah	2015
2505	Flood susceptibility analysis and its verification using a novel ensemble support vector machine and frequency ratio method.	Tehrany, Mahyat; Pradhan, Biswajeet; Jebur, Mustafa	2015
2507	Early Warning System for Disasters within Health Organisations: A Mandatory System for Developing Countries.	Zaboli, Rouhollah; Seyedin, Seyed Hesam; Malmoon, Zainab	2013
2528	Preparation and scheduling system of emergency supplies in disasters.	Jia, Liu; Kefan, Xie	2015
2541	Near optimal allocation strategy for making a staged evacuation plan with multiple exits.	Xie, Jun; Li, Qiang; Wan, Qing; Li, Xiang	2014
2547	14. Implementation, execution, and completion of projects.	Nordic Societies of Public Health	2014
2552	Methodological aspects of the implementation of the new ICRP recommendations.	Raskob, W.; Landman, C.	2013

ID	*Title*	*Author*	*Year*
2553	Modeling and simulation of stranded passengers' transferring decision-making on the basis of herd behaviors.	Shen, Yang; Liu, Sifeng; Fang, Zhigeng; Hu, Mingli	2012
2575	A Public Health Academic-Practice Partnership to Develop Capacity for Exercise Evaluation and Improvement Planning.	Wright, Kate S.; Thomas, Michael W.; Durham Jr, Dennis P.; Jackson, Lillie M.; Porth, Leslie L.; Buxton, Mark	2010
2580	Distributed Building Evacuation Simulator for Smart Emergency Management.	Dimakis, Nikolaos; Filippoupolitis, Avgoustinos; Gelenbe, Erol	2010
2583	Web-Design Evaluation of the Crisis Map of the Czech Republic Using Eye-Tracking.	Brychtova, Alzbeta; Paszto, Vit; Marek, Lukas; Panek, Jiri	2013
2595	A new hybrid evolutionary based RBF networks method for forecasting time series: A case study of forecasting emergency supply demand time series.	Mohammadi, Reza; Fatemi Ghomi, S. M. T.; Zeinali, Farzad	2014
2604	Intersection Group Dynamic Subdivision and Coordination at Intraregional Boundaries in Sudden Disaster.	Lin, Ciyun; Gong, Bowen; Yang, Zhaosheng; Qu, Xin	2015
2608	Lessons from Hurricane Sandy: a Community Response in Brooklyn, New York.	Schmeltz, Michael; GonzÃƒÂ¡lez, Sonia; Fuentes, Liza; Kwan, Amy; Ortega-Williams, Anna; Cowan, Lisa	2013
2611	Assessing Public Health Capabilities During Emergency Preparedness Tabletop Exercises: Reliability and Validity of a Measurement Tool.	Savoia, Elena; Testa, Marcia A.; Biddinger, Paul D.; Cadigan, Rebecca O.; Koh, Howard; Campbell, Paul; Stoto, Michael A.	2009
2624	Leadership in complex, stressful rescue operations: A quantitative test of a qualitatively developed model.	Sjoberg, Misa; Wallenius, Claes; Larsson, Gerry	2011
2628	Operationalizing "resilience" for disaster risk reduction in mountainous Nepal.	Sudmeier, Karen I.; Jaboyedoff, Michel; Jaquet, Stephanie	2013
2634	FireGrid: An e-infrastructure for next-generation emergency response support	Han, Liangxiu; Potter, Stephen; Beckett, George; Pringle, Gavin; Welch, Stephen; Koo, Sung-Han; Wickler, Gerhard; Usmani, Asif; Torero, José L.; Tate, Austin	2010
2636	A model for a multi-agency response management system (MARMS) for South Africa.	Pat Reid; Dewald van Niekerk	2008

ID	Title	Author	Year
2653	Decentralized Dynamic Task Allocation Using Overlapping Potential Games.	Chapman, Archie C.; Micillo, Rosa Anna; Kota, Ramachandra; Jennings, Nicholas R.	2010
2657	Multiobjective Model for Emergency Resources Allocation.	Zhaosheng Yang; Huxing Zhou; Xueying Gao; Songnan Liu	2013
2662	Simulation-assisted burn disaster planning.	Nilsson, Heléne; Jonson, Carl-Oscar; Vikström, Tore; BenTGTsson, Eva; Thorfinn, Johan; Huss, Fredrik; Kildal, Morten; Sjöberg, Folke	2013
2674	Nato's New Strategic Concept: A Critical View.	Cervera, Rafael Calduch	2011
2675	Developing Disaster Preparedness Competence: An Experiential Learning Exercise for Multiprofessional Education.	Silenas, Rasa; Akins, Ralitsa; Parrish, Alan R.; Edwards, Janine C.	2008
2687	Rallying the Troops: A Four-Step Guide to Preparing a Residency Program for Short-Term Weather Emergencies.	Chow, Grant V.; Hayashi, Jennifer; Hirsch, Glenn A.; Christmas, Colleen	2011
2691	Situation awareness and virtual globes: Applications for disaster management	Tomaszewski, Brian	2011
2719	Task force deployment for big events	Drechsel, J.; Kimms, A.	2008
2738	Space-enabled information environment for crisis management. Scenario-based analysis and evaluation in an operational environment	Ryzenko, Jakub; Smolarkiewicz, Marcin	2010
2746	Evaluating the effectiveness of an emergency preparedness training programme for public health staff in China.	Chongjian Wang; Sheng Wei; Hao Xiang; Yihua Xu; Shenghong Han; Baaliy Mkangara, Ommari; Shaofa Nie	2008
2782	Earthquake relief: Iranian nurses' responses in Bam, 2003, and lessons learnt.	Nasrabadi, A. N.; Naji, H.; Mirzabeigi, G.; Dadbakhs, M.	2007
2813	Failure Prevention in Design Through Effective Catalogue Utilization of Historical Failure Events.	Grantham Lough, K.; Stone, R.; Tumer, I.	2008
2815	A contingency model of decision-making involving risk of accidental loss	Rosness, Ragnar	2009

ID	*Title*	*Author*	*Year*
2823	Teaching the NIATx Model of Process Improvement as an Evidence-Based Process.	Evans, Alyson C.; Rieckmann, Traci; Fitzgerald, Maureen M.; Gustafson, David H.	2007
2836	Resolving crises through automated bilateral negotiations	Kraus, Sarit; Hoz-Weiss, Penina; Wilkenfeld, Jonathan; Andersen, David R.; Pate, Amy	2008
2858	Disaster Readiness: A Community – University Partnership.	Adams, Lavonne M.; Canclini, Sharon B.	2008
2554	Sustainable Measures in Rebuilding After Disasterpaper.	Veronescu, Otilia; Szitar, Mirela	2012
2592	Advances in Drama Theory for Managing Global Hazards and Disasters. Part I: Theoretical Foundation.	Levy, Jason K.; Hipel, Keith W.; Howard, N.	2009

5.3 Where to Find Further Information

Project website: www.driver-project.eu
TGM website: https://tgm.ercis.org
GitHub: https://github.com/DRIVER-EU/test-bed
PoS: https://pos.driver-project.eu
Training module: www.sisekaitse.ee/en/training-module
CMINE: www.cmine.eu

INDEX

Page locators in **bold** indicate a table

FSC
www.fsc.org
MIX
Papier | Fördert
gute Waldnutzung
FSC® C083411

Kolibri 360 GmbH
Ferdinand-Jühlke-Straße 7
99095 Erfurt, Deutschland
produktsicherheit@kolibri360.de